VOLUME SIXTY ONE

# ADVANCES IN
# ECOLOGICAL RESEARCH

Mechanisms underlying the relationship between biodiversity and ecosystem function

# ADVANCES IN ECOLOGICAL RESEARCH

*Series Editors*

**DAVID A. BOHAN**
*Directeur de Recherche*
*UMR Agroécologie, AgroSup Dijon*
*INRA, University of Bourgogne Franche-Comté*
*Dijon, France*

**ALEX J. DUMBRELL**
*School of Life Sciences*
*University of Essex*
*Wivenhoe Park, Colchester*
*Essex, United Kingdom*

VOLUME SIXTY ONE

# Advances in
# ECOLOGICAL RESEARCH

Mechanisms underlying the relationship between biodiversity and ecosystem function

Edited by

**NICO EISENHAUER**

*German Centre for Integrative Biodiversity Research (iDiv)
Halle-Jena-Leipzig; Institute of Biology,
Leipzig University, Leipzig, Germany*

**DAVID A. BOHAN**

*Directeur de Recherche
UMR Agroécologie, AgroSup Dijon
INRA, University of Bourgogne Franche-Comté
Dijon, France*

**ALEX J. DUMBRELL**

*School of Life Sciences
University of Essex
Wivenhoe Park, Colchester Essex,
United Kingdom*

ACADEMIC PRESS
An imprint of Elsevier

Academic Press is an imprint of Elsevier
125 London Wall, London, EC2Y 5AS, United Kingdom
The Boulevard, Langford Lane, Kidlington, Oxford OX5 1GB, United Kingdom
525 B Street, Suite 1650, San Diego, CA 92101, United States
50 Hampshire Street, 5th Floor, Cambridge, MA 02139, United States

First edition 2019

© 2019 Elsevier Ltd. All rights reserved.

No part of this publication may be reproduced or transmitted in any form or by any means, electronic or mechanical, including photocopying, recording, or any information storage and retrieval system, without permission in writing from the publisher. Details on how to seek permission, further information about the Publisher's permissions policies and our arrangements with organizations such as the Copyright Clearance Center and the Copyright Licensing Agency, can be found at our website: www.elsevier.com/permissions.

This book and the individual contributions contained in it are protected under copyright by the Publisher (other than as may be noted herein).

**Notices**
Knowledge and best practice in this field are constantly changing. As new research and experience broaden our understanding, changes in research methods, professional practices, or medical treatment may become necessary.

Practitioners and researchers must always rely on their own experience and knowledge in evaluating and using any information, methods, compounds, or experiments described herein. In using such information or methods they should be mindful of their own safety and the safety of others, including parties for whom they have a professional responsibility.

To the fullest extent of the law, neither the Publisher nor the authors, contributors, or editors, assume any liability for any injury and/or damage to persons or property as a matter of products liability, negligence or otherwise, or from any use or operation of any methods, products, instructions, or ideas contained in the material herein.

ISBN: 978-0-08-102912-1
ISSN: 0065-2504

For information on all Academic Press publications
visit our website at https://www.elsevier.com/books-and-journals

*Publisher:* Zoe Kruze
*Acquisition Editor:* Jason Mitchell
*Editorial Project Manager:* Joanna Collett
*Production Project Manager:* Abdulla Sait
*Cover Designer:* Mark Rogers

Typeset by SPi Global, India

# Contents

Contributors   xi
Preface: Mechanistic links between biodiversity and ecosystem functioning   xix

1. **A multitrophic perspective on biodiversity–ecosystem functioning research**   1
   Nico Eisenhauer, Holger Schielzeth, Andrew D. Barnes, Kathryn E. Barry, Aletta Bonn, Ulrich Brose, Helge Bruelheide, Nina Buchmann, François Buscot, Anne Ebeling, Olga Ferlian, Grégoire T. Freschet, Darren P. Giling, Stephan Hättenschwiler, Helmut Hillebrand, Jes Hines, Forest Isbell, Eva Koller-France, Birgitta König-Ries, Hans de Kroon, Sebastian T. Meyer, Alexandru Milcu, Jörg Müller, Charles A. Nock, Jana S. Petermann, Christiane Roscher, Christoph Scherber, Michael Scherer-Lorenzen, Bernhard Schmid, Stefan A. Schnitzer, Andreas Schuldt, Teja Tscharntke, Manfred Türke, Nicole M. van Dam, Fons van der Plas, Anja Vogel, Cameron Wagg, David A. Wardle, Alexandra Weigelt, Wolfgang W. Weisser, Christian Wirth, and Malte Jochum

   1. What are the key achievements of BEF research?   3
   2. What are the key challenges of future BEF research?   12
   3. Concluding remarks   33
   Acknowledgements   34
   References   34

2. **Above- and belowground overyielding are related at the community and species level in a grassland biodiversity experiment**   55
   Kathryn E. Barry, Alexandra Weigelt, Jasper van Ruijven, Hans de Kroon, Anne Ebeling, Nico Eisenhauer, Arthur Gessler, Janneke M. Ravenek, Michael Scherer-Lorenzen, Natalie J. Oram, Anja Vogel, Cameron Wagg, and Liesje Mommer

   1. Introduction   56
   2. Methods   60

|   |   |   |
|---|---|---|
| 3. Results | | 67 |
| 4. Discussion | | 77 |
| 5. Conclusions | | 83 |
| Authorship statement | | 83 |
| Acknowledgements | | 83 |
| References | | 84 |

## 3. Lost in trait space: species-poor communities are inflexible in properties that drive ecosystem functioning — 91

Anja Vogel, Peter Manning, Marc W. Cadotte, Jane Cowles, Forest Isbell, Alexandre L.C. Jousset, Kaitlin Kimmel, Sebastian T. Meyer, Peter B. Reich, Christiane Roscher, Michael Scherer-Lorenzen, David Tilman, Alexandra Weigelt, Alexandra J. Wright, Nico Eisenhauer, and Cameron Wagg

|   |   |
|---|---|
| 1. Introduction | 92 |
| 2. Methods | 97 |
| 3. Results | 104 |
| 4. Discussion | 116 |
| 5. Conclusions | 121 |
| Acknowledgements | 122 |
| References | 122 |

## 4. Terrestrial laser scanning reveals temporal changes in biodiversity mechanisms driving grassland productivity — 133

Claudia Guimarães-Steinicke, Alexandra Weigelt, Anne Ebeling, Nico Eisenhauer, Joaquín Duque-Lazo, Björn Reu, Christiane Roscher, Jens Schumacher, Cameron Wagg, and Christian Wirth

|   |   |
|---|---|
| 1. Introduction | 134 |
| 2. Material and methods | 138 |
| 3. Results | 144 |
| 4. Discussion | 149 |
| 5. Conclusions | 155 |
| Acknowledgements | 155 |
| References | 155 |

5. **Plant functional trait identity and diversity effects on soil meso- and macrofauna in an experimental grassland**    **163**

Rémy Beugnon, Katja Steinauer, Andrew D. Barnes, Anne Ebeling, Christiane Roscher, and Nico Eisenhauer

    1. Introduction    164
    2. Material and methods    167
    3. Results    171
    4. Discussion    174
    5. Conclusions    179
    Acknowledgements    180
    References    180
    Further reading    184

6. **How plant diversity impacts the coupled water, nutrient and carbon cycles**    **185**

Markus Lange, Eva Koller-France, Anke Hildebrandt, Yvonne Oelmann, Wolfgang Wilcke, and Gerd Gleixner

    1. Introduction    186
    2. Plant diversity effects on the soil microbial community and soil processes and functions    187
    3. Consequences of the element and water cycles and their coupling for the BEF relationships    207
    Acknowledgements    211
    References    211

7. **A new experimental approach to test why biodiversity effects strengthen as ecosystems age**    **221**

Anja Vogel, Anne Ebeling, Gerd Gleixner, Christiane Roscher, Stefan Scheu, Marcel Ciobanu, Eva Koller-France, Markus Lange, Alfred Lochner, Sebastian T. Meyer, Yvonne Oelmann, Wolfgang Wilcke, Bernhard Schmid, and Nico Eisenhauer

    1. Introduction    222
    2. Methods    227
    3. Results    236

| | |
|---|---:|
| 4. Discussion | 250 |
| 5. Conclusions | 256 |
| Acknowledgements | 256 |
| References | 257 |

## 8. Linking local species coexistence to ecosystem functioning: a conceptual framework from ecological first principles in grassland ecosystems     265

Kathryn E. Barry, Hans de Kroon, Peter Dietrich, W. Stanley Harpole, Anna Roeder, Bernhard Schmid, Adam T. Clark, Margaret M. Mayfield, Cameron Wagg, and Christiane Roscher

| | |
|---|---:|
| 1. Introduction | 266 |
| 2. Jointly emerging local coexistence and ecosystem functioning from ecological first principles | 269 |
| 3. Population level effects of abiotic and biotic conditions on fecundity, growth, and survival | 273 |
| 4. How ecological first principles influence trade-offs between fecundity, growth, and survival and in turn influence local coexistence and ecosystem functioning | 279 |
| 5. Conclusion | 285 |
| Author contributions | 286 |
| Acknowledgements | 286 |
| References | 286 |
| Further reading | 296 |

## 9. Mapping change in biodiversity and ecosystem function research: food webs foster integration of experiments and science policy     297

Jes Hines, Anne Ebeling, Andrew D. Barnes, Ulrich Brose, Christoph Scherber, Stefan Scheu, Teja Tscharntke, Wolfgang W. Weisser, Darren P. Giling, Alexandra M. Klein, and Nico Eisenhauer

| | |
|---|---:|
| 1. Topic networks as a way to visualize global conversation about biodiversity and ecosystem functioning | 298 |
| 2. Divisions among research domains: influences on food webs | 307 |
| 3. Summary and outlook: towards integrative food-web ecology | 312 |
| Acknowledgements | 314 |
| References | 315 |

## 10. Transferring biodiversity-ecosystem function research to the management of 'real-world' ecosystems 323

Peter Manning, Jacqueline Loos, Andrew D. Barnes, Péter Batáry,
Felix J.J.A. Bianchi, Nina Buchmann, Gerlinde B. De Deyn, Anne Ebeling,
Nico Eisenhauer, Markus Fischer, Jochen Fründ, Ingo Grass,
Johannes Isselstein, Malte Jochum, Alexandra M. Klein,
Esther O.F. Klingenberg, Douglas A. Landis, Jan Lepš, Regina Lindborg,
Sebastian T. Meyer, Vicky M. Temperton, Catrin Westphal,
and Teja Tscharntke

| | |
|---|---|
| 1. Introduction | 324 |
| 2. Small-grain and highly-controlled experiments (Cluster A) | 326 |
| 3. Small-grain studies with low experimental control (Cluster B) | 337 |
| 4. Large-grain studies without experimental control (Cluster C) | 340 |
| 5. Conclusion | 344 |
| Acknowledgements | 345 |
| References | 346 |

# Contributors

**Andrew D. Barnes**
German Centre for Integrative Biodiversity Research (iDiv) Halle-Jena-Leipzig; Institute of Biology, Leipzig University, Leipzig; Institute of Ecology and Evolution, Freidrich Schiller University Jena, Jena, Germany; School of Science, University of Waikato, Hamilton, New Zealand

**Kathryn E. Barry**
Systematic Botany and Functional Biodiversity, Institute of Biology, Leipzig University; German Centre for Integrative Biodiversity Research (iDiv) Halle-Jena-Leipzig, Leipzig, Germany

**Péter Batáry**
MTA Centre for Ecological Research, Institute of Ecology and Botany, Lendület Landscape and Conservation Ecology Research Group, Pest, Hungary

**Rémy Beugnon**
German Centre for Integrative Biodiversity Research (iDiv) Halle-Jena-Leipzig; Institute of Biology, Leipzig University, Leipzig, Germany

**Felix J.J.A. Bianchi**
Farming Systems Ecology, Wageningen University, Wageningen, Netherlands

**Aletta Bonn**
UFZ—Helmholtz Centre for Environmental Research, Soil Ecology Department, Halle (Saale); German Centre for Integrative Biodiversity Research (iDiv) Halle-Jena-Leipzig, Leipzig; Institute of Biodiversity, Friedrich Schiller University Jena, Jena, Germany

**Ulrich Brose**
German Centre for Integrative Biodiversity Research (iDiv) Halle-Jena-Leipzig, Leipzig; EcoNetLab, Institute of Biodiversity; Institute of Ecology and Evolution, Friedrich Schiller University Jena, Jena, Germany

**Helge Bruelheide**
German Centre for Integrative Biodiversity Research (iDiv) Halle-Jena-Leipzig, Leipzig; Institute of Biology/Geobotany and Botanical Garden, Martin Luther University Halle-Wittenberg, Halle (Saale), Germany

**Nina Buchmann**
Institute of Agricultural Sciences; Department of Environmental Systems Science, ETH Zürich, Zürich, Switzerland

**François Buscot**
UFZ—Helmholtz Centre for Environmental Research, Soil Ecology Department, Halle (Saale); German Centre for Integrative Biodiversity Research (iDiv) Halle-Jena-Leipzig, Leipzig, Germany

**Marc W. Cadotte**
Deptartment of Biological Sciences, University of Toronto Scarborough, Toronto, ON, Canada

**Marcel Ciobanu**
Institute of Biological Research, Branch of the National Institute of Research and Development for Biological Sciences, Cluj-Napoca, Romania

**Adam T. Clark**
German Centre for Integrative Biodiversity Research (iDiv) Halle-Jena-Leipzig; Synthesis Centre for Biodiversity Sciences (sDiv); Department of Physiological Diversity, Helmholtz Centre for Environmental Research (UFZ), Leipzig, Germany

**Jane Cowles**
Department of Ecology, Evolution, and Behavior, University of Minnesota, St. Paul, MN, United States

**Gerlinde B. De Deyn**
Soil Biology Group, Wageningen University, Wageningen, Netherlands

**Hans de Kroon**
Department of Experimental Plant Ecology, Institute for Water and Wetland Research, Radboud University, Nijmegen, The Netherlands

**Peter Dietrich**
Department of Physiological Diversity, Helmholtz Centre for Environmental Research (UFZ); German Centre for Integrative Biodiversity Research (iDiv) Halle-Jena-Leipzig, Leipzig, Germany

**Joaquín Duque-Lazo**
Department of Forestry, School of Agriculture and Forestry, University of Córdoba, Córdoba, Spain

**Anne Ebeling**
Institute of Ecology and Evolution, Friedrich Schiller University Jena, Jena, Germany

**Nico Eisenhauer**
German Centre for Integrative Biodiversity Research (iDiv) Halle-Jena-Leipzig; Institute of Biology, Leipzig University, Leipzig, Germany

**Olga Ferlian**
German Centre for Integrative Biodiversity Research (iDiv) Halle-Jena-Leipzig; Institute of Biology, Leipzig University, Leipzig, Germany

**Markus Fischer**
Institute of Plant Sciences, University of Bern, Bern, Switzerland

**Grégoire T. Freschet**
Centre d'Ecologie Fonctionnelle et Evolutive, UMR 5175 (CNRS—Université de Montpellier—Université Paul-Valéry Montpellier—EPHE), Montpellier, France

**Jochen Fründ**
Department of Biometry and Environmental System Analysis, Albert-Ludwigs-University Freiburg, Freiburg, Germany

**Arthur Gessler**
Institute for Landscape Biochemistry, Leibniz Centre for Agricultural Landscape Research (ZALF), Müncheberg; Berlin-Brandenburg Institute of Advanced Biodiversity Research (BBIB), Berlin, Germany; Swiss Federal Research Institute for Forest, Snow and Landscape Research WSL, Birmensdorf, Switzerland

**Darren P. Giling**
German Centre for Integrative Biodiversity Research (iDiv) Halle-Jena-Leipzig; Institute of Biology, Leipzig University, Leipzig; Institute of Ecology and Evolution, Friedrich Schiller University Jena, Jena, Germany

**Gerd Gleixner**
Department of Biogeochemical Processes, Max Planck Institute for Biogeochemistry, Jena, Germany

**Ingo Grass**
Department of Crop Sciences, Division of Agroecology, University of Göttingen, Göttingen, Germany

**Claudia Guimarães-Steinicke**
Systematic Botany and Functional Biodiversity, Institute of Biology, Leipzig University, Leipzig, Germany

**W. Stanley Harpole**
Department of Physiological Diversity, Helmholtz Centre for Environmental Research (UFZ); German Centre for Integrative Biodiversity Research (iDiv) Halle-Jena-Leipzig, Leipzig; Institute of Biology, Martin Luther University Halle-Wittenberg, Halle (Saale), Germany

**Stephan Hättenschwiler**
Centre d'Ecologie Fonctionnelle et Evolutive, UMR 5175 (CNRS—Université de Montpellier—Université Paul-Valéry Montpellier—EPHE), Montpellier, France

**Anke Hildebrandt**
Max Planck Institute for Biogeochemistry, Department of Biogeochemical Processes; Institute of Geosciences, Friedrich-Schiller-University Jena, Jena; UFZ-Helmholtz Centre for Environmental Research; German Centre for Integrative Biodiversity Research (iDiv) Halle-Jena-Leipzig, Leipzig, Germany

**Helmut Hillebrand**
German Centre for Integrative Biodiversity Research (iDiv) Halle-Jena-Leipzig, Leipzig; Institute for Chemistry and Biology of Marine Environments [ICBM], Carl-von-Ossietzky University Oldenburg, Wilhelmshaven; Helmholtz-Institute for Functional Marine Biodiversity at the University of Oldenburg (HIFMB), Oldenburg, Germany

**Jes Hines**
German Centre for Integrative Biodiversity Research (iDiv) Halle-Jena-Leipzig; Institute of Biology, Leipzig University, Leipzig, Germany

**Forest Isbell**
Department of Ecology, Evolution, and Behavior, University of Minnesota, St. Paul, MN, United States

**Johannes Isselstein**
Institute of Grassland Science, Georg-August-University Göttingen, Göttingen, Germany

**Malte Jochum**
German Centre for Integrative Biodiversity Research (iDiv) Halle-Jena-Leipzig; Institute of Biology, Leipzig University, Leipzig, Germany; Institute of Plant Sciences, University of Bern, Bern, Switzerland

**Alexandre L.C. Jousset**
Institute for Environmental Biology, Utrecht University, Utrecht, The Netherlands

**Kaitlin Kimmel**
Department of Ecology, Evolution, and Behavior, University of Minnesota, St. Paul, MN, United States

**Alexandra M. Klein**
Nature Conservation and Landscape Ecology, Albert-Ludwigs-University Freiburg; Faculty of Environment and Natural Resources, University of Freiburg, Freiburg, Germany

**Esther O.F. Klingenberg**
Department of Plant Ecology and Ecosystem Research, Georg-August University Göttingen, Göttingen, Germany

**Eva Koller-France**
Institute of Geography and Geoecology, Karlsruhe Institute of Technology (KIT), Karlsruhe; Geoecology, University of Tuebingen, Tuebingen, Germany

**Birgitta König-Ries**
Institute of Computer Science, Friedrich Schiller Universität Jena, Jena; German Centre for Integrative Biodiversity Research (iDiv) Halle-Jena-Leipzig, Leipzig, Germany

**Douglas A. Landis**
Department of Entomology and Great Lakes Bioenergy Research Center, 204 Center for Integrated Plant System, Michigan State University, East Lansing, MI, United States

**Markus Lange**
Department of Biogeochemical Processes, Max Planck Institute for Biogeochemistry, Jena, Germany

**Jan Lepš**
Department of Botany, Faculty of Science, University of South Bohemia, Ceske Budejovice, Czech Republic

**Regina Lindborg**
Deptartment of Physical Geography, Stockholm University, Stockholm, Sweden

**Alfred Lochner**
German Centre for Integrative Biodiversity Research (iDiv) Halle-Jena-Leipzig; Institute of Biology, Leipzig University, Leipzig, Germany

**Jacqueline Loos**
Department of Crop Sciences, Division of Agroecology, University of Göttingen, Göttingen; Faculty of Sustainability Science, Institute of Ecology, Leuphana University, Lüneburg, Germany

**Peter Manning**
Senckenberg Biodiversity and Climate Research Centre (SBiK-F), Frankfurt am Main, Germany

**Margaret M. Mayfield**
The University of Queensland, School of Biological Sciences, Brisbane, QLD, Australia

**Sebastian T. Meyer**
Department of Ecology and Ecosystem Management, Terrestrial Ecology Research Group, School of Life Sciences Weihenstephan, Technical University of Munich, Freising, Germany

**Alexandru Milcu**
Centre d'Ecologie Fonctionnelle et Evolutive, UMR 5175 (CNRS—Université de Montpellier—Université Paul-Valéry Montpellier—EPHE), Montpellier; Ecotron Européen de Montpellier, Centre National de la Recherche Scientifique (CNRS), Montferrier-sur-Lez, France

**Liesje Mommer**
Plant Ecology and Nature Conservation Group, Wageningen University, Wageningen, The Netherlands

**Jörg Müller**
Field Station Fabrikschleichach, Department of Animal Ecology and Tropical Biology, Biocenter, University of Würzburg, Rauhenebrach; Bavarian Forest National Park, Grafenau, Germany

**Charles A. Nock**
Geobotany, Faculty of Biology, University of Freiburg, Freiburg, Germany; Department of Renewable Resources, University of Alberta, Edmonton, AB, Canada

**Yvonne Oelmann**
Geoecology, University of Tübingen, Tübingen, Germany

**Natalie J. Oram**
Plant Ecology and Nature Conservation Group, Wageningen University, Wageningen, The Netherlands

**Jana S. Petermann**
Department of Biosciences, University of Salzburg, Salzburg, Austria

**Janneke M. Ravenek**
Department of Experimental Plant Ecology, Institute for Water and Wetland Research, Radboud University, Nijmegen, The Netherlands

**Peter B. Reich**
Department of Forest Resources, University of Minnesota, St. Paul, MN, United States

**Björn Reu**
School of Biology, Industrial University of Santander, Bucaramanga, Colombia

**Anna Roeder**
Department of Physiological Diversity, Helmholtz Centre for Environmental Research (UFZ); German Centre for Integrative Biodiversity Research (iDiv) Halle-Jena-Leipzig, Leipzig, Germany

**Christiane Roscher**
Department of Physiological Diversity, Helmholtz Centre for Environmental Research (UFZ); German Centre for Integrative Biodiversity Research (iDiv) Halle-Jena-Leipzig, Leipzig, Germany

**Christoph Scherber**
Institute of Landscape Ecology, University of Münster, Münster, Germany

**Michael Scherer-Lorenzen**
Institute of Biology/Geobotany, University of Freiburg, Freiburg, Germany

**Stefan Scheu**
J.F. Blumenbach Institute of Zoology and Anthropology, University of Göttingen, Göttingen, Germany

**Holger Schielzeth**
Institute of Ecology and Evolution, Friedrich Schiller University Jena; German Centre for Integrative Biodiversity Research (iDiv) Halle-Jena-Leipzig, Jena, Germany

**Bernhard Schmid**
Department of Geography, University of Zürich, Zürich, Switzerland; Institute of Ecology, College of Urban and Environmental Sciences, Peking University, Beijing, China

**Stefan A. Schnitzer**
Department of Biology, Marquette University, Milwaukee, WI, United States

**Andreas Schuldt**
Forest Nature Conservation, Faculty of Forest Sciences and Forest Ecology, University of Göttingen, Göttingen, Germany

**Jens Schumacher**
Institute of Mathematics, Friedrich Schiller University Jena, Jena, Germany

**Katja Steinauer**
German Centre for Integrative Biodiversity Research (iDiv) Halle-Jena-Leipzig; Institute of Biology, Leipzig University, Leipzig, Germany; Department of Terrestrial Ecology, Netherlands Institute of Ecology, Wageningen, The Netherlands

**Vicky M. Temperton**
Faculty of Sustainability Science, Institute of Ecology, Leuphana University, Lüneburg, Germany

**David Tilman**
Department of Ecology, Evolution, and Behavior, University of Minnesota, St. Paul, MN, United States

**Teja Tscharntke**
Department of Crop Sciences, Division of Agroecology; Centre of Biodiversity and Sustainable Land Use (CBL), University of Göttingen, Göttingen, Germany

**Manfred Türke**
German Centre for Integrative Biodiversity Research (iDiv) Halle-Jena-Leipzig; Institute of Biology, Leipzig University, Leipzig; Institute of Biological and Medical Imaging (IBMI), Helmholtz Zentrum München (HMGU)—German Research Center for Environmental Health, Neuherberg, Germany

**Nicole M. van Dam**
German Centre for Integrative Biodiversity Research (iDiv) Halle-Jena-Leipzig, Leipzig; Institute of Biodiversity, Friedrich Schiller University Jena, Jena, Germany

**Jasper van Ruijven**
Plant Ecology and Nature Conservation Group, Wageningen University, Wageningen, The Netherlands

**Fons van der Plas**
Institute of Biology, Leipzig University, Leipzig, Germany

**Anja Vogel**
Institute of Biology, Leipzig University; German Centre for Integrative Biodiversity Research (iDiv) Halle-Jena-Leipzig, Leipzig; Institute of Ecology and Evolution, Friedrich Schiller University Jena, Jena, Germany

**Cameron Wagg**
Fredericton Research and Development Centre, Agriculture and Agri-Food Canada, Fredericton, NB, Canada; Department of Evolutionary Biology and Environmental Studies, University of Zürich, Zürich, Switzerland

**David A. Wardle**
Asian School of the Environment, Nanyang Technological University, Singapore, Singapore

**Alexandra Weigelt**
Systematic Botany and Functional Biodiversity, Institute of Biology, Leipzig University; German Centre for Integrative Biodiversity Research (iDiv) Halle-Jena-Leipzig, Leipzig, Germany

**Wolfgang W. Weisser**
Department of Ecology and Ecosystem Management, Terrestrial Ecology Research Group, School of Life Sciences Weihenstephan, Technical University of Munich, Freising, Germany

**Catrin Westphal**
Functional Agrobiodiversity, Department of Crop Sciences, University of Göttingen, Göttingen, Germany

**Wolfgang Wilcke**
Institute of Geography and Geoecology, Karlsruhe Institute of Technology (KIT), Karlsruhe, Germany

**Christian Wirth**
Systematic Botany and Functional Biodiversity, Institute of Biology, Leipzig University; German Centre for Integrative Biodiversity Research (iDiv) Halle-Jena-Leipzig, Leipzig; Max Planck Institute for Biogeochemistry, Jena, Germany

**Alexandra J. Wright**
Department of Biological Sciences, California State University Los Angeles, Los Angeles, CA, United States

# Preface: Mechanistic links between biodiversity and ecosystem functioning

Biodiversity–ecosystem functioning (BEF) research is currently one of the most vibrant research areas in Ecology, as ecosystem functioning is critical for human well-being, and the time available to make decisions for preserving global biodiversity is short. Since the first experiments in the early 1990s, BEF research has evolved, as the knowledge gained has inspired new questions, ideas, and approaches. By combining case studies, data syntheses, reviews, and perspectives papers, this special issue on the *Mechanisms underlying the relationship between biodiversity and ecosystem functioning* provides an overview of the key achievements, current research frontiers, and future perspectives in BEF research, with a strong focus on terrestrial, mostly grassland, ecosystems. Moreover, this special issue is mostly based on research in the framework of the Jena Experiment, a large-scale and long-term grassland BEF experiment. Since its establishment in 2002, the Jena Experiment has focused on grassland biodiversity effects on multitrophic interactions and carbon and nutrient cycling and has utilized trait-based approaches to understand the mechanisms underlying BEF relationships. At the same time, researchers of the Jena Experiment have synthesized data and information from other experiments and natural ecosystems—this plethora of approaches is reflected by the diversity of chapters of this special issue. The first chapter provides a short history of the key advances and future challenges of BEF research. The subsequent chapters fall into two broad categories: Chapters 2–6 cover case studies, reviews, and syntheses that explore different mechanisms underlying BEF relationships; and Chapters 7–10 review current BEF knowledge to provide recommendations for future research. Together, this special issue represents a synthesis of the *status quo* in terrestrial BEF research as well as stimulating perspectives of the most pressing challenges and important future research directions in this field.

In Chapter 1, Eisenhauer et al. (2019a) review the key achievements of hundreds of experiments that have manipulated biodiversity as an independent variable and present a multitrophic, eco-evolutionary perspective on future biodiversity–ecosystem functioning research. The authors summarize the history of research that inspired BEF experiments. For instance, one important step was to appreciate that biodiversity is not only a dependent variable of the environmental context, but that biodiversity itself is important for the functioning of ecosystems. Based on the extensive scientific debate, the design and focus of BEF experiments changed over time, and the authors explain how this field of research became more integrative in terms of scientific disciplines. As a result, in addition to a general exploration of BEF relationships, the underlying mechanisms came into focus. A whole-ecosystem perspective allowed an exploration of the implications of BEF relationships for ecosystem services, contributing to the establishment of and discussions in The Intergovernmental Science-Policy Platform on Biodiversity and Ecosystem Services (IPBES, 2018). In this chapter, Eisenhauer and colleagues argue that a multitrophic and eco-evolutionary perspective of biotic interactions, in random- and non-random biodiversity change scenarios, will be key to further advances in the field. They discuss how multitrophic interactions and their eco-evolutionary underpinnings will help to understand the context dependency of BEF relationships and to scale up BEF to management-relevant spatial scales across ecosystem boundaries, potentially allowing implementation of this knowledge for ecosystem management, society, and policy.

In order to better understand BEF relationships, the key role of plants needs to be explored, such as competition for above- and belowground resources. In Chapter 2, Barry et al. (2019b) present the results of a case study that explores plant species richness effects on above- and belowground plant productivity in the Trait-Based Experiment (Ebeling et al., 2014) in the Jena Experiment (Roscher et al., 2004). Specifically, the authors explore potential relationships between above- and belowground overyielding at the community and species level. Barry et al. (2019b) find support for positive correlations between above- and belowground overyielding at the community level and at the species level for 8 out of 13 investigated species. Notably, plants tended to invest more in overyielding aboveground than belowground, which leads to the conclusion that aboveground competition (i.e. for light) might be stronger than for belowground resources like nutrients. Moreover, these results indicate that the competitive environment changes along the plant diversity gradient, which may lead to differential biomass allocation into plant organs and potentially different selection pressures (Tilman and Snell-Rood, 2014; Zuppinger-Dingley et al., 2014). Related changes in plant traits over time might contribute to increasing biodiversity effects in long-term experiments (Eisenhauer et al., 2019a, Chapter 1; Tilman and Snell-Rood, 2014).

Another mechanism underlying positive BEF relationships may be that high-diversity plant communities have a higher capacity to adapt to environmental changes through shifts in dominance structure and community composition. In Chapter 3, Vogel et al. (2019b) synthesized data from five long-term biodiversity experiments in grassland. These experiments manipulated different drivers of environmental change related to stress (e.g. drought) and nutrient availability (e.g. nitrogen addition). Vogel et al. (2019b, Chapter 3) used species-specific plant traits to calculate functional diversity and determined the phylogenetic diversity of the plant communities. Overall, high-diversity plant communities indeed showed a greater capacity to shift in functional trait space over time, supporting the notion that they are more flexible at adapting to different environments (Isbell et al., 2017) and maintaining high and stable levels of ecosystem functioning (Isbell et al., 2015).

Empirical evidence suggests that short-term variation of plant productivity between seasons is important for understanding the stability of ecosystem functions (Wagg et al., 2017), and that studying the fine temporal-scale dynamics of ecosystem processes might provide additional insights into species interactions and biodiversity effects. In Chapter 4, Guimarães-Steinicke et al. (2019) used terrestrial laser scanning to study the effects of different plant diversity facets on an approximated measure of plant growth every 2 weeks across two growing seasons in the Trait-Based Experiment (Ebeling et al., 2014). Terrestrial laser scanning was found to be an appropriate method to quantify plant growth in a nondestructive way. Moreover, they report that different diversity facets varied in their explanatory power for plant productivity across the seasons, with functional identity being of higher importance after disturbance (e.g. mowing) and functional dispersion being of higher significance when competition for light was high (e.g., close to peak biomass). These results indicate that in order to understand how plant interactions contribute to BEF relationships, the temporal dynamics of interactions and activity patterns need to be explored in detail, as the functions of plant interactions can range between facilitation and competition depending on the environmental context, such as along the plant diversity gradient (Wright et al., 2017).

In the same experiment, Beugnon et al. (2019) report in Chapter 5 the community responses of soil animals to variations in plant species richness, functional diversity, and the community-weighted means of different plant traits related to spatial and temporal resource use (Ebeling et al., 2014). While significant plant diversity effects on soil organisms have been reported before (e.g., Eisenhauer et al., 2013; Scherber et al., 2010), the underlying mechanisms are not well understood, and exploring the role of plant traits may be a promising approach to gain better insight (Eisenhauer and Powell, 2017; Laliberté, 2016). Beugnon et al. (2019, Chapter 5) classified different groups of soil fauna by considering their body size (meso- and macrofauna) and feeding strategy (prey [i.e. mostly detritivores and herbivores] and predators). Plant traits had a higher predictive power than plant species richness in explaining variation in soil animal abundances and diversity. However, the significance of different traits varied across the soil animal groups, contributing further evidence to the working hypothesis that different plant diversity facets are key to driving consumer communities (Schuldt et al., 2019), ecosystem processes (Guimarães-Steinicke et al., 2019, Chapter 4), and their stability (Craven et al., 2018).

In Chapter 6, Lange et al. (2019) summarized recent findings for the effects of plant diversity on nutrient, carbon, and water cycles in the Jena Experiment. In line with previous studies (e.g. Fornara and Tilman, 2008; Lange et al., 2015), carbon and nitrogen stocks accumulated over the course of the experiment, and this increase was more pronounced at higher than at lower plant diversity. However, they observed the opposite temporal trend for nitrogen and phosphorous concentrations in soil solution, an effect that they relate to the constant removal of plant biomass from the experimental plots during harvest. Moreover, Lange et al. (2019, Chapter 6) discuss the important role of soil microbial communities in soil nutrient dynamics, which may primarily be driven by rhizodeposition (Lange et al., 2015). This effect of root-derived inputs was found to be more pronounced at high plant diversity, with effects also reaching down to deeper soil layers. The authors conclude that tightening interactions between plants and soil microbes over time may not only be important for altered nutrient cycling and stocks along the plant diversity gradient but also influence other ecosystem functions, such as plant productivity.

Soil feedback effects on plant productivity will be studied in the new $\Delta$BEF Experiment, which is introduced by Vogel et al. (2019a) in Chapter 7. This experiment was inspired by observations of a strengthening biodiversity effect on ecosystem functioning over time (e.g., Guerrero-Ramírez et al., 2017; Reich et al., 2012). Vogel et al. (2019a, Chapter 7) present the rationale and design of the experiment to explore soil history effects on ecosystem functioning and to separate plant diversity-induced gradual changes in abiotic and abiotic conditions from environmental effects. This will be achieved by comparing soil treatments, with and without plant community-specific history, to the long-term plots of the Jena

Experiment, set up in 2002 (Roscher et al., 2004). The first results show significant differences among the experimental treatments for different ecosystem functions like plant productivity and soil nematodes. Differences in soil conditions and communities are likely to have even stronger feedback effects on multiple ecosystem functions in the future (Lange et al., 2019, Chapter 6). However, the authors note that future studies should also comanipulate plant and soil history in order to investigate any potential interactions.

In Chapter 8, Barry et al. (2019a) present a unifying framework based on ecological first principles of resources, and abiotic and biotic conditions. The authors review the ways in which ecosystem functioning may be linked to coexistence in plant communities and the implications for BEF relationships. They explore population-level effects of resources and abiotic and biotic conditions on fecundity, growth, and survival of plants, and how the ecological first principles influence trade-offs among fecundity, growth, and survival, which in turn affect plant coexistence and ecosystem functioning. This combination of theory and empirical results exemplifies how complementary theories from otherwise unconnected research fields can be brought together to study the reciprocal relationships between different coexistence processes and ecosystem functioning. While a simple mapping of biological processes that promote coexistence to those generating BEF relationships might not be feasible (Turnbull et al., 2016), coexistence theory can help understanding the context dependency of BEF relationships (Eisenhauer et al., 2019a, Chapter 1) and developing working hypotheses for future experiments.

In Chapter 9, Hines et al. (2019) use a keyword co-occurrence analysis to study the past three decades of BEF research. They document the rapid growth of the field and define four core research domains that have emerged over time. Notably, the authors also identify elements that connect associated research domains, such as the integrative domain of food web research that link BEF experiments to science policy. Given the important role of plant–consumer interactions in multiple ecosystem functions and that many consumer species at higher trophic levels of food webs may be particularly sensitive to environmental change (Eisenhauer et al., 2019a, Chapter 1; Hines et al., 2015), there is concern that plant–consumer interactions will suffer from biodiversity declines (Dirzo et al., 2014; Eisenhauer et al., 2019a). That multitrophic biodiversity research integrates different BEF research domains suggests that food web research must have a high research priority in the future.

Despite these now obvious links between multitrophic biodiversity and ecosystem functioning and the recent establishment of IPBES, knowledge from BEF research has, to date, had little influence on policy and the management of 'real-world' ecosystems in agriculture and managed grasslands (Eisenhauer et al., 2019a, Chapter 1). In Chapter 10, Manning et al. (2019) try to bridge this gap by classifying BEF research into three clusters by considering the level of human control over species composition and the spatial scale of the BEF studies. They suggest how the research of each cluster can inform particular fields of ecosystem management, discuss the barriers to and challenges of knowledge transfer, and present means to overcome these challenges. Notably, 1:1 comparisons between BEF experiments and real-world ecosystems can be misleading (Wardle, 2016), but important mechanisms have been identified that play a role in real-world ecosystems and that have advanced our understanding of ecosystem responses to species gains and losses (Eisenhauer et al., 2016; Isbell et al., 2017). Manning et al. (2019, Chapter 10) recommend an intensification of transdisciplinary research that considers the social–ecological context of the ecosystems in order to address challenges of sustainable land use. They conclude that the social and economic value of biodiversity for ecosystem services needs to be recognized and communicated to land managers and policy makers. The clustering approach may be a first critical step towards operationalizing BEF insights for ecosystem management, society, and policy (Eisenhauer et al., 2019a, Chapter 1).

Taken together, the chapters of this special issue show that there is no one single mechanism that can explain all BEF relationships, but that multiple, nonmutually exclusive mechanisms are likely to link the multitrophic biodiversity of ecosystems to the multiple functions provided by ecosystems (Fig. 1; e.g. Eisenhauer et al., 2019a, Chapter 1; Hines et al., 2019, Chapter 9). The field has moved away from a purely plant-centred perspective that mainly focused on resource use complementarity and competition. The list of potential BEF mechanisms presented here is not supposed to be comprehensive. Yet, the chapters of this special issue suggest that in addition to studying competitive interactions (Barry et al., 2019b, Chapter 2) and the consequences for and feedback effects of plant coexistence (Barry et al., 2019a, Chapter 8), changes in soil abiotic (Lange et al., 2019, Chapter 6) and biotic conditions (Eisenhauer et al., 2019a, Chapter 1; Vogel et al., 2019a, Chapter 7), multitrophic interactions in complex food webs (Eisenhauer et al., 2019a, Chapter 1; Hines et al., 2019, Chapter 9), and changes in plant traits and their composition (Beugnon et al., 2019, Chapter 5; Vogel et al., 2019b, Chapter 3) are likely to contribute to different biodiversity facets driving the functioning of

**Fig. 1** Schematic illustration of researchers with different expertise exploring several nonmutually exclusive mechanisms underlying BEF relationships (as indicated by the multiple ends of the Gordian knot). Depicted are broad categories of mechanisms, including mutualism of plants with multitrophic interaction partners above and below the ground (top left), above- and belowground complementarity in habitat space use (bottom left), the role of plant pathogens (top right), and plant community-induced changes in nutrient dynamics (bottom right). Apart from testing the different hypotheses presented in this special issue, a future challenge will be to integrate the respective mechanisms in order to understand in which contexts they may prevail. Ultimately, by addressing and integrating these different hypotheses, i.e., cutting the Gordian knot, BEF research will become a predictive science.

ecosystems (Beugnon et al., 2019, Chapter 5; Guimarães-Steinicke et al., 2019, Chapter 4) (Fig. 1). Accordingly, complementary future research directions are proposed to develop a more comprehensive understanding of the mechanisms underlying BEF relationships across environmental contexts as well as spatial and temporal scales. Ultimately, such knowledge will be integral to information for land managers and policy makers about implications of BEF research for the sustainable management of ecosystems (Eisenhauer et al., 2019a, Chapter 1; Manning et al., 2019, Chapter 10).

NICO EISENHAUER
German Centre for Integrative Biodiversity
Research (iDiv) Halle-Jena-Leipzig, Leipzig, Germany
Institute of Biology, Leipzig University, Leipzig, Germany

DAVID A. BOHAN
Directeur de Recherche, UMR Agroécologie, AgroSup Dijon, INRA,
University of Bourgogne Franche-Comté Dijon, France

ALEX J. DUMBRELL
School of Life Sciences, University of Essex, Colchester, United Kingdom

## Acknowledgements

The Jena Experiment is funded by the Deutsche Forschungsgemeinschaft (DFG, German Research Foundation; FOR 1451), with additional support by the Friedrich Schiller University Jena, the Max Planck Institute for Biogeochemistry, and the Swiss National Science Foundation (SNF). N.E. acknowledges the support by the German Centre for Integrative Biodiversity Research Halle–Jena–Leipzig, funded by the German Research Foundation (FZT 118).

## References

Barry, K.E., de Kroon, H., Dietrich, P., Harpole, W.S., Roeder, A., Schmid, B., Clark, A.T., Mayfield, M.M., Wagg, C., Roscher, C., 2019a. Linking local species coexistence to ecosystem functioning: a conceptual framework from ecological first principles in grassland ecosystems. Adv. Ecol. Res. 61, 265–296.

Barry, K.E., Weigelt, A., van Ruijven, J., de Kroon, H., Ebeling, A., Eisenhauer, N., Gessler, A., Ravenek, J.M., Scherer-Lorenzen, M., Oram, N.J., Vogel, A., Wagg, C., Mommer, L., 2019b. Above- and belowground overyielding are related at the community and species level in a grassland biodiversity experiment. Adv. Ecol. Res. 61, 55–89.

Beugnon, R., Steinauer, K., Barnes, A.D, Ebeling, A., Roscher, C., Eisenhauer, N., 2019. Plant functional trait identity and diversity effects on soil meso- and macrofauna in an experimental grassland. Adv. Ecol. Res. 61, 163–184.

Craven, D., Eisenhauer, N., et al., 2018. Multiple facets of biodiversity drive the diversity–stability relationship. Nat. Ecol. Evol. 2, 1579–1587. https://doi.org/10.1038/s41559-018-0647-7.

Dirzo, R., Young, H.S., Galetti, M., Ceballos, G., Isaac, N.J.B., Collen, B., 2014. Defaunation in the anthropocene. Science 345, 401–406.

Ebeling, A., Pompe, S., Baade, J., Eisenhauer, N., Hillebrand, H., Proulx, R., Roscher, C., Schmid, B., Wirth, C., Weisser, W.W., 2014. A trait-based experimental approach to understand the mechanisms underlying biodiversity-ecosystem functioning relationships. Basic Appl. Ecol. 15, 229–240.

Eisenhauer, N., Powell, J.R., 2017. Plant trait effects on soil organisms and functions. Pedobiologia 65, 1–4.

Eisenhauer, N., Dobies, T., Cesarz, S., Hobbie, S.E., Meyer, R.J., Worm, K., Reich, P.B., 2013. Plant diversity effects on soil food webs are stronger than those of elevated $CO_2$ and N deposition in a long-term grassland experiment. Proc. Natl. Acad. Sci. U. S. A. 110, 6889–6894.

Eisenhauer, N., Barnes, A.D., Cesarz, S., Craven, D., Ferlian, O., Gottschall, F., Hines, J., Sendek, A., Siebert, J., Thakur, M.P., Türke, M., 2016. Biodiversity–ecosystem function experiments reveal the mechanisms underlying the consequences of biodiversity change in real world ecosystems. J. Veg. Sci. 27, 1061–1070.

Eisenhauer, N., Schielzeth, H., Barnes, A.D., Barry, K.E., Bonn, A., Brose, U., Bruelheide, H., Buchmann, N., Buscot, F., Ebeling, A., Ferlian, O., Freschet, G.T., Giling, D.P., Hättenschwiler, S., Hillebrand, H., Hines, J., Isbell, F., Koller-France, E., König-Ries, B., de Kroon, H., Meyer, S.T., Milcu, A., Müller, J., Nock, C.A., Petermann, J.S., Roscher, C., Scherber, C., Scherer-Lorenzen, M., Schmid, B., Schnitzer, S.A., Schuldt, A., Tscharntke, T., Türke, M., van Dam, N.M., van der Plas, F., Vogel, A., Wagg, C., Wardle, D.A., Weigelt, A., Weisser, W.W., Wirth, C., Jochum, M., 2019a. A multitrophic perspective on biodiversity–ecosystem functioning research. Adv. Ecol. Res. 61, 1–54.

Fornara, D.A., Tilman, D., 2008. Plant functional composition influences rates of soil carbon and nitrogen accumulation. J. Ecol. 96, 314–322.

Guerrero-Ramírez, N.R., Craven, D., Reich, P.B., Ewel, J.J., Koricheva, J., Parrotta, J.A., Auge, H., Erickson, H.E., Forrester, D.I., Hector, A., Joshi, J., Montagnini, F., 2017. Diversity-dependent temporal divergence of ecosystem functioning in experimental ecosystems. Nat. Ecol. Evol. 1, 1639–1642.

Guimarães-Steinicke, C., Weigelt, A., Ebeling, A., Eisenhauer, N., Duque-Lazo, J., Reu, B., Roscher, C., Schumacher, J., Wagg, C., Wirth, C., 2019. Terrestrial laser scanning reveals temporal changes in biodiversity mechanisms driving grassland productivity. Adv. Ecol. Res. 61, 133–161.

Hines, J., van der Putten, W.H., De Deyn, G.B., Wagg, C., Voigt, W., Mulder, C., Weisser, W.W., Engel, J., Melian, C., Scheu, S., Birkhofer, K., Ebeling, A., et al., 2015. Towards an integration of biodiversity-ecosystem functioning and food web theory to evaluate relationships between multiple ecosystem services. Adv. Ecol. Res. 53, 161–199.

Hines, J., Ebeling, A., Barnes, A.D., Brose, U., Scherber, C., Scheu, S., Tscharntke, T., Weisser, W.W., Giling, D.P., Klein, A.M., Eisenhauer, N., 2019. Mapping change in biodiversity and ecosystem function research: food webs foster integration of experiments and science policy. Adv. Ecol. Res. 61, 297–322.

IPBES, 2018. Summary for policymakers of the regional assessment report on biodiversity and ecosystem services for Europe and Central Asia of the Intergovernmental Science-Policy Platform on Biodiversity and Ecosystem Services. In: Fischer, M., Rounsevell, M., Torre-Marin Rando, A., Mader, A., Church, A., Elbakidze, M., Elias, V., Hahn, T., Harrison, P.A., Hauck, J., Martín-López, B., Ring, I., Sandström, C., Sousa Pinto, I., Visconti, P., Zimmermann, N.E., Christie, M. (Eds.), IPBES secretariat. Bonn, Germany, p. 48 pages.

Isbell, F., Craven, D., Connolly, J., Loreau, M., Schmid, B., Beierkuhnlein, C., Bezemer, T.M., Bonin, C., Bruelheide, H., De Luca, E., Ebeling, A., et al., 2015. Biodiversity increases the resistance of ecosystem productivity to climate extremes. Nature 526, 574–577.

Isbell, F., Adler, P.R., Eisenhauer, N., Fornara, D., Kimmel, K., Kremen, C., Letourneau, D.K., Liebman, M., Polley, H.W., Quijas, S., Scherer-Lorenzen, M., 2017. Benefits of increasing plant diversity in sustainable agroecosystems. J. Ecol. 105, 871–879.

Laliberté, E., 2016. Below-ground frontiers in trait-based plant ecology. New Phytol. 213, 1597–1603.

Lange, M., Eisenhauer, N., Sierra, C.A., Bessler, H., Engels, C., Griffiths, R.I., Mellado-Vázquez, P.G., Malik, A.A., Roy, J., Scheu, S., Steinbeiss, S., Thomson, B.C., et al., 2015. Plant diversity increases soil microbial activity and soil carbon storage. Nat. Commun. 6, 6707.

Lange, M., Koller-France, E., Hildebrandt, A., Oelmann, Y., Wilcke, W., Gleixner, G., 2019. How plant diversity impacts the coupled water, nutrient and carbon cycles. Adv. Ecol. Res. 61, 185–219.

Manning, P., Loos, J., Barnes, A.D., Batàry, P., Bianchi, F.J.J.A., Buchmann, N., De Deyn, G.B., Ebeling, A., Eisenhauer, N., Fischer, M., Fründ, J., Grass, I., Isselstein, J., Jochum, M., Klein, A.M., Klingenberg, E.O.F., Landis, D.A., Lepš, J., Lindborg, R., Meyer, S.T., Temperton, V.M., Westphal, C., Tscharntke, T., 2019. Transferring biodiversity-ecosystem function research to the management of 'real-world' ecosystems. Adv. Ecol. Res. 61, 323–356.

Reich, P.B., Tilman, D., Isbell, F., Mueller, K., Hobbie, S.E., Flynn, D.F.B., Eisenhauer, N., 2012. Impacts of biodiversity loss escalate through time as redundancy fades. Science 336, 589–592.

Roscher, C., Schumacher, J., Baade, J., Wilcke, W., Gleixner, G., Weisser, W.W., Schmid, B., Schulze, E.D., 2004. The role of biodiversity for element cycling and trophic interactions: an experimental approach in a grassland community. Basic Appl. Ecol. 5, 107–121.

Scherber, C., Eisenhauer, N., Weisser, W.W., Schmid, B., Voigt, W., Fischer, M., Schulze, E.D., Roscher, C., Weigelt, A., Allan, E., Beler, H., Bonkowski, M., et al., 2010. Bottom-up effects of plant diversity on multitrophic interactions in a biodiversity experiment. Nature 468, 553–556.

Schuldt, A., Ebeling, A., Kunz, M., Staab, M., Guimarães-Steinicke, C., Bachmann, D., Buchmann, N., Durka, W., Fichtner, A., Fornoff, F., Härdtle, W., Hertzog, L., et al., 2019. Multiple plant diversity components drive consumer communities across ecosystems. Nat. Commun. 10, 1460.

Tilman, D., Snell-Rood, E.C., 2014. Diversity breeds complementarity. Nature 515, 44.

Turnbull, L.A., Isbell, F., Purves, D.W., Loreau, M., Hector, A., 2016. Understanding the value of plant diversity for ecosystem functioning through niche theory. Proc. R. Soc. B 283, 20160536.

Vogel, A., Ebeling, A., Gleixner, G., Roscher, C., Scheu, S., Ciobanu, M., Koller-France, E., Lange, M., Lochner, A., Meyer, S.T., Oelmann, Y., Wilcke, W., Schmid, B., Eisenhauer, N., 2019a. A new experimental approach to test why biodiversity effects strengthen as ecosystems age. Adv. Ecol. Res. 61, 221–264.

Vogel, A., Manning, P., Cadotte, M.W., Cowles, J., Isbell, F., Jousset, A.L.C., Kimmel, K., Meyer, S.T., Reich, P.B., Roscher, C., Scherer-Lorenzen, M., Tilman, D., Weigelt, A., Wright, A.J., Eisenhauer, N., Wagg, C., 2019b. Lost in trait space: species-poor communities are inflexible in properties that drive ecosystem functioning. Adv. Ecol. Res. 61, 91–131.

Wagg, C., O'Brien, M.J., Vogel, A., Scherer-Lorenzen, M., Eisenhauer, N., Schmid, B., Weigelt, A., 2017. Plant diversity maintains long-term ecosystem productivity under frequent drought by increasing short-term variation. Ecology 98, 2952–2961.

Wardle, D.A., 2016. Do experiments exploring plant diversity-ecosystem functioning relationships inform how biodiversity loss impacts natural ecosystems? J. Veg. Sci. 27, 646–653.

Wright, A.J., Wardle, D.A., Callaway, R., Gaxiola, A., 2017. The overlooked role of facilitation in biodiversity experiments. Trends Ecol. Evol. 32, 383–390.

Zuppinger-Dingley, D., Schmid, B., Petermann, J.S., Yadav, V., De Deyn, G.B., Flynn, D.F.B., 2014. Selection for niche differentiation in plant communities increases biodiversity effects. Nature 515, 108–111.

## Further reading

Eisenhauer, N., Bonn, A., Guerra, C.A., 2019b. Recognizing the quiet extinction of invertebrates. Nat. Commun. 10, 50.

Wilkinson, M.D., Dumontier, M., Aalbersberg, I.J., Appleton, G., Axton, M., Baak, A., et al., 2016. The FAIR Guiding Principles for scientific data management and stewardship. Sci. Data 3, 160018.

CHAPTER ONE

# A multitrophic perspective on biodiversity–ecosystem functioning research

Nico Eisenhauer[a,b,*], Holger Schielzeth[a,c], Andrew D. Barnes[a,b], Kathryn E. Barry[a,b], Aletta Bonn[a,g,ac], Ulrich Brose[a,d], Helge Bruelheide[a,e], Nina Buchmann[f], François Buscot[a,g], Anne Ebeling[c], Olga Ferlian[a,b], Grégoire T. Freschet[h], Darren P. Giling[a,b,c], Stephan Hättenschwiler[h], Helmut Hillebrand[a,i,ah], Jes Hines[a,b], Forest Isbell[j], Eva Koller-France[k], Birgitta König-Ries[a,l], Hans de Kroon[m], Sebastian T. Meyer[n], Alexandru Milcu[h,o], Jörg Müller[p,q], Charles A. Nock[r,s], Jana S. Petermann[t], Christiane Roscher[a,u], Christoph Scherber[v], Michael Scherer-Lorenzen[r], Bernhard Schmid[w], Stefan A. Schnitzer[x], Andreas Schuldt[y], Teja Tscharntke[z,aa], Manfred Türke[a,b,ab], Nicole M. van Dam[a,ac], Fons van der Plas[b], Anja Vogel[a,b,c], Cameron Wagg[ad,ae], David A. Wardle[af], Alexandra Weigelt[a,b], Wolfgang W. Weisser[n], Christian Wirth[a], Malte Jochum[a,b,ag]

[a]German Centre for Integrative Biodiversity Research (iDiv) Halle-Jena-Leipzig, Leipzig, Germany
[b]Institute of Biology, Leipzig University, Leipzig, Germany
[c]Institute of Ecology and Evolution, Friedrich Schiller University Jena, Jena, Germany
[d]EcoNetLab, Institute of Biodiversity, Friedrich Schiller University Jena, Jena, Germany
[e]Institute of Biology/Geobotany and Botanical Garden, Martin Luther University Halle-Wittenberg, Halle (Saale), Germany
[f]Institute of Agricultural Sciences, ETH Zürich, Zürich, Switzerland
[g]UFZ—Helmholtz Centre for Environmental Research, Soil Ecology Department, Halle (Saale), Germany
[h]Centre d'Ecologie Fonctionnelle et Evolutive, UMR 5175 (CNRS—Université de Montpellier—Université Paul-Valéry Montpellier—EPHE), Montpellier, France
[i]Institute for Chemistry and Biology of Marine Environments [ICBM], Carl-von-Ossietzky University Oldenburg, Wilhelmshaven, Germany
[j]Department of Ecology, Evolution and Behavior, University of Minnesota, St. Paul, MN, United States
[k]Karlsruher Institut für Technologie (KIT), Institut für Geographie und Geoökologie, Karlsruhe, Germany
[l]Institute of Computer Science, Friedrich Schiller Universität Jena, Jena, Germany
[m]Radboud University, Institute for Water and Wetland Research, Animal Ecology and Physiology & Experimental Plant Ecology, Nijmegen, The Netherlands
[n]Terrestrial Ecology Research Group, Technical University of Munich, School of Life Sciences Weihenstephan, Freising, Germany
[o]Ecotron Européen de Montpellier, Centre National de la Recherche Scientifique (CNRS), Montferrier-sur-Lez, France
[p]Field Station Fabrikschleichach, Department of Animal Ecology and Tropical Biology, Biocenter, University of Würzburg, Rauhenebrach, Germany
[q]Bavarian Forest National Park, Grafenau, Germany
[r]Geobotany, Faculty of Biology, University of Freiburg, Freiburg, Germany
[s]Department of Renewable Resources, University of Alberta, Edmonton, AB, Canada

[t]Department of Biosciences, University of Salzburg, Salzburg, Austria
[u]UFZ—Helmholtz Centre for Environmental Research, Department Physiological Diversity, Leipzig, Germany
[v]Institute of Landscape Ecology, University of Münster, Münster, Germany
[w]Department of Geography, University of Zürich, Zürich, Switzerland
[x]Department of Biology, Marquette University, Milwaukee, WI, United States
[y]Forest Nature Conservation, Faculty of Forest Sciences and Forest Ecology, University of Göttingen, Göttingen, Germany
[z]Department of Crop Sciences, Division of Agroecology, University of Göttingen, Göttingen, Germany
[aa]Centre of Biodiversity and Sustainable Land Use (CBL), University of Göttingen, Göttingen, Germany
[ab]Institute of Biological and Medical Imaging (IBMI), Helmholtz Zentrum München (HMGU)—German Research Center for Environmental Health, Neuherberg, Germany
[ac]Institute of Biodiversity, Friedrich Schiller University Jena, Jena, Germany
[ad]Fredericton Research and Development Centre, Agriculture and Agri-Food Canada, Fredericton, NB, Canada
[ae]Department of Evolutionary Biology and Environmental Studies, University of Zürich, Zürich, Switzerland
[af]Asian School of the Environment, Nanyang Technological University, Singapore, Singapore
[ag]Institute of Plant Sciences, University of Bern, Bern, Switzerland
[ah]Helmholtz-Institute for Functional Marine Biodiversity at the University of Oldenburg (HIFMB), Oldenburg, Germany
*Corresponding author: e-mail address: nico.eisenhauer@idiv.de

## Contents

1. What are the key achievements of BEF research? 3
   1.1 A short history of BEF research 5
   1.2 A new BEF era provides novel insights 8
   1.3 Identification of BEF mechanisms 9
   1.4 BEF in multitrophic communities 10
   1.5 BEF implications for ecosystem services 11
2. What are the key challenges of future BEF research? 12
   2.1 Non-random biodiversity change across trophic levels 13
   2.2 Predicting the strength of BEF relationships across environmental contexts 19
   2.3 Spatial scaling of BEF relationships 23
   2.4 Eco-evolutionary implications of multitrophic BEF 26
   2.5 FAIR data and beyond 29
   2.6 Operationalizing BEF insights for ecosystem management, society, and decision making 31
3. Concluding remarks 33
Acknowledgements 34
References 34

## Abstract

Concern about the functional consequences of unprecedented loss in biodiversity has prompted biodiversity–ecosystem functioning (BEF) research to become one of the most active fields of ecological research in the past 25 years. Hundreds of experiments have manipulated biodiversity as an independent variable and found compelling support that the functioning of ecosystems increases with the diversity of their ecological communities. This research has also identified some of the mechanisms underlying BEF relationships, some context-dependencies of the strength of relationships, as well as implications for various ecosystem services that humankind depends upon. In this chapter, we argue that a multitrophic perspective of biotic interactions in random and non-random biodiversity change scenarios is key to advance future BEF research and to address some of its most important remaining challenges. We discuss that the study and the quantification of multitrophic interactions in space and time facilitates scaling up from small-scale biodiversity manipulations and ecosystem function assessments to management-relevant spatial scales across ecosystem boundaries. We specifically consider multitrophic conceptual frameworks to understand and predict the context-dependency of BEF relationships. Moreover, we highlight the importance of the eco-evolutionary underpinnings of multitrophic BEF relationships. We outline that FAIR data (meeting the standards of findability, accessibility, interoperability, and reusability) and reproducible processing will be key to advance this field of research by making it more integrative. Finally, we show how these BEF insights may be implemented for ecosystem management, society, and policy. Given that human well-being critically depends on the multiple services provided by diverse, multitrophic communities, integrating the approaches of evolutionary ecology, community ecology, and ecosystem ecology in future BEF research will be key to refine conservation targets and develop sustainable management strategies.

## 1. What are the key achievements of BEF research?

*"The community is indeed the hierarchical level where the basic characteristics of life—its diversity, complexity, and historical nature—are perhaps the most daunting and challenging. [...] however, most of the theoretical insights that have been gained about the effects of biodiversity on ecosystem functioning come from approaches developed in community ecology."*

**Loreau (2010)**

Human activities influence virtually all ecosystems around the globe through a large variety of environmental alterations (MEA, 2005). Habitat destruction (Maxwell et al., 2016), changing and intensified land use (Gossner et al., 2016; Newbold et al., 2015), climate change (Urban et al., 2016), and invasion of exotic species (Murphy and Romanuk, 2014; van Kleunen et al., 2015; Vitousek et al., 1997; Wardle et al., 2011) are some of the most significant drivers of biodiversity change (Maxwell et al., 2016). Subsequent changes in ecological communities raise substantial ethical and aesthetic concerns as well as questions regarding the functioning of altered ecosystems (Hooper et al., 2005; Isbell et al., 2017a; Naeem et al., 2012). Biodiversity–ecosystem functioning (BEF) research has revealed strong positive effects of biodiversity on various ecosystem functions, and has linked these effects to underlying mechanisms. Positive BEF relationships can be observed at different spatial (Cardinale et al., 2012; Hautier et al., 2018; Isbell et al., 2011; Roscher et al., 2005; Thompson et al., 2018; van der Plas et al., 2016a,b) and temporal scales (Guerrero-Ramírez et al., 2017; Reich et al., 2012; Zavaleta et al., 2010), and can be multi-dimensional on both the predictor (i.e., multidiversity) and response side (multifunctionality) (e.g., Hector and Bagchi, 2007; Meyer et al., 2018; Schuldt et al., 2018; Soliveres et al., 2016a). Accordingly, one of the most important conclusions of BEF research is that the strength of BEF relationships is strongly context-dependent. BEF relationships have been shown to depend on climatic conditions (Maestre et al., 2012; Ratcliffe et al., 2017), local site conditions (Allan et al., 2015; Eisenhauer et al., 2018; Fridley, 2002; Reich et al., 2001), and disturbance and management regimes (Guerrero-Ramírez et al., 2017; Kardol et al., 2018; Weigelt et al., 2009), which interact with biodiversity (Guerrero-Ramírez and Eisenhauer, 2017; but see Craven et al., 2016). Accordingly, mechanisms underlying biodiversity effects have been found to differ from one community to the next. Before discussing how the *status quo* can inspire future research to address some of the most important challenges in BEF research and ecology in general, we provide an overview of key achievements of past BEF work.

This chapter is based on a survey among researchers in the Jena Experiment, of a workshop on the "Future of BEF research" organized in the framework of the Jena Experiment, and of the German Centre for Integrative Biodiversity Research (iDiv) and thus has a bias towards BEF research in terrestrial ecosystems and in controlled experiments. Rather than proving a comprehensive picture of all important research directions in BEF and how these directions may have developed since past reviews (e.g., Cardinale et al., 2012; Hillebrand and Matthiessen, 2009; Hooper et al., 2005; Loreau et al., 2001; Naeem et al., 2012; Scherer-Lorenzen, 2014; Tilman et al., 2014; van der Plas, 2019; Weisser et al., 2017), we focus on the key aspects that materialized from the survey. In December 2016, all researchers were asked to answer the following two questions:
- What are the key achievements of past BEF research?
- What are the key challenges/topics of future BEF research? Where should the field move?

Contributions were synthesized by N.E. and discussed at the "Future of BEF research"-workshop in Jena, Germany in February 2017. As an outcome, we highlight six priority areas of future BEF research, namely non-random biodiversity change across trophic levels; predicting the strength of BEF relationships across environmental contexts; spatial scaling of BEF relationships; eco-evolutionary implications of multitrophic BEF; FAIR data and reproducible processing; and operationalizing BEF insights for ecosystem management, society, and decision making.

## 1.1 A short history of BEF research

Prior to the era of BEF research, nature conservation efforts targeted biodiversity separately from ecosystem functioning. On the one hand, the goal of conservation was to prevent species extinctions (Mace, 2014). On the other hand, ecosystems were protected and managed to conserve and maximize their functions and services (such as forests for groundwater recharge, erosion control, or recreation), but without explicit consideration of their diversity (Costanza et al., 1997). Conservation had mostly been ethically motivated, while BEF research moved the argument to take a utilitarian view of biodiversity to convince target groups like politicians and land managers. Although, there was a consensus that it was important to protect different species and certain functions, these aims were, and still are in many places of the world, regarded as poorly connected, as well as insufficiently linked to ecological theory. BEF research helped to provide an empirical underpinning for these inherently related objects, thus adding an important justification for conserving biodiversity that went beyond ethical and aesthetic motivations (Dallimer et al., 2012; Potthast, 2014).

Early observations of natural communities inspired the notion that biodiversity may be a key determinant of the functioning of ecosystems (Darwin and Wallace, 1858; Elton, 1958; McNaughton, 1977; Schulze and Mooney, 1994). This idea was supported by theoretical models (Loreau, 1998; Tilman et al., 1997a; Yachi and Loreau, 1999) and experiments (Hooper et al., 2005; O'Connor et al., 2017). In fact, over the past 25 years, BEF research has led us to recognize that the identity and combinations of species are powerful drivers of ecosystem processes (Hooper et al., 2005; Isbell et al., 2017a; Schulze and Mooney, 1994; Tilman et al., 2014; Weisser et al., 2017).

More specifically, prior to the mid-1990s, ecologists focused more on abiotic factors driving variation in biodiversity, such as geology and climate, than biotic factors, such as species diversity and species interactions (Hobbie, 1992). Early topical questions were related to the environmental determinants of biodiversity (Fig. 1; van der Plas, 2019). The search for

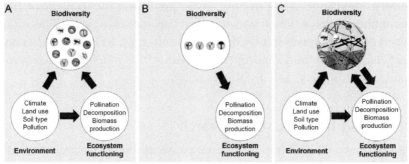

**Fig. 1** The evolution of biodiversity research. Main foci of biodiversity–ecosystem functioning research over time (Chapin et al., 2000; De Laender et al., 2016; Eisenhauer et al., 2016; Isbell et al., 2013; van der Plas, 2019). While studying example environmental drivers of different facets of biodiversity and ecosystem functioning has been an important subdiscipline in ecological research for many decades (i.e., community ecology) (A), in the mid-1990s, researchers started to manipulate biodiversity (mostly at the producer level; mostly random biodiversity loss scenarios) as an independent variable (functional biodiversity research or BEF research) (B). More recently, ecologists started focusing on the complex interplay between anthropogenically driven environmental gradients, non-random biodiversity change across trophic levels in food webs (C) (see also Fig. 2), and the consequences for ecosystem function (e.g., Barnes et al., 2018; De Laender et al., 2016; Hines et al., 2019; Mori et al., 2013; Sobral et al., 2017; Soliveres et al., 2016a) (C). *Figure modified after van der Plas, F., 2019.*

answers to these fundamental questions in biodiversity yielded major scientific achievements, such as Darwin's theory of evolution (Darwin, 1859), Hutchinson's concept of the ecological niche (Hutchinson, 1957), and MacArthur and Wilson's theory of island biogeography (MacArthur and Wilson, 1967; summarized in Craven et al., 2019). Still today, the exploration of the determinants of biodiversity is a crucial field in ecology (e.g., Adler et al., 2011), which is important to some of the most pressing challenges of humankind, particularly given the unprecedented rate of anthropogenic environmental change.

While the importance of species diversity to ecosystem functioning was recognized more than 150 years ago; e.g., Darwin and Wallace (1858) stated "… it has been experimentally shown that a plot of land will yield a greater weight if sown with several species and genera of grasses, than if sown with only two or three species", this recognition of the importance of biodiversity took a back seat (Hector and Hooper, 2002). In fact, one of the first experiments of the 20th century reporting on BEF relationships was originally designed to study how different concentrations of nitrogen drive plant diversity (Tilman and Downing, 1994). However, when these plant communities were unexpectedly hit by an extreme drought, it became evident that the response to the extreme event and the stability of the ecosystem function "plant productivity" depended on the species richness of the community (Tilman and Downing, 1994). This study showed a positive biodiversity-stability relationship, but was criticized because it did not manipulate biodiversity as an independent factor, meaning that stability of plant biomass production was likely (co-)determined by the nitrogen treatment (Givnish, 1994; Huston, 1997). After the first "wave" of scientific debate, Grime (1997) concluded that "…neither evolutionary theory nor empirical studies have presented convincing evidence that species diversity and ecosystem function are consistently and causally connected."

This debate stimulated a series of controlled experiments that directly manipulated biodiversity aiming to quantify the effect of plant species richness on ecosystem functioning under controlled environmental conditions (e.g., Díaz et al., 2003; Ebeling et al., 2014; Hector et al., 1999; Hooper et al., 2005; Naeem et al., 1994; Niklaus et al., 2001; O'Connor et al., 2017; Roscher et al., 2004; Tilman et al., 1997b; Wardle and Zackrisson, 2005; Fig. 1). The results were surprisingly clear: community biomass production, in particular, increased with an increasing number of plant species (Hooper et al., 2005). Subsequent debates (e.g., Eisenhauer et al., 2016; Wardle, 2016) and adjustments of experimental designs stimulated the collection of evidence that BEF relationships could occur irrespective of the inclusion of certain species, functional groups, or combinations of species (Eisenhauer et al., 2016; Huang et al., 2018; van Ruijven and Berendse, 2003; Wilsey and Polley, 2004).

The focus on the manipulation of plant diversity and productivity, however, led to calls, and actions, to study a wider range of taxa and functions. Subsequently, BEF research became more integrative in terms of scientific disciplines by realizing that a whole-ecosystem perspective, including, e.g., multitrophic interactions and element cycles, is required to explore the mechanistic underpinnings and implications of biodiversity change (Roscher et al., 2004; Schuldt et al., 2018). Nonetheless, these experiments have also provoked debate over their realism. Randomly-assembled communities may not mirror real-world assembly and disassembly (Lepš, 2004; Wardle, 2016), which are determined by the simultaneous interplay of abiotic and biotic filters in time and space (Götzenberger et al., 2012). Some recent experiments thus shifted their focus from the number of species to the functional and phylogenetic dissimilarity of species assemblages (Cadotte, 2013; Dias et al., 2013; Ebeling et al., 2014; Ferlian et al., 2018; Scherer-Lorenzen et al., 2007) or have implemented non-random biodiversity loss scenarios (e.g., Bracken et al., 2008; Bruelheide et al., 2014; Schläpfer et al., 2005).

Non-random changes in biodiversity and the notion that the strength of BEF relationships is context-dependent (Baert et al., 2018; Guerrero-Ramírez et al., 2017; Ratcliffe et al., 2017) have led contemporary BEF research to re-introduce non-random and indirect manipulations of biodiversity using environmental change drivers, such as various climate variables, management intensity, chemical pollutants, and nutrient enrichment, as well as observations along environmental gradients (De Laender et al., 2016; Everwand et al., 2014; Grace et al., 2016; Isbell et al., 2013; Fig. 1). Although empirical evidence is limited to date, the findings of, e.g., Duffy et al. (2017) and Isbell et al. (2013) substantiate the general predictions from BEF experiments by demonstrating that the repeatedly-reported discrepancies in results between experimental and real-world BEF studies may, in fact, be due to multiple interacting or unrecognized drivers typically operating in real-world systems (De Laender et al., 2016; Eisenhauer et al., 2016; Loreau, 1998).

## 1.2 A new BEF era provides novel insights

In the last ~10 years, multiple review papers on BEF relationships have comprehensively summarized the major achievements and novel insights by BEF research (e.g., Balvanera et al., 2006; Cardinale et al., 2012; Dirzo et al., 2014; Hooper et al., 2005; Isbell et al., 2017a; Loreau et al., 2001; Naeem et al., 2012; Scherer-Lorenzen, 2014; Tilman et al., 2014; van der Plas, 2019; Weisser et al., 2017). Briefly, this research has shown the importance of biodiversity (from microorganisms to trees, but mostly of primary producers) in driving the functioning of ecosystems, with functions ranging from very specific ones, such as the molecular transformation of organic compounds, to highly integrated ones, such as primary productivity. Positive BEF relationships arise from phenotypically- and genetically-based differences or trade-offs in species characteristics that drive the evolutionary diversification of niches (and the niches created by other species) through selective pressures, such that there is no single species or few species that perform(s) the different functions in exactly the same way or contribute(s) to all of the different functions (Turnbull et al., 2016). Consequently, it has been shown that the conservation of species diversity is necessary to sustain long-term functioning (Guerrero-Ramírez et al., 2017; Meyer et al., 2016; Reich et al., 2012) and multi-functionality of ecosystems (Allan et al., 2013; Hector and Bagchi, 2007; Isbell et al., 2011; Lefcheck et al., 2015; Meyer et al., 2018; Schuldt et al., 2018).

While BEF research has mostly focused on uncovering the links between species richness and ecosystem function, showing that some particular species or functional groups have a disproportionately strong contribution to BEF relationships, variation at different levels of ecological organization (genetic diversity, phylogenetic species diversity, functional diversity) can have comparable effects on ecosystem functioning (e.g., Hughes et al., 2008). In contrast to earlier assumptions (Cardinale et al., 2011), there seems to be low functional redundancy of coexisting species (Reich et al., 2012), particularly so across environmental contexts (Isbell et al., 2011), and therefore, at larger spatial scales that may cover more different environmental conditions (Isbell et al., 2017a). Thus, there is increasing awareness of the mechanistic links between traits involved in coexistence and resource use and traits affecting emerging properties and processes in ecosystems (Bannar-Martin et al., 2018; Chesson et al., 2001; Mori et al., 2018; Mouquet et al., 2002; Turnbull et al., 2013, 2016); although empirical evidence for the role of response and effect traits in ecosystem functioning still is limited (e.g., Beugnon et al., 2019; Paine et al., 2015; Yang et al., 2018).

## 1.3 Identification of BEF mechanisms

BEF research has identified a list of (non-mutually exclusive) mechanisms that contribute to enhancing ecosystem functioning with increasing biodiversity (e.g., increased biotope space describing the number of different ecological niches, more efficient resource use, multitrophic interactions, facilitation; Hooper et al., 2005; Weisser et al., 2017; reviewed by Barry et al., 2019a). Mathematical approaches and experimental treatments were established to disentangle different facets of biodiversity effects (e.g., complementarity effect, selection effect, and species asynchrony; Fox, 2005; de Mazancourt et al., 2013; Isbell et al., 2018; Loreau and Hector, 2001). More recent research has provided insights into niche dynamics. This means that species' realized niches change over time according to their competitive environment and their interaction network that are both dynamic in time and space (Hofstetter et al., 2007). As a consequence, this might lead to increasing biodiversity effects on certain ecosystem functions over time

(Allan et al., 2011; Huang et al., 2018; Isbell et al., 2011; Lange et al., 2019; Meyer et al., 2016; Reich et al., 2012; Zuppinger-Dingley et al., 2014).

Previous studies, particularly short-term studies, may have underestimated the strength of biodiversity–ecosystem functioning relationships by missing these longer-term effects (Eisenhauer et al., 2012; Finn et al., 2013; Schmid et al., 2008). Among those is the important finding of strengthening complementarity effects (calculated based on Loreau and Hector, 2001) of species-rich communities over time (Cardinale et al., 2007; Huang et al., 2018; Reich et al., 2012; but see Kardol et al., 2018). These complementarity effects may be driven by several underlying mechanisms. For example, at low biodiversity, negative density-dependent effects of pests and pathogens may contribute to the deterioration of community functions in comparison to more diverse communities (Eisenhauer et al., 2012; Guerrero-Ramírez et al., 2017; Maron et al., 2011; Schnitzer et al., 2011; Schuldt et al., 2017b; Weisser et al., 2017). In contrast, species-rich communities may support more mutualistic interactions (e.g., Schuldt et al., 2017b; Wright et al., 2014), which may increase ecosystem functioning over time (Eisenhauer et al., 2012). These two mechanisms are not mutually exclusive (Guerrero-Ramírez et al., 2017), and different ecosystem functions show varying relative importance of the two mechanisms at the same time (Meyer et al., 2016). Despite these first promising insights into potential explanations of complementarity effects, the underlying ecological and evolutionary mechanisms remain elusive.

## 1.4 BEF in multitrophic communities

BEF research has demonstrated that biodiversity change at one trophic level cascades to other trophic levels. For example, plant diversity increases the diversity of above- and belowground consumer communities ("biodiversity begets biodiversity"; e.g., Ebeling et al., 2018; Eisenhauer et al., 2013; Haddad et al., 2009; Hines et al., 2019; Scherber et al., 2010; Thebault and Loreau, 2003), and independent biodiversity changes at more than one trophic level interactively affect ecosystem functions (e.g., Coulis et al., 2015; Eisenhauer et al., 2012; Gessner et al., 2010; Handa et al., 2014). Relatedly, it has been shown that complex, multitrophic communities affect the relationship between biodiversity and multiple ecosystem functions (Naeem et al., 1994; Schuldt et al., 2018; Soliveres et al., 2016a; van der Heijden et al., 1998; Wang et al., 2019). For instance, across a land-use intensity gradient in German grasslands, the diversity of primary producers, herbivorous insects, and microbial decomposers were particularly important predictors of plant biomass and forage quality (Soliveres et al., 2016a). For Chinese subtropical forests, it was shown that individual ecosystem functions central to energy and nutrient flows across trophic levels are more strongly related to the diversity of heterotrophs promoting decomposition and nutrient cycling, and affected by plant functional-trait diversity and composition, than by tree species richness (Schuldt et al., 2018). In managed Inner Mongolian grasslands, diversifying livestock by mixing both sheep and cattle promoted multidiversity (including the diversity of plants, insects, soil microbes, and nematodes) and multifunctionality (including plant biomass, insect abundance, nutrient cycling, and soil carbon) (Wang et al., 2019).

Perspectives papers have suggested to integrate BEF- and food-web theory to advance the understanding of causal relationships between complex communities and multiple ecosystem functions (Barnes et al., 2018; Duffy et al., 2007; Hines et al., 2015b, 2019; Thompson et al., 2012). Moreover, multitrophic interactions may play a decisive role in shaping BEF relationships *via* diversity-induced species plasticity in physiology, morphology, and micro-evolutionary processes (Mraja et al., 2011; Zuppinger-Dingley et al., 2014). However, even though one of the first biodiversity experiments manipulated multitrophic biodiversity in terrestrial ecotrons (Naeem et al., 1994), multitrophic BEF research in terrestrial ecosystems is still in its infancy, and the majority of existing studies focus on aquatic systems (Lefcheck et al., 2015; O'Connor et al., 2017; Seibold et al., 2018; Stachowicz et al., 2007, 2008a).

## 1.5 BEF implications for ecosystem services

Beyond its focus on ecosystem functioning, BEF research has also shown that biodiversity is important for a wide range of potential ecosystem services (Allan et al., 2015; Balvanera et al., 2006, 2014; Cardinale et al., 2012; Isbell et al., 2017a,b). These include provisioning, regulating, and also cultural services, underpinned by supporting services and includes, e.g., forage production (Binder et al., 2018; Finn et al., 2013), wood production (Isbell et al., 2017b), soil carbon storage for climate regulation (Fornara and Tilman, 2008; Lange et al., 2015), soil erosion control (Berendse et al., 2015; Pérès et al., 2013), water quality regulation (Scherer-Lorenzen et al., 2003), natural attenuation of pollutants in soil (Bandowe et al., 2019), pollination (Ebeling et al., 2008), and pest control (Hertzog et al., 2017) or herbivory reduction (Civitello et al., 2015; Schuldt et al., 2017b).

Moreover, BEF research has stressed the role of multifunctionality, including the simultaneous provisioning of many functions at one location (e.g., Lefcheck et al., 2015; Schuldt et al., 2018) and across environmental contexts (Eisenhauer et al., 2018; Isbell et al., 2015a), as well as single functions in different settings (Isbell et al., 2011). However, this research has also highlighted that biodiversity does not necessarily enhance all ecosystem functions at the same time (Cardinale et al., 2012; van der Plas et al., 2016a,b), and trade-offs have been observed among different functions (Allan et al., 2015; Meyer et al., 2018). Moreover, studies simultaneously exploring a range of functions remain scarce, poorly represent the whole range of services provided by ecosystems, and are often disconnected from the utilitarian value of the (agro-)ecosystem (Manning et al., 2018; Swift et al., 2004; van der Plas et al., 2018). Nonetheless, these assessments of multifunctional ecosystems represent first important steps towards operationalizing BEF insights for society and policy makers (Manning et al., 2018) and will help to incorporate the importance of biodiversity for ecosystem-service provision in political discussions around the globe (including, e.g., halting biodiversity loss is included among sustainable development goals, changes to the European Common Agricultural Policy; IPBES reports, https://www.ipbes.net/).

## 2. What are the key challenges of future BEF research?

*"The central problem in understanding and measuring biological diversity is that we still have a lot of work to do. And while we are taking inventory, the shelves are already being cleared."*

**Christian Wirth (2013)**

Congruent to the statement above, biodiversity research is a field under time pressure. Biodiversity change can alter the functioning of ecosystems in dramatic ways and at an unprecedented pace, which will have important consequences for the provision of ecosystem services (Balvanera et al., 2006; Cardinale et al., 2012) and human health (Civitello et al., 2015; Lozupone et al., 2012; Wall et al., 2015). Some of the related key challenges of BEF research have been described in previous review papers (e.g., Cardinale et al., 2012; Hooper et al., 2005; Isbell et al., 2017a), and the plethora of (meta-)studies and mechanistic insights that were derived in the last years has helped to refine existing and ask novel questions in BEF research. Here, we argue that taking a multitrophic (Eisenhauer, 2017; Seibold et al., 2018) and eco-evolutionary perspective (Tilman and Snell-Rood, 2014; Zuppinger-Dingley et al., 2014) of biotic interactions will advance this field of research by identifying previously unknown mechanisms. Despite the broad consensus on the significance of BEF relationships, the underlying ecological and evolutionary mechanisms are not well understood, which impedes the transition from a description of patterns to a predictive science. Importantly, the focus should now not only be on generalizable patterns, but more on the context-dependency of BEF relationships (Baert et al., 2018; Craven et al., 2016; Eisenhauer et al., 2018; Fridley, 2002; Guerrero-Ramírez et al., 2017; Jousset et al., 2011; Kardol et al., 2018; Ratcliffe et al., 2017; Schuldt et al., 2017a). Understanding why and how the strength of biodiversity effects varies with environmental conditions and at which spatial scales different mechanisms operate will be key to operationalizing BEF insights for ecosystem management, society, and decision making. We will discuss these research frontiers in the following sections.

## 2.1 Non-random biodiversity change across trophic levels

*"What escapes the eye... is a much more insidious kind of extinction: the extinction of ecological interactions."*

*Janzen (1974)*

Real-world biodiversity change (both invasions and extinctions) can be highly non-random (Haddad et al., 2008; Wardle, 2016). Thus, future BEF research has to investigate how non-random biodiversity loss affects ecosystem functioning in real-world ecosystems (Isbell et al., 2017a,b). Addressing this question is particularly important in order to facilitate the application of BEF results to agriculture, forestry, and biodiversity conservation. At the same time, this is a very challenging task as biodiversity change and species turnover may be hard to predict due to multiple co-occurring and interacting global-change drivers (Scherber, 2015; Tylianakis et al., 2008) and their context-dependent effects on species and their interactions (Bowler et al., 2018; Schmid and Hector, 2004). Global change experiments, particularly those that manipulate multiple global change drivers, may be particularly valuable to study biodiversity changes and subsequent ecosystem responses (Giling et al., 2019; Vogel et al., 2019a). Furthermore, it might be promising to look more closely into the many published studies using random extinction scenarios, as some of them might by chance resemble extinction patterns that are actually observed in nature and provide opportunities for re-analysis (Manning et al., 2019). At the same time, the existing literature needs to be synthesized to derive a better understanding of trait-specific extinction risks of different taxonomic groups (Cardillo et al., 2005; Kotiaho et al., 2005; Seibold et al., 2015).

Another aspect of non-random species loss that has attracted increasing scientific attention over the last years is the role of rare species for community functioning. Across ecosystems, the large majority of species are rare and thus prone to extinction (Jousset et al., 2017; Soliveres et al., 2016b; Suding et al., 2005). In contrast to the mass-ratio hypothesis, which assumes that locally abundant species drive ecosystem functioning (Grime, 1998), many studies have shown that rare species can have disproportionately strong impacts on ecosystems (Allan et al., 2013; Connolly et al., 2013; Klein et al., 2003; Lyons et al., 2005; Mouillot et al., 2013; Soliveres et al., 2016b). Future experiments thus need to investigate the role of rare species and their interactions with common species, and compare "real-world," non-random extinction scenarios with random extinction scenarios. Such an experiment was, for example, established in the subtropical BEF-China experiment, where two non-random extinction scenarios were included: one is based on local rarity and one on specific leaf area (SLA) of tree species, mimicking habitat loss through fragmentation and climate change, two current and likely future key drivers of change in Chinese subtropical forest communities (Bruelheide et al., 2014).

Both high trophic level and high body mass have been associated with vulnerability to extinction (with many related traits; Fig. 2; e.g., Dirzo et al., 2014; Voigt et al., 2003), but vulnerability to environmental change occurs at all trophic levels according to species' life history traits. Thus, the focus of previous terrestrial BEF experiments on manipulating the primary producer level does not necessarily reflect that this is the most vulnerable trophic level to environmental change. Although early BEF research already considered multiple trophic levels (e.g., Naeem and Li, 1997; Naeem et al., 1994), the understanding of how multitrophic communities change their diversity and how this affects their functioning in terrestrial ecosystems remains limited (Eisenhauer et al., 2013; Haddad et al., 2009; Scherber et al., 2010). Moreover, terrestrial BEF research so far has virtually neglected the fact that primary producers do not function in isolation, but in a complex network of multitrophic, and also non-trophic interactions (Fig. 2; Duffy, 2002; Hines et al., 2015b; Seabloom et al., 2017; Sobral et al., 2017; Tiede et al., 2016).

In complex food webs, the magnitude or rates of different ecosystem functions are tightly coupled to the community size structure describing how the body masses of species and individuals are distributed across trophic levels (Brose et al., 2017; Dossena et al., 2012). For instance, subtle shifts in the body mass structure of top consumer populations can induce strong trophic cascades with pronounced effects on primary production

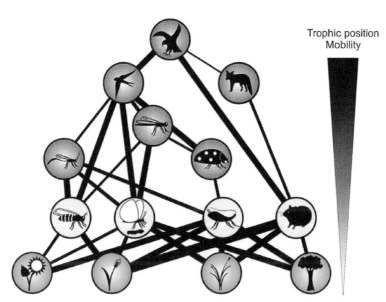

**Fig. 2** A multitrophic perspective on biodiversity–ecosystem functioning research. Mobility tends to increase with increasing trophic position in ecological networks, and some work suggests that the vulnerability to environmental change does so, too (Hines et al., 2015a; Voigt et al., 2003), although species at all trophic levels may be vulnerable to changing environments based on their specific life-history traits. This means that the previous focus of BEF experiments on the primary producer level does not necessarily reflect that this is the most vulnerable trophic level to environmental change. This simple aboveground food web serves as the basis for other figures in this chapter. It illustrates that species within complex communities are connected by trophic links that can represent ecosystem functions and services (see also Fig. 3); although not shown here, the same concept applies to belowground food webs and ecosystem functions.

(Jochum et al., 2012). Consistently, analyses of complex food-web models demonstrated that primary production may be more tightly coupled to the trophic level and body mass of the top consumer than to total or plant diversity (Wang and Brose, 2018). Thus, ecological networks are an important tool that can be used to evaluate links that drive trade-offs between multiple ecosystem functions (Fig. 3; Brose et al., 2017; Hines et al., 2015b).

Across ecosystems, there is strong empirical evidence that the diversity at higher trophic levels is important for providing multiple ecosystem functions and services (Barnes et al., 2018; Bruno et al., 2006, 2008; Gessner et al., 2010; Hines et al., 2015b; Lefcheck et al., 2015; Schneider et al., 2012, 2016; Schuldt et al., 2018; Soliveres et al., 2016a; Wang et al., 2019). This was, for example, shown by manipulating stream-living macroinvertebrates and investigating their effect on decomposition (Cardinale et al., 2002; Handa

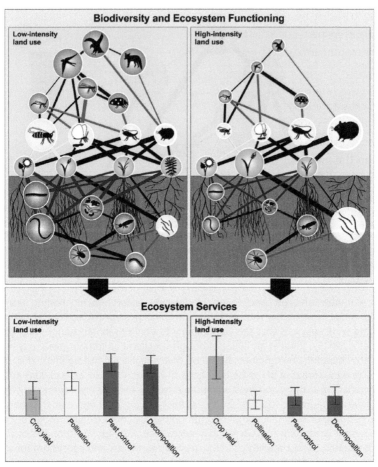

**Fig. 3** Multitrophic communities drive ecosystem multifunctionality. This scheme depicts relationships between the diversity of species in aboveground-belowground networks and the management of multiple ecosystem services across adjacent agricultural ecosystems. Management decisions, such as intensifying agricultural practices (right part of the figure), that focus on locally maximizing one ecosystem service, such as crop yield, can limit the other ecosystem services provided in complex food webs in a given area (e.g., pest control is reduced, indicated by higher biomass of aphid and vole). Note that the stability of delivering the focal service decreases in this example (larger error bar in crop yield) at high land-use intensity (Isbell et al., 2017b). Socio-political context related to human population density and stakeholder interests can influence feedbacks between ecosystem services and the management of complex ecosystems. Importantly, ecosystem services are not solely provided by single nodes in the food web and at a single location, but by the interaction among multiple nodes (colours of example links between nodes in upper part, correspond to ecosystem service bar colours in lower part) across adjacent ecosystems. *Redrawn after Hines et al., 2015b.*

et al., 2014), or by manipulating the diversity of aphid natural enemies and investigating pest control (Cardinale et al., 2003). Biodiversity changes at higher trophic levels of aquatic ecosystems have been shown to exert cascading effects on the biomass production at lower levels (Duffy et al., 2007; Worm and Duffy, 2003). This finding was generalized by models of complex food webs, in which increased animal diversity led not only to higher herbivory but also, counter-intuitively, to higher primary production by plants (Schneider et al., 2016). This surprising finding is explained by systematic trait shifts in the plant communities that are induced by the increased top-down pressure (Schneider et al., 2016). These results contribute to the general notion that biodiversity changes across trophic levels can have complex indirect effects, which strongly calls for a multitrophic whole-ecosystem perspective for mechanistically understanding BEF relationships (Barnes et al., 2018; Brose and Hillebrand, 2016; Eisenhauer, 2017; Hines et al., 2015b; Seibold et al., 2018; Thompson et al., 2012; Worm and Duffy, 2003).

Ultimately, the understanding of real-world BEF relationships requires coupling multitrophic biodiversity change and indirect effects among species addressed at local habitat scales with spatio-temporal upscaling to the landscape level. However, research on multitrophic interactions and quantitative food-web changes in space and time is little developed so far (but see, e.g., Grass et al., 2018; Tscharntke et al., 2012). Across ecosystems, the increase in the number of interactions between species is predictably linked to the simultaneous increase in the number of species (Brose et al., 2004). This connection between species-area and link-area relationships facilitates the prediction of food-web complexity at the landscape level, but upscaling of BEF relationships would also require integrating the identities or traits of species and their interactions into models. In this vein, behavior- and trait-based allometric random walk models (Hirt et al., 2018), as well as extensions of the classic theory of island biogeography that account for effects of the species' trophic levels (Gravel et al., 2011), body masses (Jacquet et al., 2017), and network-area relationships (Galiana et al., 2018), have great potential to become important cornerstones of novel BEF upscaling approaches (see also section "Spatial scaling of BEF relationships").

In order to account for the finding of substantial species turnover and biotic homogenization due to human activities (Dornelas et al., 2014; Gossner et al., 2016), future BEF experiments may also include both species gains and losses (Mori et al., 2018; Wardle et al., 2011) across different trophic levels. Integrating trophic complexity will be key to account for cascading, facilitative, and competitive effects in order to understand how biodiversity affects whole-ecosystem functioning (Barnes et al., 2018), regardless of the direction of biodiversity change (loss or gain; Wardle, 2016). Moreover, biotic homogenization across trophic levels may have important implications for the stable provisioning of multiple ecosystem services (Hautier et al., 2018; Pasari et al., 2013; van der Plas et al., 2016a,b) as synchrony in responses across species may compromise ecosystem functioning (Craven et al., 2018; de Mazancourt et al., 2013). Higher synchrony among species in space and time may be particularly deleterious for ecosystems with ongoing global change as predicted by the temporal and spatial insurance hypotheses of biodiversity (Loreau et al., 2003a; Yachi and Loreau, 1999).

The explicit quantification of fluxes of energy and matter in BEF experiments would greatly facilitate the integration of different trophic levels (Barnes et al., 2014, 2018; Lindeman, 1942; Stocker et al., 1999; Wilsey and Polley, 2004). Flux rates may be more sensitive and may show faster responses to variations in biodiversity than pools (Meyer et al., 2016; but see Liu et al., 2018 for a counter example). Evidence for this, however, is scarce (but see Allan et al., 2013; Niklaus et al., 2016), but this deserves further attention, particularly in long-term (Huang et al., 2018; Meyer et al., 2016) and multitrophic experiments (Eisenhauer, 2017). An Ecotron study with intact soil monoliths from the Jena Experiment (Milcu et al., 2014) under controlled conditions allowed for the quantification of the effects of plant diversity on ecosystem carbon fluxes and uptake efficiency of plants. Indeed, it was observed that increasing plant species and functional diversity led to higher gross and net ecosystem carbon uptake rates, and effects were partly mediated by the leaf area index and the diversity of leaf nitrogen concentrations of the plant community (Milcu et al., 2014). While the consideration of multitrophic interaction partners in such studies is still in its infancy, new research infrastructures have been established to explore the role of above- and belowground food webs in fluxes of energy and matter (Eisenhauer and Türke, 2018).

Assessing energy flux dynamics in ecological networks provides the mechanistic underpinning of multitrophic BEF relationships, which is why the quantification of energy fluxes in food webs may be a powerful tool for studying ecosystem functioning in multitrophic systems ranging from biodiversity experiments to real-world ecosystems (Barnes et al., 2018). By combining food-web theory with BEF theory (Hines et al., 2015b), whole community energy-flux assessment enables investigators to quantify many different types of multitrophic ecosystem processes without having to measure them all separately (Barnes et al., 2018). Energy flux can then be used as an integrated measure and a common currency to compare certain types of processes (e.g., herbivory or predation) across different ecosystem types (Barnes et al., 2018); however, energy-flux calculations need to be validated by actual process measurements (e.g., Schwarz et al., 2017), which in complex ecosystems such as grasslands and forests poses a serious challenge.

## 2.2 Predicting the strength of BEF relationships across environmental contexts

*"The idea that the mechanisms underpinning species coexistence are the same as those that link biodiversity with ecosystem functioning can be traced all the way back to Darwin's principle of divergence..."*

**Turnbull et al. (2013)**

The strength and sign of BEF relationships have been reported to differ among studies as well as among biotic and environmental contexts (e.g., Baert et al., 2018; Fridley, 2002; Guerrero-Ramírez et al., 2017; Jousset et al., 2011; Jucker et al., 2016; Kardol et al., 2018; Ratcliffe et al., 2017; Steudel et al., 2012; but see Craven et al., 2016). We still have scant knowledge about how and why effects of the diversity and composition of communities on ecosystem functions vary. How context-dependent are BEF relationships, and what biotic and abiotic factors drive this context-dependency?

There have been several empirical attempts to study BEF relationships under different environmental contexts, such as the BIODEPTH experiment across eight European countries (Hector et al., 1999), the COST Agrodiversity experimental network across 31 sites in Europe and Canada (Finn et al., 2013; Kirwan et al., 2007), the global network of tree diversity experiments in TreeDivNet (Grossman et al., 2018; Paquette et al., 2018), the global Nutrient Network (Borer et al., 2014, 2017), the global meta-analyses in drylands (Maestre et al., 2012) and forests (Guerrero-Ramírez et al., 2017), the BioCON experiment in Cedar Creek studying effects of elevated $CO_2$ concentrations and N deposition (Reich et al., 2001), the BAC experiment in Cedar Creek exploring warming effects (Cowles et al., 2016; Pennekamp et al., 2018; Thakur et al., 2017), the two sites of the BEF-China experiment (Huang et al., 2018), and the Jena drought experiment (Vogel et al., 2012). Moreover, in the Jena Experiment (Roscher et al., 2004), researchers have applied a large number of subplot treatments to study if plant diversity effects are contingent upon management intensity (Weigelt et al., 2009), above- and belowground consumers (Eisenhauer et al., 2011), and plant invasion (Petermann et al., 2010;

Roscher et al., 2009; Steinauer et al., 2016). Although some studies report BEF relationships in plant diversity experiments to be consistent across abiotic and biotic contexts (e.g., Craven et al., 2016; O'Connor et al., 2017; Thakur et al., 2015), there is substantial variability within and across studies depending on the point in time of the measurement (Kardol et al., 2018; Reich et al., 2012; Wright et al., 2015), the biodiversity facet investigated (Craven et al., 2016), and the trophic level and complexity of the studied community (Beugnon et al., 2019; Mulder et al., 1999; O'Connor et al., 2017; Seabloom et al., 2017).

In response to some of the initial debates regarding the validity of BEF relationships across environmental contexts (e.g., Givnish, 1994; Tilman and Downing, 1994), previous BEF research focused heavily on completely removing any "confounding" effects of abundance, biomass, and environmental gradients, in order to isolate and quantify "true" biodiversity effects. It is, however, important to understand biodiversity effects in the context of other co-varying factors to better predict scenarios of ecosystem function given species gains or losses (which covary with many other factors; Wardle, 2016). Future research should thus aim at understanding the functional role of biodiversity in dynamic ecosystems that are not at competitive equilibrium (Brose and Hillebrand, 2016; Leibold et al., 2004) as well as in affecting multiple dimensions of stability under changing environmental conditions (Donohue et al., 2016; Pennekamp et al., 2018). Such information is, for instance, urgently needed to inform predictive BEF models and to provide tailored management recommendations that account for local environmental conditions (Guerrero-Ramírez et al., 2017).

Conceptual advances are likely to be achieved by utilizing niche and coexistence theory to understand the context-dependency of BEF relationships (Barry et al., 2019; Turnbull et al., 2016). Environmental change often affects the composition of communities by altering the environmental conditions, modifying available niche space directly (niche destruction; Harpole et al., 2016) and/or indirectly through altered biotic interactions (Turnbull et al., 2016). For instance, the addition of nutrients has been repeatedly shown to favour the growth of certain plant species with high nutritional demands and fast uptake strategies (Clark et al., 2007; Harpole and Tilman, 2007; Harpole et al., 2016; Vogel et al., 2019a). Increased plant growth of some species, in turn, induces the shading of other species, which then disappear because their niche requirements are no longer met (Hautier et al., 2009). The resulting loss of species then undermines ecosystem functions of the depauperate plant communities (Isbell et al., 2013).

The same mechanisms that permit the coexistence of different species, namely niche differences, also are the key for the complementary resource use and resultant overyielding (Barry et al., 2019c; Loreau, 2004; Tilman et al., 1997b; Turnbull et al., 2013, 2016; Vandermeer, 1981) and transfer of energy across trophic levels (Barnes et al., 2014). Niche differentiation and facilitation within (Cardinale et al., 2007; Reich et al., 2012; Wright et al., 2017) and across trophic levels (Ferlian et al., 2018; Poisot et al., 2013) are often found to be the main mechanisms behind positive BEF relationships. As a consequence, changes of the environmental conditions that influence the co-existence of species are also likely to affect the strength of BEF relationships (Barry et al., 2019b). In support of this notion, positive BEF relationships have been shown to be strongest in complex resource environments (Fig. 4) and to become non-significant or even negative in homogenous resource environments (Eisenhauer et al., 2013; Hodapp et al., 2016; Jousset et al., 2011; Mouquet et al., 2002; Norberg et al., 2001). Hodapp et al. (2016) generalized this to resource supply heterogeneity landscapes and showed that strongly positive effects of richness on ecosystem function occur only if (1) species differ in traits, (2) environments show heterogeneity, and (3) dispersal allows effective species sorting. Research on algal model communities in relatively structured environments (flow habitats and disturbance regimes) has shown that communities with more species take greater advantage of the niche opportunities in a given environment, and this allows diverse systems to better perform ecosystem functions (Cardinale, 2011; Stachowicz et al., 2007, 2008a). Taken together, these results indicate that environmental heterogeneity promotes complementarity effects (see, e.g., Wacker et al., 2008) and thus steeper BEF relationships (Fig. 4), suggesting that habitat homogenization may compromise positive biodiversity effects on ecosystems.

To study the context-dependence of BEF relationships, different site-specific conditions for biodiversity effects, including environmental stress and resource availability (Fig. 4), will need to be disentangled (Baert et al., 2018; Guerrero-Ramírez et al., 2017). Global networks of experiments using standardized methods (Grossman et al., 2018; Lefcheck et al., 2016; Meyer et al., 2015) and syntheses of data are needed and have proven to be extremely powerful in detecting global biodiversity(-function) patterns and underlying mechanisms (e.g., Nutrient Network; Borer et al., 2014, 2017; Grace et al., 2016). Notably, such standardized assessments are particularly important for quantifying multitrophic interactions across environmental gradients (Kambach et al., 2016; Roslin et al., 2017) that are

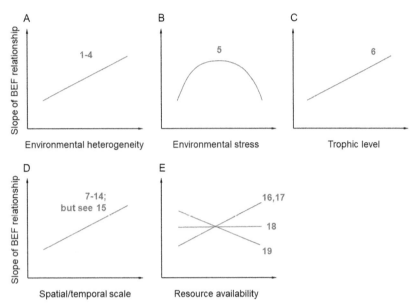

**Fig. 4** Context-dependent biodiversity–ecosystem functioning (BEF) relationships; examples include (A) environmental heterogeneity, (B) environmental stress, (C) trophic level, (D) spatial and temporal scale, and (E) resource availability. Although the proposed relationships are supported by some studies (examples given, no comprehensive list of studies), a thorough understanding of the context-dependency of BEF and the underlying mechanisms is elusive. Thus, the depicted relationships should be regarded as working hypotheses for future research. See also Bardgett and Wardle (2010) (Fig. 5.3 and references therein) for a similar conceptualization of the context-dependency of BEF relationships that are mostly based on observational studies and removal experiments, rather than on random biodiversity manipulation experiments, as done here. For panel (B), we followed the definition by Chase and Leibold (2003), stating that "stressful niche factors limit the per capita population growth rate of the focal population, but are not influenced by changes in the population size." *1*: Stachowicz et al. (2008b), *2*: Griffin et al. (2009), *3*: Cardinale (2011), *4*: Jousset et al. (2011), *5*: Baert et al. (2018), *6*: Lefcheck et al. (2015), *7*: Cardinale et al. (2007), *8*: Eisenhauer et al. (2010), *9*: Cardinale et al. (2011), *10*: Isbell et al. (2011), *11*: Reich et al. (2012), *12*: Thakur et al. (2015), *13*: Meyer et al. (2016), *14*: Guerrero-Ramírez et al. (2017), *15*: Kardol et al. (2018), *16*: Reich et al. (2001), *17*: Fridley (2002), *18*: Craven et al. (2016), *19*: Zhang and Zhang (2006).

intimately linked with ecosystem function (Eisenhauer et al., 2019). For instance, different tree diversity experiments around the globe collaborate in the framework of TreeDivNet (Paquette et al., 2018; Verheyen et al., 2016) and allow for countering criticisms related to realism, generality, and lack of mechanistic explanation in their work (Grossman et al., 2018; Paquette et al., 2018). However, empirical work and syntheses should not be restricted to certain ecosystems, but should span across ecosystem types (e.g., aquatic and terrestrial; Balvanera et al., 2006; Cardinale et al., 2011; Handa et al., 2014; Lefcheck et al., 2015; Ruiz-González et al., 2018; Schuldt et al., 2019). Recent modelling (e.g., Baert et al., 2018) and empirical work (e.g., Guerrero-Ramírez et al., 2017) provided exciting working hypotheses for future research (Fig. 4).

## 2.3 Spatial scaling of BEF relationships

*"Biodiversity loss substantially diminishes several ecosystem services by altering ecosystem functioning and stability, especially at the large temporal and spatial scales that are most relevant for policy and conservation."*

*Isbell et al. (2017a)*

To date, BEF relationships have mostly been investigated at small scales (e.g., in microcosms, mesocosms, or small plots; Cardinale et al., 2011), raising the question "How does the BEF relationship change with spatial scale?" (Barnes et al., 2016; Isbell et al., 2017a; Manning et al., 2019; Thompson et al., 2018). Accordingly, Mori et al. (2018) recently stressed the need for unification of beta-diversity and among-patch ecosystem-function theory. The focus on small-scale studies may also be one reason for described mismatches between local-scale observational and experimental BEF studies and conclusions drawn for management-relevant scales in non-experimental settings (Oehri et al., 2017; van der Plas et al., 2016a,b). Thus, future research needs to bridge the gap between results from local-scale BEF experiments and real-world relevant scales in order to understand whether and how biodiversity effects are important at the landscape scale (Cardinale et al., 2012; Isbell et al., 2017a; Thompson et al., 2018). There is empirical evidence suggesting that the importance of biodiversity in driving ecosystem functions increases as more spatial contexts, i.e., different environmental conditions, are considered (Grace et al., 2016; Hautier et al., 2018; Isbell et al., 2011; Mori et al., 2016; Thompson et al., 2018), stressing the role of environmental heterogeneity in driving the strength and mechanisms of BEF relationships (Cardinale, 2011; Griffin et al., 2009). One solution may be the development of spatial upscaling algorithms to relate local BEF findings to patterns at the landscape scale. Using such an approach, Barnes et al. (2016), however, showed that the relative importance of biodiversity for ecosystem functions decreased with increasing spatial scale. Such contradicting findings are also observed in fragmentation-biodiversity studies when focusing on patches or landscapes (Fahrig et al., 2019), and integrating the ecosystem function aspect in fragmentation studies may help bridging this field of research to BEF (Fahrig, 2017). Hence, the mechanisms dominating biodiversity and functions might differ between small and large spatial scales (Loreau et al., 2003a,b). This indicates the need for future research on this topic, particularly if we are to integrate knowledge from BEF experiments in ecosystem service modelling and other spatial mapping exercises.

While BEF experiments have been "stuck" in plots and buckets, meta-community theory has been dealing with species appearance and disappearance without an explicit link to ecosystem functioning (Bannar-Martin et al., 2018; Leibold and Chase, 2018; Leibold et al., 2004, 2017; but see Loreau et al., 2003b). Thus, species pools and their turnover and dynamics need to be incorporated into BEF research (Bannar-Martin et al., 2018; Wardle, 2016) to consider the links between community assembly/coexistence mechanisms (e.g., dispersal, demographic stochasticity, niches/traits) and ecosystem functioning (Hillebrand et al., 2018). One step towards this goal may be to identify trade-offs in spatial and temporal scales at which diversity maximizes single and multiple ecosystem functions. In fact, considering multitrophic consumer networks that link different landscape patches and ecosystem compartments through the flux of energy across trophic levels (Barnes et al., 2014) might be a promising approach to facilitate the upscaling of local processes to landscape-level function (Fig. 5; Barnes et al., 2018). For instance, future research efforts on land-use change and restoration could be targeted towards manipulating biodiversity at different spatial scales and exploring whole-ecosystem consequences within and across different patches and compartments. Another option are disturbances acting at the landscape scale. They offer excellent options for BEF studies at larger spatial scales, but research plans have to be made long before such disturbances happen (Lindenmayer et al., 2010).

Dispersal may promote the functioning of ecosystems in two ways (Leibold et al., 2017; Loreau et al., 2003a; Thompson and Gonzalez, 2016). First, species dispersal and community assembly processes may allow species to track local environmental changes by shifting in space, which may then preserve biodiversity and ensure high ecosystem functioning (Leibold et al., 2017; Loreau et al., 2003a; Thompson and Gonzalez, 2016). Second, source–sink dynamics may allow species to persist in suboptimal environments, thus increasing local biodiversity over time, although this does not necessarily promote functioning (Leibold et al., 2017). Species-sorting dynamics also provide spatial insurance, so that compensatory dynamics

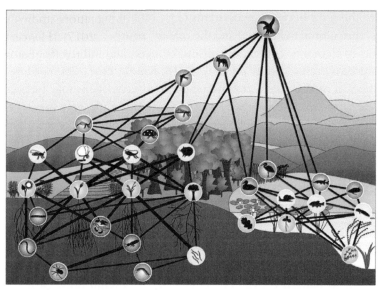

**Fig. 5** Complex communities link different habitats, a consideration that may facilitate the upscaling of BEF. Conceptual illustration of how multitrophic interactions across ecosystem boundaries can link different ecosystem types and compartments, including above- and belowground compartments, forests and grasslands, as well as terrestrial and aquatic ecosystems. Links between different network modules in these subsystems provide stability of trophic dynamics, matter and energy flow across system boundaries as well as stability of ecosystem function and service delivery (Barnes et al., 2018).

stabilize the fluctuations of each function through time at the regional, but not necessarily at the local scale (Loreau et al., 2003a; Thompson and Gonzalez, 2016). Relatedly, spatial network modularity has a buffering effect in perturbed experimental metapopulations, protecting some local subpopulations from the perturbation (Gilarranz et al., 2017) or providing empty patches for recolonization, ultimately stabilizing the metapopulation (Fox et al., 2017). Adding to this complex picture, there is experimental evidence suggesting that also habitat isolation and matrix quality influence biodiversity and ecosystem functioning (Spiesman et al., 2018). Thus, the fragmentation and connectivity of habitat patches as well as the mobility of taxa driving meta-community dynamics are likely to be of great importance, yet understudied in BEF research (Gonzalez et al., 2017).

Most terrestrial ecosystems have soft boundaries that are highly permeable for many species. Accordingly, different ecosystems should not be regarded in isolation but in the context of surrounding ecosystems. For instance, biodiversity effects on adjacent ecosystems should be studied to explore the links between ecosystems (e.g., by linking aquatic ecosystems, forests, agricultural fields, etc.) and the role of "source" and "sink" dynamics in fluxes of elements, energy, organisms, biomass, and information between adjacent ecosystems (Gounand et al., 2018; Knight et al., 2005). Here, the investigation of key organisms linking different ecosystems may be particularly relevant to move from the plot scale to the landscape scale (Barnes et al., 2018; Fig. 5) as well as to define conservation priorities and corresponding management practices.

## 2.4 Eco-evolutionary implications of multitrophic BEF

*"Nothing in evolution or ecology makes sense except in the light of the other."*
**Pelletier et al. (2009)**

Ecology and evolution are sometimes thought of as acting at different timescales, which might explain why evolutionary processes have rarely been considered in past BEF research. However, a growing body of literature shows that evolutionary processes can be rapid and of relevance at what is commonly considered ecological timescales (Carroll et al., 2007; Hendry, 2016), such that a strict time-scale separation is no longer useful. Furthermore, the study of the molecular basis of adaptation has experienced a boost due to recent technological developments (Bosse et al., 2017; Savolainen et al., 2013; Stapley et al., 2010; Wuest and Niklaus, 2018). BEF research may greatly benefit from embracing the rich and growing body of knowledge on micro-evolutionary processes, population genetics, and the molecular basis of adaptation, because adaptation and evolutionary processes are likely to contribute to the dynamic nature of BEF relationships (e.g., Tilman and Snell-Rood, 2014; van Moorsel et al., 2018; Zuppinger-Dingley et al., 2014). Such eco-evolutionary processes can ideally be studied in the few long-term experiments worldwide that have been run for multiple generations of the organisms studied.

Undoubtedly, members of an ecological community impose selection pressures onto each other. For example, changing phenotypes have been reported in a number of plant species in response to manipulated biodiversity gradients (Lipowsky et al., 2011, 2012; Schöb et al., 2018; Zuppinger-Dingley et al., 2014, 2016). Phenotypic changes may allow different coexisting species to use resources in more dissimilar and complementary ways, thereby reducing competition, maximizing growth, and favouring stable coexistence (Tilman and Snell-Rood, 2014; Zuppinger-Dingley et al., 2014). Yet, we know too little about the relative importance of phenotypic plasticity, transgenerational epigenetic processes (Schmid et al., 2018), and genuine evolutionary adaptation that simultaneously contribute to phenotypic changes (Hoffmann and Sgrò, 2011; Zuppinger-Dingley et al., 2014). Such knowledge is important, however, in order to estimate how lasting and/or reversible the effects are.

There are a number of ways in which micro-evolutionary processes may help to understand and predict BEF relationships. For example, a significant role of the comparatively slow process of evolutionary adaptation may explain the observation of strengthening BEF relationships over time in grassland experiments (Tilman and Snell-Rood, 2014; Vogel et al., 2019b; Zuppinger-Dingley et al., 2014). Furthermore, micro-evolutionary dynamics may lead to positive feedback loops that can affect ecosystem functioning. Natural selection is usually expected to reduce genetic variance, but genetic variation provides the raw material for future adaptation (Mousseau and Roff, 1987). Frequent changes in the selective regime may thus jeopardize populations' persistence (Hoffmann and Sgrò, 2011). Phenotypic plasticity, in contrast, may buffer populations against changing selection regimes (Charmantier et al., 2008). Taking into account the relative importance of phenotypic plasticity and micro-evolutionary adaptation will be essential for the understanding of how adaptation processes affect BEF relationships.

Members of a community mutually influence each other during the selection process (Jousset et al., 2016; Tilman and Snell-Rood, 2014). Two aspects seem to be particularly relevant in the context of community assembly. First, functionally similar and/or related species will be selected for character displacement and niche differentiation, thereby promoting specialization, coexistence, and ecosystem processes (Harmon et al., 2009; Tilman and Snell-Rood, 2014). The genetic and evolutionary mechanisms of such processes have rarely been studied in BEF research. Second, species may co-evolve together with their antagonists, e.g., pathogens (Vogel et al., 2019b). Here, the species involved can differ substantially in generation time. Pathogens may adapt and change quickly, imposing persistent and likely fluctuating selection pressure on host species. Indeed, several studies showed that negative plant-soil feedback effects can induce a decrease in plant growth in monoculture (e.g., Hendriks et al., 2013; Maron et al., 2011; Schnitzer et al., 2011). Deteriorating monocultures over time indicates that Janzen-Connell effects, the accumulation of species-specific plant antagonists, may play an important role in BEF relationships (Petermann et al., 2008). Zuppinger-Dingley et al. (2014) proposed that a respective selection pressure should be particularly pronounced in low-diversity plant communities (see also van Moorsel et al., 2018). In contrast, accumulation of such species-specific plant antagonists in high-diversity plant communities would be impeded because of lower host densities (Civitello et al., 2015; Hantsch et al., 2013, 2014; Rottstock et al., 2014). On the other hand, prolonged time in monocultures in the Jena Experiment has converted negative into positive net plant-soil feedback effects (Zuppinger-Dingley et al., 2016), which could be partly due to evolved resistance of the plants and/or a slower build-up of communities of mutualists like the accumulation of plant growth promoting rhizobacteria (Latz et al., 2012) in comparison to antagonists. Taken together, these lines of evidence suggest that dissimilar host-pathogen interactions at low versus high biodiversity may impose different selection pressures on community members, both at the level of plant species and genotypes (Roscher et al., 2007).

Genetic diversity within species offers the raw material for future adaptations (Jousset et al., 2016), even if some of the variation may not be utilized under current conditions (Paaby and Rockman, 2014). Genetic variation, thus, serves as a genetic insurance for population persistence and ultimately for sustained ecosystem functioning. It is vital to understand the processes that affect intra-specific diversity in communities differing in species diversity (Vellend and Geber, 2005). Genetic diversity depends on the effective population size, which in turn is determined by census size, reproductive system, spatial structure, and the intensity and shape of natural selection. Strong directional and stabilizing selection both tend to reduce genetic diversity. The potential cascading effects of community diversity on population diversity and eventually intra-specific and phylogenetic diversity as well as consequences for ecosystem functioning are poorly studied (but see Crutsinger et al., 2006; Hughes et al., 2008; Zeng et al., 2017). In fact, there has been a recent interest in how populations assemble with respect to functional diversity, but also phylogenetic diversity, and the underlying mechanisms are relevant in the BEF context as community assembly and disassembly processes have implications for the long-term functioning of plant communities (Vogel et al., 2019a). Species differ partly due to divergent directional selection. Under the premise that phylogenetic distance contains a signal of divergent selection for (unknown) functional traits, phylogenetic distance can be used as a proxy for functional diversity (Cadotte et al., 2008; Vogel et al., 2019a). However, such genetic conservatism may be highly variable among traits, for instance among leaf and root traits of plants (e.g., Valverde-Barrantes et al., 2017). Therefore a combination of traits and stepwise phylogeny has been proposed (Cadotte, 2013) and successfully applied in multitaxon studies (Thorn et al., 2016). Translating these challenges that have mostly been addressed for herbaceous plants to higher trophic levels, it is also relevant to explore how much phylogenetic diversity is represented within multitrophic communities for applied conservation aspects (Eisenhauer et al., 2019).

The field of ecological genetics has seen a great expansion in opportunities by the rapid development of next-generation sequencing technologies (Savolainen et al., 2013). It is now possible to sequence and assemble the genome of just about any species at manageable cost, which allows the study of the genomics of previous non-model organisms in natural conditions (Ellegren, 2014; Savolainen et al., 2013; Stapley et al., 2010). Genotyping-by-sequencing techniques allow the study of genetic polymorphisms without much cost- and labor-intensive development of genetic markers and gives an unbiased view on population-wide genetic diversity (Narum et al., 2013). There are many ways how these new technologies can be employed in a BEF context. A particularly exciting avenue is the study of co-evolutionary dynamics in multi-species systems up to the community level. Genomic and transcriptomic methods may allow to uncover the genetic architecture of functional trait variation (Schielzeth and Husby, 2014). Moreover, population genomics allows studying the population structure and inbreeding patterns at high resolution across multiple species. Ultimately, such knowledge will help to link the diversity at the genome level to ecosystem-level processes (Wuest and Niklaus, 2018) and to explore the role of species interactions driving these interlinkages.

## 2.5 FAIR data and beyond

*"The grand challenge for biodiversity informatics is to develop an infrastructure to allow the available data to be brought into a coordinated coupled modelling environment able to address questions relating to our use of the natural environment that captures the variety, distinctiveness and complexity of all life on Earth."*

**Hardisty and Roberts (2013)**

Data plays an increasingly important role for BEF research (Trogisch et al., 2017). As for other subdisciplines of biodiversity research, this results in a need for improved biodiversity informatics along all steps of the data lifecycle from data collection to data analysis and publication (Hardisty and Roberts, 2013). Due to the availability of novel methods like high-throughput sequencing, automatic monitoring, and remote sensing, more and more data are being produced in BEF research. Thus, the resulting data are likely to play an important role in future BEF research, as high-throughput sequencing has the potential to help identifying potential microbial drivers of BEF relationships (e.g., Laforest-Lapointe et al., 2017), automatic monitoring may be key to link behavioural ecology of animals to multitrophic BEF (e.g., Dell et al., 2014; Eisenhauer and Türke, 2018), and remote sensing is likely to help scaling up BEF research to the landscape scale (e.g., Cabello et al., 2012). Often, the amount of such data collected exceeds available resources for manual processing. Recently established methods in machine learning, in particular deep neural networks, have the potential to alleviate this problem (see Brust et al., 2018 and Ryo and Rillig, 2017 for successful examples). Currently, however, the applicability of these methods is restricted by their need for large sets of labelled training data. Further development of methods to reduce the need for training data and/or semi-automatically label data are needed. Additionally, better tools for data quality assurance and improvement are needed, such as comprehensive data quality frameworks (Morris et al., 2018; Veiga et al., 2017). These are not yet part of commonly used data management platforms though.

Answering important questions in BEF research often requires data that covers large temporal and spatial scales. Few projects run long enough or cover a wide enough geographical range to be able to collect all the data needed themselves. Thus, BEF research relies on data reuse and sharing—both within and across projects. This necessitates BEF data being preserved following the FAIR principles (Wilkinson et al., 2016): data should be findable, accessible, interoperable, and reusable. This urgent need was described even before the term FAIR was coined (Hampton et al., 2013). For data to be findable, it needs to be described with rich metadata. While suitable annotation schemes exist for some types of biodiversity data (e.g., ABCD for collection data or Darwin Core for occurrence data), they are still lacking for more complex BEF data. The Easy Annotation Scheme for Ecology (Pfaff et al., 2017) or BioSchemas (http://bioschemas.org; Gray et al., 2017), for instance, aim to alleviate this problem. In addition to better annotation schemes, better tools to reduce the human effort in creating these annotations are needed. First examples for such approaches in other disciplines show the general feasibility (Rodrigo et al., 2018).

The best described dataset will not be found, if no appropriate search engines exist. Even though Google recently launched a dataset search, in general, this topic is not yet well researched and poses a number of fundamental challenges (Chapman et al., 2019). Besides finding relevant data, integrating this data is a challenging and labor-intense task. Both tasks could be made considerably easier by the usage of semantic web technologies, in particular the usage of ontologies (Gruber, 1993) and compliance to the linked open data principles (Bizer et al., 2008). This is also addressed in the parts of the FAIR principles related to interoperability and reusability. Finally, there is growing awareness, that preserving data alone is not sufficient for reproducibility. Rather, analysis tools and workflows need to be preserved as well (Hardisty et al., 2019). Culturally, a shift is needed to incentivize proper data management and sharing. Although there are warnings stating that a raise in openly available datasets might create the illusion of "a free lunch for all" and that this system will collapse, if the considerable effort that goes into providing datasets is not properly incentivized (Escribano et al., 2018), we believe that open data are absolutely necessary to facilitate fast scientific progress.

## 2.6 Operationalizing BEF insights for ecosystem management, society, and decision making

*"A mix of governance options, policies and management practices is available for public and private actors in Europe and Central Asia, but further commitment is needed to adopt and effectively implement them to address the drivers of change, to safeguard biodiversity and to ensure nature's contributions to people for a good quality of life."*

*IPBES (2018)*

With the rising human population size, *per capita* consumption, and subsequent ecosystem service demands, there is an increasing need for bringing the ecological, fundamental BEF knowledge into action in order to develop applications for the sustainable management of ecosystems, such as agroecosystems (Isbell et al., 2017a,b). Will ecosystems be managed in an ecologically sustainable way or will increasing demands be temporarily compensated by higher management intensity only to be followed by long-term depletion of agriculturally used soils? Indeed, recent studies have pinpointed many potential benefits of increased biodiversity in agroecosystems and production forests (Isbell et al., 2017b; Gérard et al., 2017; Martin-Guay et al., 2018; Pretty, 2018). These conclusions are supported by a long history of intercropping literature that highlights the importance of increasing biodiversity in space and time to maintain crop yields (e.g., Darwin, 1859; Trenbath, 1974; Vandermeer, 1990). In this context, BEF research has the potential to apply the multifunctionality concept (Byrnes et al., 2014; Hector and Bagchi, 2007) to move beyond considering multifunctionality a suite of independent functions, but rather to consider synergies and trade-offs among different ecosystem services

(Fig. 3; Allan et al., 2015; Binder et al., 2018; Giling et al., 2019; Hines et al., 2015b; Manning et al., 2018; Meyer et al., 2018; see also Manning et al., 2019 for an in-depth discussion of this topic). Biodiversity potentially provides a partial substitute for many costly and non-sustainable agricultural management practices, such as the application of fertilizers, pesticides, imported pollinators, and irrigation (Finger and Buchmann, 2015; Isbell et al., 2017b; Tilman et al., 2006; Weigelt et al., 2009).

There is increasing concern that the ongoing loss of biodiversity may affect and diminish the provision of ecosystem services in the future (Cardinale et al., 2012; IPBES, 2018; Manning et al., 2018; Ricketts et al., 2016; Wall et al., 2015). While some key ecological processes may be well understood, such patterns can be difficult to translate into quantitative relationships suitable for use in an ecosystem service context. There is a need to derive quantitative "pressure-response functions" linking anthropogenic pressures with ecosystem functions that underpin key climate, water-quality, and food-regulating services. This requires the joint analysis of the complex, sometimes conflicting or interactive, effects of multiple anthropogenic pressures on different ecosystem functions and the role of biodiversity as a mediating factor determining how anthropogenic pressures translate into changes in ecosystem services. Challenges relate to the differing spatial scales and configuration of anthropogenic pressures and ecosystem service beneficiaries, and uncertainties associated with the time lags between anthropogenic pressures and ecosystem responses (Isbell et al., 2015b). Accordingly, future research needs to employ a quantitative, multi-parameter approach to assess the nature of linkages between biodiversity, ecosystem processes, and ecosystem services (Giling et al., 2019; Manning et al., 2018) within and across ecosystem boundaries (Barnes et al., 2018). This involves the effects of anthropogenic pressures on these linkages, including reversal of pressures through conservation and restoration management, and likely threshold or hysteresis functions (Isbell et al., 2013).

Results from the last decade of BEF research tend to suggest that we need to conserve a large proportion of existing species, rather than few selected species, to maximize ecosystem service provisioning across spatial and temporal scales (Isbell et al., 2011; Meyer et al., 2018; Reich et al., 2012; Winfree et al., 2018; but see Kleijn et al., 2015). BEF research has to accept the challenge to embrace socio-ecological systems with their different drivers and interaction networks (e.g., including humans; Bohan et al., 2016; Dee et al., 2017). This means, for instance, building BEF experiments based on communities realized under (more) realistic land-use regimes regarding current and future stakeholder priorities. Here, e.g., disturbances, restoration projects, and changes in management due to different financial incentives may offer real-world replicated experiments. Scientists will have to more deliberately collaborate with national or federal agencies to develop strategies to become engaged in such projects early enough.

Fully embracing socio-ecological processes can only happen at larger scales and adds several layers of complexity to research projects (Thompson et al., 2018). For operationalizing this goal, food web network theory can meet social network theory to develop combined assessments (Dee et al., 2017). It will be important to identify vulnerabilities in the network(s) and critical bottlenecks to perform opportunity and risk assessments. Knowledge about risk factors can then inform where and when to best employ management interventions. Ultimately, BEF outcomes have to be translated to show policymakers and the general public the value of biodiversity, including consequences of biodiversity decline for human well-being and health, as well as economic aspects, such as jobs, revenues, and the global climate and economy. Moreover, to date, few biodiversity studies have expressed the impact of biodiversity loss on the global warming potential (Isbell et al., 2015b)—a metric accessible to policy makers and commonly used in the IPCC reports to compare whether the greenhouse gas balance of ecosystems has a net warming or cooling effect on climate (IPCC, 2014). Thus, studies linking biodiversity change with global warming potential would not only be of great fundamental value, but could also lead to insights that are of great value for the society at large, and that could be disseminated through e.g. IPBES discussions and reports.

## 3. Concluding remarks

The BEF research field faces the critical challenge to simultaneously develop a more mechanistic understanding of BEF relationships and their context-dependencies as well as to scale up from the plot-level mechanisms and processes to management-relevant spatial and temporal scales in order to operationalize BEF insights for ecosystem management, society, and decision making. Here, we argue that further exploring trophic (e.g., Barnes et al., 2018) and non-trophic interactions (e.g., competition, facilitation; Ferlian et al., 2018) in multitrophic communities will be key to investigate the consequences of non-random biodiversity change as well as the eco-evolutionary underpinnings and implications of BEF relationships. As a consequence, the study of biotic interactions needs to consider the interaction history of the involved organisms (Zuppinger-Dingley et al., 2014). Evolutionary history may integrate information about past trophic and non-trophic interactions and thus determine the functioning of species in complex communities. As such, this knowledge may not only be essential to mechanistically understand BEF relationships, but also to develop applications for sustainable agroecosystems (Isbell et al., 2017a; Wang et al., 2019), advance ecological restoration to maintain ecosystem functioning (Kettenring et al., 2014), and sustain the integrity of Earth's ecosystems.

## Acknowledgements

The Jena Experiment is funded by the Deutsche Forschungsgemeinschaft (DFG, German Research Foundation; FOR 1451), with additional support by the Friedrich Schiller University Jena, the Max Planck Institute for Biogeochemistry, and the Swiss National Science Foundation (SNF). This project received additional support from the European Research Council (ERC) under the European Union's Horizon 2020 research and innovation program (grant agreement no. 677232 to NE) and the German Centre for Integrative Biodiversity Research Halle–Jena–Leipzig, funded by the German Research Foundation (FZT 118). MJ was additionally supported by the Swiss National Science Foundation. HB, FB, BS, MSL and CW acknowledge the funding for the BEF-China experiment by the German Research Foundation (DFG FOR 891) and the Swiss National Science Foundation (SNSF nos. 130720 and 147092). FI acknowledges funding support from the US National Science Foundation's Long-Term Ecological Research program (LTER) (DEB-1234162), as well as the LTER Network Communications Office (DEB-1545288). HB, SH, MSL, FvdP and CW acknowledge funding support from the European Union Seventh Framework Programme (FP7/2007–2013) under grant agreement no 265171, project FunDivEUROPE. We acknowledge comments by Tiffany Knight, Jonathan Levine, Yvonne Oelmann, Henrique Pereira, Wolfgang Wilcke, and Elizabeth Wolkovich during the "15 Years of the Jena Experiment" workshop or on earlier versions of this manuscript. Figs 1, 2, 3, and 5 were prepared by Thomas Fester (Scivit).

## References

Adler, P.B., Seabloom, E.W., Borer, E.T., Hillebrand, H., Hautier, Y., Hector, A., Harpole, W.S., Halloran, L.R.O., Grace, J.B., Anderson, T.M., Bakker, J.D., Biederman, L., et al., 2011. Productivity is a poor predictor of plant species richness. Science 1750, 1750–1754.

Allan, E., Weisser, W., Weigelt, A., Roscher, C., Fischer, M., Hillebrand, H., 2011. More diverse plant communities have higher functioning over time due to turnover in complementary dominant species. Proc. Natl. Acad. Sci. U. S. A. 108, 17034–17039.

Allan, E., Weisser, W.W., Fischer, M., Schulze, E.D., Weigelt, A., Roscher, C., Baade, J., Barnard, R.L., Beßler, H., Buchmann, N., Ebeling, A., Eisenhauer, N., et al., 2013. A comparison of the strength of biodiversity effects across multiple functions. Oecologia 173, 223–237.

Allan, E., Manning, P., Alt, F., Binkenstein, J., Blaser, S., Blüthgen, N., Böhm, S., Grassein, F., Hölzel, N., Klaus, V.H., Kleinebecker, T., Morris, E.K., et al., 2015. Land use intensification alters ecosystem multifunctionality via loss of biodiversity and changes to functional composition. Ecol. Lett. 18, 834–843.

Baert, J.M., Eisenhauer, N., Janssen, C.R., De Laender, F., 2018. Biodiversity effects on ecosystem functioning respond unimodally to environmental stress. Ecol. Lett. 21, 1191–1199.

Balvanera, P., Pfisterer, A.B., Buchmann, N., He, J.-S., Nakashizuka, T., Raffaelli, D., Schmid, B., 2006. Quantifying the evidence for biodiversity effects on ecosystem functioning and services. Ecol. Lett. 9, 1146–1156.

Balvanera, P., Siddique, I., Dee, L., Paquette, A., Isbell, F., Gonzalez, A., Byrnes, J., O'Connor, M.I., Hungate, B.A., Griffin, J.N., 2014. Linking biodiversity and ecosystem services: current uncertainties and the necessary next steps. Bioscience 64, 49–57.

Bandowe, B.A.M., Leimer, S., Meusel, H., Velescu, A., Dassen, S., Eisenhauer, N., Hoffmann, T., Oelmann, Y., Wilcke, W., 2019. Plant diversity enhances the natural attenuation of polycyclic aromatic compounds (PAHs and oxygenated PAHs) in grassland soils. Soil Biol. Biochem. 129, 60–70.

Bannar-Martin, K.H., Kremer, C.T., Ernest, S.K.M., Leibold, M.A., Auge, H., Chase, J., Declerck, S.A.J., Eisenhauer, N., Harpole, S., Hillebrand, H., Isbell, F., Koffel, T., et al., 2018. Integrating community assembly and biodiversity to better understand ecosystem function: the Community Assembly and the Functioning of Ecosystems (CAFE) approach. Ecol. Lett. 21, 167–180.

Bardgett, R.D., Wardle, D.A., 2010. Aboveground-Belowground Linkages: Biotic Interactions, Ecosystem Processes, and Global Change. Oxford University Press, Oxford.

Barnes, A.D., Jochum, M., Mumme, S., Haneda, N.F., Farajallah, A., Widarto, T.H., Brose, U., 2014. Consequences of tropical land use for multitrophic biodiversity and ecosystem functioning. Nat. Commun. 5, 1–7.

Barnes, A.D., Weigelt, P., Jochum, M., Ott, D., Hodapp, D., Haneda, N.F., Brose, U., 2016. Species richness and biomass explain spatial turnover in ecosystem functioning across tropical and temperate ecosystems. Philos. Trans. R. Soc. B 371 (1694) 20150279.

Barnes, A.D., Jochum, M., Lefcheck, J.S., Eisenhauer, N., Scherber, C., O'Connor, M.I., de Ruiter, P., Brose, U., 2018. Energy flux: the link between multitrophic biodiversity and ecosystem functioning. Trends Ecol. Evol. 33, 186–197.

Barry, K.E., Mommer, L., van Ruijven, J., Wirth, C., Wright, A.J., Bai, Y., Connolly, J., De Deyn, G.B., de Kroon, H., Isbell, F., Milcu, A., Roscher, C., et al., 2019a. The future of complementarity: disentangling causes from consequences. Trends Ecol. Evol. 34, 167–180.

Barry, K.E., de Kroon, H., Dietrich, P., Harpole, W.S., Roeder, A., Schmid, B., Clark, A.T., Mayfield, M.M., et al., 2019b. Linking local species coexistence to ecosystem functioning: a conceptual framework from ecological first principles in grassland ecosystems. Adv. Ecol. Res. 61, 265–296.

Barry, K.E., Weigelt, A., van Ruijven, J., de Kroon, H., Ebeling, A., Eisenhauer, N., Gessler, A., Ravenek, J.M., et al., 2019c. Above- and belowground overyielding are related at the community and species level in a grassland biodiversity experiment. Adv. Ecol. Res. 61, 55–89.

Berendse, F., van Ruijven, J., Jongejans, E., Keesstra, S., 2015. Loss of plant species diversity reduces soil erosion resistance. Ecosystems 18, 881–888.

Beugnon, R., Steinauer, K., Barnes, A.D., Ebeling, A., Roscher, C., Eisenhauer, N., 2019. Plant functional trait identity and diversity effects on soil meso- and macrofauna in an experimental grassland. Adv. Ecol. Res. 61, 163–184.

Binder, S., Isbell, F., Polasky, S., Catford, J.A., Tilman, D., 2018. Grassland biodiversity can pay. Proc. Natl. Acad. Sci. U. S. A. 115, 3876–3881.

Bizer, C., Heath, T., Berners-Lee, T., 2008. Linked data: principles and state of the art. In: Proceedings of the 17th International Conference on World Wide Web—WWW'08, pp. 1265–1266.

Bohan, D., Landuyt, D., Ma, A., Macfadyen, S., Martinet, V., Massol, F., et al., 2016. Networking our way to better ecosystem service provision. Trends Ecol. Evol. 31, 105–115.

Borer, E.T., Harpole, W.S., Adler, P.B., Lind, E.M., Orrock, J.L., Seabloom, E.W., Smith, M.D., 2014. Finding generality in ecology: a model for globally distributed experiments. Methods Ecol. Evol. 5, 65–73.

Borer, E.T., Grace, J.B., Harpole, W.S., MacDougall, A.S., Seabloom, E.W., 2017. A decade of insights into grassland ecosystem responses to global environmental change. Nat. Ecol. Evol. 1, 0118.

Bosse, M., Spurgin, L.G., Laine, V.N., Cole, E.F., Firth, J.A., Gienapp, P., Gosler, A.G., Mcmahon, K., Poissant, J., Verhagen, I., Groenen, M.A.M., Van Oers, K., et al., 2017. Recent natural selection causes adaptive evolution of an avian polygenic trait. Science 358, 365–368.

Bowler, D., Bjorkmann, A., Dornelas, M., Myers-Smith, I., Navarro, L., Niamir, A., Supp, S., Waldock, C., Vellend, M., Blowes, S., Boehning-Gaese, K., Bruelheide, H., et al., 2018. The geography of the anthropocene differs between the land and the sea. BioRxiv. https://doi.org/10.1101/432880.

Bracken, M.E.S., Friberg, S.E., Gonzalez-Dorantes, C.A., Williams, S.L., 2008. Functional consequences of realistic biodiversity changes in a marine ecosystem. Proc. Natl. Acad. Sci. U. S. A. 105, 924–928.

Brose, U., Hillebrand, H., 2016. Biodiversity and ecosystem functioning in soil. Philos. Trans. R. Soc. B 371, 20150267.

Brose, U., Ostling, A., Harrison, K., Martinez, N.D., 2004. Unified spatial scaling of species and their trophic interactions. Nature 428, 167–171.

Brose, U., Blanchard, J.L., Eklöf, A., Galiana, N., Hartvig, M., Hirt, M.R., Kalinkat, G., Nordström, M.C., O'Gorman, E.J., Rall, B.C., Schneider, F.D., Thébault, E., Jacob, U., 2017. Predicting the consequences of species loss using size-structured biodiversity approaches. Biol. Rev. 92, 684–697.

Bruelheide, H., Nadrowski, K., Assmann, T., Bauhus, J., Both, S., Buscot, F., Chen, X.Y., Ding, B., Durka, W., Erfmeier, A., Gutknecht, J.L.M., Guo, D., et al., 2014. Designing forest biodiversity experiments: general considerations illustrated by a new large experiment in subtropical China. Methods Ecol. Evol. 5, 74–89.

Bruno, J.F., Lee, S.C., Kertesz, J.S., Carpenter, R.C., Long, Z.T., Duffy, J.E., 2006. Partitioning the effects of algal species identity and richness on benthic marine primary production. Oikos 115, 170–178.

Bruno, J.F., Boyer, K.E., Duffy, E., Lee, S.C., 2008. Relative and interactive effects of plant and grazer richness in a benthic marine community. Ecology 89, 2518–2528.

Brust, C.A., Burghardt, T., Groenenberg, M., Käding, C., Kühl, H.S., Manguette, M.L., Denzler, J., 2018. Towards automated visual monitoring of individual gorillas in the wild. In: Proceedings—2017 IEEE International Conference on Computer Vision Workshops, ICCVW 2017, pp. 2820–2830.

Byrnes, J.E.K., Gamfeldt, L., Isbell, F., Lefcheck, J.S., Griffin, J.N., Hector, A., Cardinale, B.J., Hooper, D.U., Dee, L.E., Duffy, J.E., 2014. Investigating the relationship between biodiversity and ecosystem multifunctionality: challenges and solutions. Methods Ecol. Evol. 5, 111–124.

Cabello, J., Fernández, N., Alcaraz-Segura, D., Oyonarte, C., Pineiro, G., Altesor, A., Delibes, M., Paruelo, J.M., 2012. The ecosystem functioning dimension in conservation: insights from remote sensing. Biodivers. Conserv. 21, 3287–3305.

Cadotte, M.W., 2013. Experimental evidence that evolutionarily diverse assemblages result in higher productivity. Proc. Natl. Acad. Sci. U. S. A. 110, 8996–9000.

Cadotte, M.W., Cardinale, B.J., Oakley, T.H., 2008. Evolutionary history and the effect of biodiversity on plant productivity. Proc. Natl. Acad. Sci. U. S. A. 105, 17012–17017.

Cardillo, M., Mace, G.M., Jones, K.E., Bielby, J., Bininda-Emonds, O.R.P., Sechrest, W., Orme, C.D.L., Purvis, A., 2005. Multiple causes of high extinction risk in large mammal species. Science 309, 1239–1241.

Cardinale, B.J., 2011. Biodiversity improves water quality through niche partitioning. Nature 472, 86–91.

Cardinale, B.J., Palmer, M.A., Collins, S.L., 2002. Species diversity enhances ecosystem functioning through interspecific facilitation. Nature 415, 426–429.

Cardinale, B.J., Harvey, C.T., Gross, K., Ives, A.R., 2003. Biodiversity and biocontrol: emergent impacts of a multi-enemy assemblage on pest suppression and crop yield in an agroecosystem. Ecol. Lett. 6, 857–865.

Cardinale, B.J., Wright, J.P., Cadotte, M.W., Carroll, I.T., Hector, A., Srivastava, D.S., Loreau, M., Weis, J.J., 2007. Impacts of plant diversity on biomass production increase through time because of species complementarity. Proc. Natl. Acad. Sci. U. S. A. 104, 18123–18128.

Cardinale, B.J., Matulich, K.L., Hooper, D.U., Byrnes, J.E., Duffy, E., Gamfeldt, L., Balvanera, P., O'Connor, M.I., Gonzalez, A., 2011. The functional role of producer diversity in ecosystems. Am. J. Bot. 98, 572–592.

Cardinale, B.J., Duffy, J.E., Gonzalez, A., Hooper, D.U., Perrings, C., Venail, P., Narwani, A., Mace, G.M., Tilman, D., Wardle, D.A., et al., 2012. Biodiversity loss and its impact on humanity. Nature 486, 59–67.

Carroll, S.P., Hendry, A.P., Reznick, D.N., Fox, C.W., 2007. Evolution on ecological timescales. Funct. Ecol. 21, 387–393.

Chapin III, F.S., Zavaleta, E.S., Eviner, V.T., Naylor, R.L., Vitousek, P.M., Reynolds, H.L., Hooper, D.U., Lavorel, S., Sala, O.E., Hobbie, S.E., et al., 2000. Consequences of changing biodiversity. Nature 405, 234–242.

Chapman, A., Simperl, E., Koesten, L., Konstantinidis, G., Ibáñez-Gonzalez, L.-D., Kacprzak, E., Groth, P., 2019. Dataset search: a survey. arXiv. 1901.00735.

Charmantier, A., Mccleery, R.H., Cole, L.R., Perrins, C., Kruuk, L.E.B., Sheldon, B.C., 2008. Adaptive phenotypic plasticity in response to climate change in a wild bird population. Science 800, 800–804.

Chase, J.M., Leibold, M.A., 2003. Ecological Niches: Linking Classical and Contemporary Approaches. University of Chicago Press, Chicago.

Chesson, P., Pacala, S., Neuhauser, C., 2001. Environmental niches and ecosystem functioning. In: Kinzig, A.P., Pacala, S., Tilman, D. (Eds.), The Functional Consequences of Biodiversity. Princeton University Press, pp. 213–245.

Civitello, D.J., Cohen, J., Fatima, H., Halstead, N.T., Liriano, J., McMahon, T.A., Ortega, C.N., Sauer, E.L., Sehgal, T., Young, S., Rohr, J.R., 2015. Biodiversity inhibits parasites: broad evidence for the dilution effect. Proc. Natl. Acad. Sci. U. S. A. 112, 8667–8671.

Clark, C.M., Cleland, E.E., Collins, S.L., Fargione, J.E., Gough, L., Gross, K.L., Pennings, S.C., Suding, K.N., Grace, J.B., 2007. Environmental and plant community determinants of species loss following nitrogen enrichment. Ecol. Lett. 10, 596–607.

Connolly, J., Bell, T., Bolger, T., Brophy, C., Carnus, T., Finn, J.A., Kirwan, L., Isbell, F., Levine, J., Lüscher, A., Picasso, V., Roscher, C., et al., 2013. An improved model to predict the effects of changing biodiversity levels on ecosystem function. J. Ecol. 101, 344–355.

Costanza, R., D'Arge, R., de Groot, R., Farber, S., Grasso, M., Hannon, B., Limburg, K., Naeem, S., O'Neill, R.V., Paruelo, J., Raskin, R.G., Sutton, P., van den Belt, M., 1997. The value of the world's ecosystem services and natural capital. Nature 387, 253–260.

Coulis, M., Fromin, N., David, J.F., Gavinet, J., Clet, A., Devidal, S., Roy, J., Hättenschwiler, S., 2015. Functional dissimilarity across trophic levels as a driver of soil processes in a Mediterranean decomposer system exposed to two moisture levels. Oikos 124, 1304–1316.

Cowles, J.M., Wragg, P.D., Wright, A.J., Powers, J.S., Tilman, D., 2016. Shifting grassland plant community structure drives positive interactive effects of warming and diversity on aboveground net primary productivity. Glob. Chang. Biol. 22, 741–749.

Craven, D., Isbell, F., Manning, P., Connolly, J., Bruelheide, H., Ebeling, A., Roscher, C., van Ruijven, J., Weigelt, A., Wilsey, B., Beierkuhnlein, C., de Luca, E., et al., 2016. Plant diversity effects on grassland productivity are robust to both nutrient enrichment and drought. Philos. Trans. R. Soc. B 371, 20150277.

Craven, D., Eisenhauer, N., et al., 2018. Multiple facets of biodiversity drive the diversity–stability relationship. Nat. Ecol. Evol. 2, 1579–1587.

Craven, D., Winter, M., Hotzel, K., Gaikwad, J., Eisenhauer, N., Hohmuth, M., König-Ries, B., Wirth, C., 2019. Evolution of interdisciplinarity in biodiversity science. Ecol. Evol. 9, 6744–6755.

Crutsinger, G.M., Collins, M.D., Fordyce, J.A., Gompert, Z., Nice, C.C., Sanders, N.J., 2006. Plant genotypic diversity predicts community structure and governs an ecosystem process. Science 313, 966–968.

Dallimer, M., Irvine, K.N., Skinner, A.M.J., Davies, Z.G., Rouquette, J.R., Maltby, L.L., Warren, P.H., Armsworth, P.R., Gaston, K.J., 2012. Biodiversity and the feel-good factor: understanding associations between self-reported human well-being and species richness. Bioscience 62, 47–55.

Darwin, C., 1859. On the Origins of Species by Means of Natural Selection. John Murray, United Kingdom.

Darwin, C., Wallace, A., 1858. On the tendency of species to form varieties; and on the perpetuation of varieties and species by natural means of selection. J. Proc. Linn. Soc. Lond. 3, 45–62.

De Laender, F., Rohr, J.R., Ashauer, R., Baird, D.J., Berger, U., Eisenhauer, N., Grimm, V., Hommen, U., Maltby, L., Pomati, F., Roessink, I., Radchuk, V., Van den Brink, P.J., 2016. Re-introducing environmental change drivers in biodiversity-ecosystem functioning research. Trends Ecol. Evol. 31, 905–915.

de Mazancourt, C., Isbell, F., Larocque, A., Berendse, F., De Luca, E., Grace, J.B., Haegeman, B., Polley, W.H., Roscher, C., Schmid, B., Tilman, D., van Ruijven, J., et al., 2013. Predicting ecosystem stability from community composition and biodiversity. Ecol. Lett. 16, 617–625.

Dee, L.E., Allesina, S., Bonn, A., Eklöf, A., Gaines, S.D., Hines, J., Jacob, U., McDonald-Madden, E., Possingham, H., Schröter, M., Thompson, R.M., 2017. Operationalizing network theory for ecosystem service assessments. Trends Ecol. Evol. 32, 118–130.

Dell, A.I., Bender, J.A., Branson, K., Couzin, I.D., de Polavieja, G.G., Noldus, L.P., Pérez-Escudero, A., Perona, P., Straw, A.D., Wikelski, M., Brose, U., 2014. Automated image-based tracking and its application in ecology. Trends Ecol. Evol. 29, 417–428.

Dias, A.T.C., Berg, M.P., de Bello, F., Van Oosten, A.R., Bílá, K., Moretti, M., 2013. An experimental framework to identify community functional components driving ecosystem processes and services delivery. J. Ecol. 101, 29–37.

Díaz, S., Symstad, A.J., Chapin, F.S., Wardle, D.A., Huenneke, L.F., 2003. Functional diversity revealed by removal experiments. Trends Ecol. Evol. 18, 140–146.

Dirzo, R., Young, H.S., Galetti, M., Ceballos, G., Isaac, N.J.B., Collen, B., 2014. Defaunation in the anthropocene. Science 345, 401–406.

Donohue, I., Hillebrand, H., Montoya, J.M., Petchey, O.L., Pimm, S.L., Fowler, M.S., Healy, K., Jackson, A.L., Lurgi, M., McClean, D., O'Connor, N.E., O'Gorman, E.J., Yang, Q., 2016. Navigating the complexity of ecological stability. Ecol. Lett. 19, 1172–1185.

Dornelas, M., Gotelli, N.J., McGill, B., Shimadzu, H., Moyes, F., Sievers, C., Magurran, A.E., 2014. Assemblage time series reveal biodiversity change but not systematic loss. Science 344, 296–299.

Dossena, M., Yvon-Durocher, G., Grey, J., Montoya, J.M., Perkins, D.M., Trimmer, M., Woodward, G., 2012. Warming alters community size structure and ecosystem functioning. Proc. R. Soc. B Biol. Sci. 279, 3011–3019.

Duffy, J.E., 2002. Biodiversity and ecosystem function: the consumer connection. Oikos 99, 201–219.

Duffy, J.E., Cardinale, B.J., France, K.E., McIntyre, P.B., Thébault, E., Loreau, M., 2007. The functional role of biodiversity in ecosystems: Incorporating trophic complexity. Ecol. Lett. 10, 522–538.

Duffy, E.J., Godwin, C.M., Cardinale, B.J., 2017. Biodiversity effects in the wild are common and as strong as key drivers of productivity. Nature 549, 261–264.

Ebeling, A., Klein, A.-M., Schumacher, J., Weisser, W.W., Tscharntke, T., 2008. How does plant richness affect pollinator richness and temporal stability of flower visits? Oikos 117, 1808–1815.

Ebeling, A., Pompe, S., Baade, J., Eisenhauer, N., Hillebrand, H., Proulx, R., Roscher, C., Schmid, B., Wirth, C., Weisser, W.W., 2014. A trait-based experimental approach to understand the mechanisms underlying biodiversity-ecosystem functioning relationships. Basic Appl. Ecol. 15, 229–240.

Ebeling, A., Rzanny, M., Lange, M., Eisenhauer, N., Hertzog, L.R., Meyer, S.T., Weisser, W.W., 2018. Plant diversity induces shifts in the functional structure and diversity across trophic levels. Oikos 127, 208–219.

Eisenhauer, N., 2017. Consumers control carbon. Nat. Ecol. Evol. 1, 1596–1597.

Eisenhauer, N., Türke, M., 2018. From climate chambers to biodiversity chambers. Front. Ecol. Environ. 16, 136–137.

Eisenhauer, N., Beßler, H., Engels, C., Gleixner, G., Habekost, M., Milcu, A., Partsch, S., Sabais, A.C.W., Scherber, C., Steinbeiss, S., Weigelt, A., Weisser, W.W., Scheu, S., 2010. Plant diversity effects on soil microorganisms support the singular hypothesis. Ecology 91, 485–496.

Eisenhauer, N., Milcu, A., Allan, E., Nitschke, N., Scherber, C., Temperton, V., Weigelt, A., Weisser, W.W., Scheu, S., 2011. Impact of above- and below-ground invertebrates on temporal and spatial stability of grassland of different diversity. J. Ecol. 99, 572–582.

Eisenhauer, N., Reich, P.B., Scheu, S., 2012. Increasing plant diversity effects on productivity with time due to delayed soil biota effects on plants. Basic Appl. Ecol. 13, 571–578.

Eisenhauer, N., Dobies, T., Cesarz, S., Hobbie, S.E., Meyer, R.J., Worm, K., Reich, P.B., 2013. Plant diversity effects on soil food webs are stronger than those of elevated $CO_2$ and N deposition in a long-term grassland experiment. Proc. Natl. Acad. Sci. U. S. A. 110, 6889–6894.

Eisenhauer, N., Barnes, A.D., Cesarz, S., Craven, D., Ferlian, O., Gottschall, F., Hines, J., Sendek, A., Siebert, J., Thakur, M.P., Türke, M., 2016. Biodiversity–ecosystem function experiments reveal the mechanisms underlying the consequences of biodiversity change in real world ecosystems. J. Veg. Sci. 27, 1061–1070.

Eisenhauer, N., Hines, J., Isbell, F., van Der Plas, F., Hobbie, S.E., Kazanski, C.E., Lehmann, A., Liu, M., Lochner, A., Rillig, M.C., Vogel, A., Worm, K., Reich, P.B., 2018. Plant diversity maintains multiple soil functions in future environments. eLife 7, e41228.

Eisenhauer, N., Bonn, A., Guerra, A.C., 2019. Recognizing the quiet extinction of invertebrates. Nat. Commun. 10, 50.

Ellegren, H., 2014. Genome sequencing and population genomics in non-model organisms. Trends Ecol. Evol. 29, 51–63.

Elton, C.S., 1958. The Ecology of Invasions by Animals and Plants. Methuen and Co. Ltd., London.

Escribano, N., Galicia, D., Ariño, A.H., 2018. The tragedy of the biodiversity data commons: a data impediment creeping nigher? Database, 1–6.

Everwand, G., Rösch, V., Tscharntke, T., Scherber, C., 2014. Disentangling direct and indirect effects of experimental grassland management and plant functional-group manipulation on plant and leafhopper diversity. BMC Ecol. 14, 1.

Fahrig, L., 2017. Ecological responses to habitat fragmentation *per se*. Annu. Rev. Ecol. Evol. Syst. 48, 1–23.

Fahrig, L., Arroyo-Rodríguez, V., Bennett, J.R., Boucher-Lalonde, V., Cazetta, E., Currie, D.J., Eigenbrod, F., Ford, A.T., Harrison, S.P., Jaeger, J.A.G., Koper, N., Martin, A.E., et al., 2019. Is habitat fragmentation bad for biodiversity? Biol. Conserv. 230, 179–186.

Ferlian, O., Cesarz, S., Craven, D., Hines, J., Barry, K.E., Bruelheide, H., Buscot, F., Haider, S., Heklau, H., Herrmann, S., Kühn, P., Pruschitzki, U., et al., 2018. Mycorrhiza in tree diversity-ecosystem function relationships: conceptual framework and experimental implementation. Ecosphere 9, e02226.

Finger, R., Buchmann, N., 2015. An ecological economic assessment of risk-reducing effects of species diversity in managed grasslands. Ecol. Econ. 110, 89–97.

Finn, J.A., Kirwan, L., Connolly, J., Sebastià, M.T., Helgadottir, A., Baadshaug, O.H., Bélanger, G., Black, A., Brophy, C., Collins, R.P., Čop, J., Dalmannsdóttir, S., et al., 2013. Ecosystem function enhanced by combining four functional types of plant species in intensively managed grassland mixtures: A 3-year continental-scale field experiment. J. Appl. Ecol. 50, 365–375.

Fornara, D.A., Tilman, D., 2008. Plant functional composition influences rates of soil carbon and nitrogen accumulation. J. Ecol. 96, 314–322.

Fox, J.W., 2005. Interpreting the "selection effect" of biodiversity on ecosystem function. Ecol. Lett. 8, 846–856.

Fox, J.W., Vasseur, D., Cotroneo, M., Guan, L., Simon, F., 2017. Population extinctions can increase metapopulation persistence. Nat. Ecol. Evol. 1, 1271–1278.

Fridley, J.D., 2002. Resource availability dominates and alters the relationship between species diversity and ecosystem productivity in experimental plant communities. Oecologia 132, 271–277.

Galiana, N., Lurgi, M., Claramunt-Lopez, B., Fortin, M.J., Leroux, S., Cazelles, K., Gravel, D., Montoya, J.M., 2018. The spatial scaling of species interaction networks. Nat. Ecol. Evol. 2, 782–790.

Gérard, A., Wollni, M., Hölscher, D., Irawan, B., Sundawati, L., Teuscher, M., Kreft, H., 2017. Oil-palm yields in diversified plantations: Initial results from a biodiversity enrichment experiment in Sumatra, Indonesia. Agric. Ecosyst. Environ. 240, 253–260.

Gessner, M.O., Swan, C.M., Dang, C.K., McKie, B.G., Bardgett, R.D., Wall, D.H., Hättenschwiler, S., 2010. Diversity meets decomposition. Trends Ecol. Evol. 25, 372–380.

Gilarranz, L.J., Rayfield, B., Liñán-Cembrano, G., Bascompte, J., Gonzalez, A., 2017. Effects of network modularity on the spread of perturbation impact in experimental metapopulations. Science 357, 199–201.

Giling, D.P., Beaumelle, L., Phillips, H.R.P., Cesarz, S., Eisenhauer, N., Ferlian, O., Gottschall, F., Guerra, C., Hines, J., Sendek, A., Siebert, J., Thakur, M.P., Barnes, A.D., 2019. A niche for ecosystem multifunctionality in global change research. Glob. Chang. Biol. 25, 763–774.

Givnish, T.J., 1994. Does diversity beget stability? Nature 371, 113–114.

Gonzalez, A., Thompson, P., Loreau, M., 2017. Spatial ecological networks: planning for sustainability in the long-term. Curr. Opin. Environ. Sustain. 29, 187–197.

Gossner, M.M., Lewinsohn, T.M., Kahl, T., Grassein, F., Boch, S., Prati, D., Birkhofer, K., Renner, S.C., Sikorski, J., Wubet, T., Arndt, H., Baumgartner, V., et al., 2016. Land-use intensification causes multitrophic homogenization of grassland communities. Nature 540, 266–269.

Götzenberger, L., de Bello, F., Bråthen, K.A., Davison, J., Dubuis, A., Guisan, A., Lepš, J., Lindborg, R., Moora, M., Pärtel, M., Pellissier, L., Pottier, J., et al., 2012. Ecological assembly rules in plant communities-approaches, patterns and prospects. Biol. Rev. 87, 111–127.

Gounand, I., Little, C.J., Harvey, E., Altermatt, F., 2018. Cross-ecosystem carbon flows connecting ecosystems worldwide. Nat. Commun. 9, 4825.

Grace, J.B., Anderson, T.M., Seabloom, E.W., Borer, E.T., Adler, P.B., Harpole, W.S., Hautier, Y., Hillebrand, H., Lind, E.M., Pärtel, M., Bakker, J.D., Buckley, Y.M., et al., 2016. Integrative modelling reveals mechanisms linking productivity and plant species richness. Nature 529, 390–393.

Grass, I., Jauker, B., Steffan-Dewenter, I., Tscharntke, T., Jauker, F., 2018. Past and potential future effects of habitat fragmentation on structure and stability of plant–pollinator and host–parasitoid networks. Nat. Ecol. Evol. 2, 1408–1417.

Gravel, D., Massol, F., Canard, E., Mouillot, D., Mouquet, N., 2011. Trophic theory of island biogeography. Ecol. Lett. 14, 1010–1016.

Gray, A.J.G., Goble, C., Jimenez, R.C., 2017. Bioschemas: From Potato Salad to Protein Annotation. CEUR Workshop Proceedings.

Griffin, J.N., Jenkins, S.R., Gamfeldt, L., Jones, D., Hawkins, S.J., Thompson, R.C., 2009. Spatial heterogeneity increases the importance of species richness for an ecosystem process. Oikos 118, 1335–1342.

Grime, J.P., 1997. Biodiversity and ecosystem function: the debate deepens. Science 277, 1260–1261.

Grime, J.P., 1998. Benefits of plant diversity to ecosystems: immediate, filter and founder effects. J. Ecol. 86, 902–910.

Grossman, J.J., Vanhellemont, M., Barsoum, N., Bauhus, J., Bruelheide, H., Castagneyrol, B., Cavender-Bares, J., Eisenhauer, N., Ferlian, O., Gravel, D., Hector, A., Jactel, H., et al., 2018. Synthesis and future research directions linking tree diversity to growth, survival, and damage in a global network of tree diversity experiments. Environ. Exp. Bot. 152, 68–89.

Gruber, T.R., 1993. A translation approach to portable ontologies. Knowl. Acquis. 5, 199–220.

Guerrero-Ramírez, N.R., Eisenhauer, N., 2017. Trophic and non-trophic interactions influence the mechanisms underlying biodiversity–ecosystem functioning relationships under different abiotic conditions. Oikos 126, 1748–1759.

Guerrero-Ramírez, N.R., Craven, D., Reich, P.B., Ewel, J.J., Koricheva, J., Parrotta, J.A., Auge, H., Erickson, H.E., Forrester, D.I., Hector, A., Joshi, J., Montagnini, F., 2017. Diversity-dependent temporal divergence of ecosystem functioning in experimental ecosystems. Nat. Ecol. Evol. 1, 1639–1642.

Haddad, N.M., Holyoak, M., Mata, T.M., Davies, K.F., Melbourne, B.A., Preston, K., 2008. Species traits predict the effects of disturbance and productivity on diversity. Ecol. Lett. 11, 348–356.

Haddad, N.M., Crutsinger, G.M., Gross, K., Haarstad, J., Knops, J.M.H., Tilman, D., 2009. Plant species loss decreases arthropod diversity and shifts trophic structure. Ecol. Lett. 12, 1029–1039.

Hampton, S.E., Strasser, C.A., Tewksbury, J.J., Gram, W.K., Budden, A.E., Batcheller, A.L., Duke, C.S., Porter, J.H., 2013. Big data and the future of ecology. Front. Ecol. Environ. 11, 156–162.

Handa, I.T., Aerts, R., Berendse, F., Berg, M.P., Bruder, A., Butenschoen, O., Chauvet, E., Gessner, M.O., Jabiol, J., Makkonen, M., McKie, B.G., Malmqvist, B., et al., 2014. Consequences of biodiversity loss for litter decomposition across biomes. Nature 509, 218–221.

Hantsch, L., Braun, U., Scherer-Lorenzen, M., Bruelheide, H., 2013. Tree diversity effects on species richness and infestation of foliar fungal pathogens in European tree diveristy experiments. Ecosphere 4, 81.

Hantsch, L., Bien, S., Radatz, S., Braun, U., Auge, H., Bruelheide, H., 2014. Tree diversity and the role of non-host neighbour tree species in reducing fungal pathogen infestation. J. Ecol. 102, 1673–1687.

Hardisty, A., Roberts, D., 2013. A decadal view of biodiversity informatics: challenges and priorities. BMC Ecol. 13, 16.

Hardisty, A., Michener, W.K., Agosti, D., Garcia, E.A., Bastin, L., Belbin, L., Bowser, A., Buttigieg, P.L., Canhos, D.A.L., Egloff, W., De Giovanni, R., Figueira, R., 2019. The Bari Manifesto: an interoperability framework for essential biodiversity variables. Eco. Inform. 49, 22–31.

Harmon, L.J., Matthews, B., Des Roches, S., Chase, J.M., Shurin, J.B., Schluter, D., 2009. Evolutionary diversification in stickleback affects ecosystem functioning. Nature 458, 1167–1170.

Harpole, W.S., Tilman, D., 2007. Grassland species loss resulting from reduced niche dimension. Nature 446, 791–793.

Harpole, W.S., Sullivan, L.L., Lind, E.M., Firn, J., Adler, P.B., Borer, E.T., Chase, J., Fay, P.A., Hautier, Y., Hillebrand, H., MacDougall, A.S., Seabloom, E.W., et al., 2016. Addition of multiple limiting resources reduces grassland diversity. Nature 537, 93–96.

Hautier, Y., Niklaus, P.A., Hector, A., 2009. Competition for light causes plant biodiversity loss after eutrophication. Science 324, 636–638.

Hautier, Y., Isbell, F., Borer, E.T., Seabloom, E.W., Harpole, W.S., Lind, E.M., MacDougall, A.S., Stevens, C.J., Adler, P.B., Alberti, J., Bakker, J.D., Brudvig, L.A., et al., 2018. Local loss and spatial homogenization of plant diversity reduce ecosystem multifunctionality. Nat. Ecol. Evol. 2, 50–56.

Hector, A., Bagchi, R., 2007. Biodiversity and ecosystem multifunctionality. Nature 448, 188–190.

Hector, A., Hooper, R., 2002. Darwin and the first ecological experiment. Science 295, 639–640.

Hector, A., Schmid, B., Beierkuhnlein, C., Caldeira, M.C., Diemer, M., Dimitrakopoulos, P.G., Finn, J.A., Freitas, H., Giller, P.S., Good, J., Harris, R., Högberg, P., et al., 1999. Plant diversity and productivity experiments in European grasslands. Science 286, 1123–1127.

Hendriks, M., Mommer, L., de Caluwe, H., Smit-Tiekstra, A.E., van der Putten, W.H., de Kroon, H., 2013. Independent variations of plant and soil mixtures reveal soil feedback effects on plant community overyielding. J. Ecol. 101, 287–297.

Hendry, A.P., 2016. Eco-Evolutionary Dynamics. Princeton University Press, Princeton.

Hertzog, L.R., Ebeling, A., Weisser, W.W., Meyer, S.T., 2017. Plant diversity increases predation by ground-dwelling invertebrate predators. Ecosphere 8 e01990.

Hillebrand, H., Matthiessen, B., 2009. Biodiversity in a complex world: consolidation and progress in functional biodiversity research. Ecol. Lett. 12, 1405–1419.

Hillebrand, H., Blasius, B., Borer, E.T., Chase, J.M., Downing, J.A., Eriksson, B.K., Filstrup, C.T., Harpole, W.S., Hodapp, D., Larsen, S., Lewandowska, A.M., Seabloom, E.W., et al., 2018. Biodiversity change is uncoupled from species richness trends: consequences for conservation and monitoring. J. Appl. Ecol. 55, 169–184.

Hines, J., Eisenhauer, N., Drake, B.G., 2015a. Inter-annual changes in detritus-based food chains can enhance plant growth response to elevated atmospheric $CO_2$. Glob. Chang. Biol. 21, 4642–4650.

Hines, J., van der Putten, W.H., De Deyn, G.B., Wagg, C., Voigt, W., Mulder, C., Weisser, W.W., Engel, J., Melian, C., Scheu, S., Birkhofer, K., Ebeling, A., et al., 2015b. Towards an integration of biodiversity-ecosystem functioning and food web theory to evaluate relationships between multiple ecosystem services. Adv. Ecol. Res. 53, 161–199.

Hines, J., Ebeling, A., Barnes, A.D., Brose, U., Scherber, C., Scheu, S., Tscharntke, T., Weisser, W.W., Giling, D.P., Klein, A.-M., Eisenhauer, N., 2019. Mapping change in biodiversity and ecosystem function research: food webs foster integration of experiments and science policy. Adv. Ecol. Res. 61, 297–322.

Hirt, M.R., Grimm, V., Li, Y., Rall, B.C., Rosenbaum, B., Brose, U., 2018. Bridging scales: allometric random walks link movement and biodiversity research. Trends Ecol. Evol. 33, 701–712.

Hobbie, S.E., 1992. Effects of Plant Species Nutrient Cycling. Trends Ecol. Evol. 7, 336–339.

Hodapp, D., Hillebrand, H., Blasius, B., Ryabov, A.B., 2016. Environmental and trait variability constrain community structure and the biodiversity-productivity relationship. Ecology 97, 1463–1474.

Hoffmann, A.A., Sgró, C.M., 2011. Climate change and evolutionary adaptation. Nature 470, 479–485.

Hofstetter, R., Dempsey, T., Klepzig, K., Ayres, M., 2007. Temperature-dependent effects on mutualistic, antagonistic, and commensalistic interactions among insects, fungi and mites. Commun. Ecol. 8, 47–56.

Hooper, D.U., Chapin III, F.S., Ewel, J.J., Hector, A., Inchausti, P., Lavorel, S., Lawton, J.H., Lodge, D.M., Loreau, M., Naeem, S., Schmid, B., Setälä, H., et al., 2005. Effects of biodiversity on ecosystem functioning: a consensus of current knowledge. Ecol. Monogr. 75, 3–35.

Huang, Y., Chen, Y., Castro-Izaguirre, N., Baruffol, M., Brezzi, M., Lang, A., Li, Y., Härdtle, W., Von Oheimb, G., Yang, X., Liu, X., Pei, K., et al., 2018. Impacts of species richness on productivity in a large-scale subtropical forest experiment. Science 362, 80–83.

Hughes, A.R., Inouye, B.D., Johnson, M.T.J., Underwood, N., Vellend, M., 2008. Ecological consequences of genetic diversity. Ecol. Lett. 11, 609–623.

Huston, M.A., 1997. Hidden treatments in ecological experiments: re-evaluating the ecosystem function of biodiversity. Oecologia 110, 449–460.

Hutchinson, G.E., 1957. Concluding remarks. In: Cold Spring Harbor Symposia on Quantitative Biology. vol. 22. The Biological Laboratory, Cold Spring Harbor, pp. 415–421.

IPBES, 2018. In: Fischer, M., Rounsevell, M., Torre-Mari, A. (Eds.), Summary for policymakers of the regional assessment report on biodiversity and ecosystem services for Europe and Central Asia of the intergovernmental science-policy platform on biodiversity and ecosystem services. IPBES Secretariat, Bonn, Germany.

IPCC, 2014. Climate Change 2014: Synthesis report. In: Core Writing Team, , Pachauri, R.K., Meyer, L.A. (Eds.), Contribution of Working Groups I, II and III to the Fifth Assessment Report of the Intergovernmental Panel on Climate Change. IPCC, Geneva, Switzerland. 151 pp.

Isbell, F., Calcagno, V., Hector, A., Connolly, J., Harpole, W.S., Reich, P.B., Scherer-Lorenzen, M., Schmid, B., Tilman, D., van Ruijven, J., Weigelt, A., Wilsey, B.J., et al., 2011. High plant diversity is needed to maintain ecosystem services. Nature 477, 199–202.

Isbell, F., Reich, P.B., Tilman, D., Hobbie, S.E., Polasky, S., Binder, S., 2013. Nutrient enrichment, biodiversity loss, and consequent declines in ecosystem productivity. Proc. Natl. Acad. Sci. U. S. A. 110, 11911–11916.

Isbell, F., Craven, D., Connolly, J., Loreau, M., Schmid, B., Beierkuhnlein, C., Bezemer, T.M., Bonin, C., Bruelheide, H., De Luca, E., Ebeling, A., et al., 2015a. Biodiversity increases the resistance of ecosystem productivity to climate extremes. Nature 526, 574–577.
Isbell, F., Tilman, D., Polasky, S., Loreau, M., 2015b. The biodiversity-dependent ecosystem service debt. Ecol. Lett. 18, 119–134.
Isbell, F., Gonzalez, A., Loreau, M., Cowles, J., Díaz, S., Hector, A., Mace, G.M., Wardle, D.A., O'Connor, M.I., Duffy, J.E., Turnbull, L.A., Thompson, P.L., Larigauderie, A., 2017a. Linking the influence and dependence of people on biodiversity across scales. Nature 546, 65–72.
Isbell, F., Adler, P.R., Eisenhauer, N., Fornara, D., Kimmel, K., Kremen, C., Letourneau, D.K., Liebman, M., Polley, H.W., Quijas, S., Scherer-Lorenzen, M., 2017b. Benefits of increasing plant diversity in sustainable agroecosystems. J. Ecol. 105, 871–879.
Isbell, F., Cowles, J., Dee, L.E., Loreau, M., Reich, P.B., Gonzalez, A., Hector, A., Schmid, B., 2018. Quantifying effects of biodiversity on ecosystem functioning across times and places. Ecol. Lett. 21, 763–778.
Jacquet, C., Mouillot, D., Kulbicki, M., Gravel, D., 2017. Extensions of island biogeography theory predict the scaling of functional trait composition with habitat area and isolation. Ecol. Lett. 20, 135–146.
Janzen, D.H., 1974. The deflowering of Central America. Nat. Hist. 83, 49–53.
Jochum, M., Schneider, F.D., Crowe, T.P., Brose, U., O'Gorman, E.J., 2012. Climate-induced changes in bottom-up and top-down processes independently alter a marine ecosystem. Philos. Trans. R. Soc. B 367, 2962–2970.
Jousset, A., Schmid, B., Scheu, S., Eisenhauer, N., 2011. Genotypic richness and dissimilarity opposingly affect ecosystem functioning. Ecol. Lett. 14, 537–545.
Jousset, A., Eisenhauer, N., Merker, M., Mouquet, N., Scheu, S., 2016. High functional diversity stimulates diversification in experimental microbial communities. Sci. Adv. 2, e1600124.
Jousset, A., Bienhold, C., Chatzinotas, A., Gallien, L., Gobet, A., Kurm, V., Küsel, K., Rillig, M.C., Rivett, D.W., Salles, J.F., van Der Heijden, M.G.A., Youssef, N.H., et al., 2017. Where less may be more: how the rare biosphere pulls ecosystems strings. ISME J. 11, 853–862.
Jucker, T., Avăcăritei, D., Bărnoaiea, I., Duduman, G., Bouriaud, O., Coomes, D.A., 2016. Climate modulates the effects of tree diversity on forest productivity. J. Ecol. 104, 388–398.
Kambach, S., Kühn, I., Castagneyrol, B., Bruelheide, H., 2016. The impact of tree diversity on different aspects of insect herbivory along a global temperature gradient—a meta-analysis. PLoS One 11, 1–14.
Kardol, P., Fanin, N., Wardle, D.A., 2018. Long-term effects of species loss on community properties across contrasting ecosystems. Nature 557, 710–713.
Kettenring, K.M., Mercer, K.L., Reinhardt Adams, C., Hines, J., 2014. Application of genetic diversity-ecosystem function research to ecological restoration. J. Appl. Ecol. 51, 339–348.
Kirwan, L., Lüscher, A., Sebastià, M.T., Finn, J.A., Collins, R.P., Porqueddu, C., Helgadottir, A., Baadshaug, O.H., Brophy, C., Coran, C., Dalmannsdóttir, S., Delgado, I., et al., 2007. Evenness drives consistent diversity effects in intensive grassland systems across 28 European sites. J. Ecol. 95, 530–539.
Kleijn, D., Winfree, R., Bartomeus, I., Carvalheiro, L.G., Henry, M., Isaacs, R., Klein, A.M., Kremen, C., M'Gonigle, L.K., Rader, R., Ricketts, T.H., Williams, N.M., et al., 2015. Delivery of crop pollination services is an insufficient argument for wild pollinator conservation. Nat. Commun. 6, 7414.

Klein, A.M., Steffan-Dewenter, I., Tscharntke, T., 2003. Fruit set of highland coffee increases with the diversity of pollinating bees. Proc. R. Soc. B Biol. Sci. 270, 955–961.

Knight, T.M., McCoy, M.W., Chase, J.M., McCoy, K.A., Holt, R.D., 2005. Trophic cascades across ecosystems. Nature 437, 880–883.

Kotiaho, J.S., Kaitala, V., Komonen, A., Päivinen, J., 2005. Predicting the risk of extinction from shared ecological characteristics. Proc. Natl. Acad. Sci. U. S. A. 102, 1963–1967.

Laforest-Lapointe, I., Paquette, A., Messier, C., Kembel, S.W., 2017. Leaf bacterial diversity mediates plant diversity and ecosystem function relationships. Nature 546, 145–147.

Lange, M., Eisenhauer, N., Sierra, C.A., Bessler, H., Engels, C., Griffiths, R.I., Mellado-V ázquez, P.G., Malik, A.A., Roy, J., Scheu, S., Steinbeiss, S., Thomson, B.C., et al., 2015. Plant diversity increases soil microbial activity and soil carbon storage. Nat. Commun. 6, 6707.

Lange, M., Koller-France, E., Hildebrandt, A., Oelmann, Y., Wilcke, W., Gleixner, G., 2019. How plant diversity impacts the coupled water, nutrient and carbon cycles. Adv. Ecol. Res. 61, 185–219.

Latz, E., Eisenhauer, N., Rall, B.C., Allan, E., Roscher, C., Scheu, S., Jousset, A., 2012. Plant diversity improves protection against soil-borne pathogens by fostering antagonistic bacterial communities. J. Ecol. 100, 597–604.

Lefcheck, J.S., Byrnes, J.E.K., Isbell, F., Gamfeldt, L., Griffin, J.N., Eisenhauer, N., Hensel, M.J.S., Hector, A., Cardinale, B.J., Duffy, J.E., 2015. Biodiversity enhances ecosystem multifunctionality across trophic levels and habitats. Nat. Commun. 6, 6936.

Lefcheck, J.S., Brandl, S.J., Reynolds, P.L., Smyth, A.R., Meyer, S.T., 2016. Extending rapid ecosystem function assessments to marine ecosystems: a reply to Meyer. Trends Ecol. Evol. 31, 251–253.

Leibold, M.A., Chase, J.M., 2018. Metacommunity ecology. In: Monographs in Population Biology. vol. 59. Princeton University Press.

Leibold, M.A., Holyoak, M., Mouquet, N., Amarasekare, P., Chase, J.M., Hoopes, M.F., Holt, R.D., Shurin, J.B., Law, R., Tilman, D., Loreau, M., Gonzalez, A., 2004. The metacommunity concept: a framework for multi-scale community ecology. Ecol. Lett. 7, 601–613.

Leibold, M.A., Chase, J.M., Ernest, S.K.M., 2017. Community assembly and the functioning of ecosystems: how metacommunity processes alter ecosystems attributes. Ecology 98, 909–919.

Lepš, J., 2004. What do the biodiversity experiments tell us about consequences of plant species loss in the real world? Basic Appl. Ecol. 5, 529–534.

Lindeman, R.L., 1942. The trophic-dynamic aspect of ecology. Ecology 23, 399–417.

Lindenmayer, D.B., Likens, G.E., Franklin, J.F., 2010. Rapid responses to facilitate ecological discoveries from major disturbances. Front. Ecol. Environ. 8, 527–532.

Lipowsky, A., Schmid, B., Roscher, C., 2011. Selection for monoculture and mixture genotypes in a biodiversity experiment. Basic Appl. Ecol. 12, 360–371.

Lipowsky, A., Roscher, C., Schumacher, J., Schmid, B., 2012. Density-independent mortality and increasing plant diversity are associated with differentiation of *Taraxacum officinale* into r- and K-strategists. PLoS One 7 e28121.

Liu, X., Trogisch, S., He, J.S., Niklaus, P.A., Bruelheide, H., Tang, Z., Erfmeier, A., Scherer-Lorenzen, M., Pietsch, K.A., Yang, B., Kühn, P., Scholten, T., et al., 2018. Tree species richness increases ecosystem carbon storage in subtropical forests. Proc. R. Soc. B Biol. Sci. 285, 20181240.

Loreau, M., 1998. Biodiversity and ecosystem functioning: a mechanistic model. Proc. Natl. Acad. Sci. U. S. A. 95, 5632–5636.

Loreau, M., 2010. Linking biodiversity and ecosystems: towards a unifying ecological theory. Philos. Trans. R. Soc. B 365 (1537), 49–60.

Loreau, M., 2004. Does functional redundancy exist? Oikos 104, 606–611.

Loreau, M., Hector, A., 2001. Partitioning selection and complementarity in biodiversity experiments. Nature 412, 72–76.
Loreau, M., Naeem, S., Inchausti, P., Bengtsson, J., Grime, J.P., Hector, A., Hooper, D.U., Huston, M.A., Raffaelli, D., Schmid, B., Tilman, D., Wardle, D.A., 2001. Biodiversity and ecosystem functioning: current knowledge and future challenges. Science 294, 804–808.
Loreau, M., Mouquet, N., Gonzalez, A., 2003a. Biodiversity as spatial insurance in heterogeneous landscapes. Proc. Natl. Acad. Sci. U. S. A. 100, 12765–12770.
Loreau, M., Mouquet, N., Holt, R.D., 2003b. Meta-ecosystems: a theoretical framework for a spatial ecosystem ecology. Ecol. Lett. 6, 673–679.
Lozupone, C.A., Stombaugh, J.I., Gordon, J.I., Jansson, J.K., Knight, R., 2012. Diversity, stability and resilience of the human gut microbiota. Nature 489, 220–230.
Lyons, K.G., Brigham, C.A., Traut, B.H., Schwartz, M.W., 2005. Rare species and ecosystem functioning. Conserv. Biol. 19, 1019–1024.
MacArthur, R.H., Wilson, E.O., 1967. The Theory of Island Biogeography. Princeton University Press, Princeton.
Mace, G.M., 2014. Whose conservation? Science 345, 1558–1560.
Maestre, F., Quero, J., Gotelli, N., Escudero, A., Ochoa, V., Delgado-Baquerizo, M., Garcia-Gomez, M., Bowker, M.A., Soliveres, S., Escolar, C., Garcia-Palacios, P., Berdugo, M., 2012. Plant species richness and ecosystem multifunctionality in global drylands. Science 335, 214–219.
Manning, P., Van Der Plas, F., Soliveres, S., Allan, E., Maestre, F.T., Mace, G., Whittingham, M.J., Fischer, M., 2018. Redefining ecosystem multifunctionality. Nat. Ecol. Evol. 2, 427–436.
Manning, P., Loos, J., Barnes, A.D., Batàry, P., Bianchi, F.J.J.A., Buchmann, N., De Deyn, G.B., Ebeling, A., et al., 2019. Transferring biodiversity-ecosystem function research to the management of 'real-world' ecosystems. Adv. Ecol. Res. 61, 323–356.
Maron, J.L., Marler, M., Klironomos, J.N., Cleveland, C.C., 2011. Soil fungal pathogens and the relationship between plant diversity and productivity. Ecol. Lett. 14, 36–41.
Martin-Guay, M.O., Paquette, A., Dupras, J., Rivest, D., 2018. The new Green Revolution: sustainable intensification of agriculture by intercropping. Sci. Total Environ. 615, 767–772.
Maxwell, S.L., Fuller, R.A., Brooks, T.M., Watson, J.E., 2016. Biodiversity: the ravages of guns, nets and bulldozers. Nature News 536 (7615), 143.
McNaughton, S.J., 1977. Diversity and stability of ecological communities: a comment on the role of empiricism in ecology. Am. Nat. 111, 515–525.
MEA, 2005. Millennium Ecosystem Assessment (MEA). www.maweb.org.
Meyer, S.T., Koch, C., Weisser, W.W., 2015. Towards a standardized Rapid Ecosystem Function Assessment (REFA). Trends Ecol. Evol. 30, 390–397.
Meyer, S.T.M., Ebeling, A., Eisenhauer, N., Hertzog, L., Hillebrand, H., Milcu, A., Pompe, S., Abbas, M., Bessler, H., Buchmann, N., De Luca, E., Engels, C., et al., 2016. Effects of biodiversity strengthen over time as ecosystem functioning declines at low and increases at high biodiversity. Ecosphere 7 e01619.
Meyer, S.T., Ptacnik, R., Hillebrand, H., Bessler, H., Buchmann, N., Ebeling, A., Eisenhauer, N., Engels, C., Fischer, M., Halle, S., Klein, A.M., Oelmann, Y., et al., 2018. Biodiversity-multifunctionality relationships depend on identity and number of measured functions. Nat. Ecol. Evol. 2, 44–49.
Milcu, A., Roscher, C., Gessler, A., Bachmann, D., Gockele, A., Guderle, M., Landais, D., Piel, C., Escape, C., Devidal, S., Ravel, O., Buchmann, N., Gleixner, G., Hildebrandt, A., Roy, J., 2014. Functional diversity of leaf nitrogen concentrations drives grassland carbon fluxes. Ecol. Lett. 17, 435–444.

Mori, A.S., Furukawa, T., Sasaki, T., 2013. Response diversity determines the resilience of ecosystems to environmental change. Biol. Rev. 88, 349–364.

Mori, A.S., Isbell, F., Fujii, S., Makoto, K., Matsuoka, S., Osono, T., 2016. Low multifunctional redundancy of soil fungal diversity at multiple scales. Ecol. Lett. 19, 249–259.

Mori, A.S., Isbell, F., Seidl, R., 2018. β-Diversity, community assembly, and ecosystem functioning. Trends Ecol. Evol. 33, 549–564.

Morris, P.J., Hanken, J., Lowery, D., Ludäscher, B., Macklin, J., McPhillips, T., Wieczorek, J., Zhang, Q., 2018. Kurator: tools for improving fitness for use of biodiversity data. Biodivers. Inform. Sci. Stand. 2, e26539.

Mouillot, D., Bellwood, D.R., Baraloto, C., Chave, J., Galzin, R., Harmelin-Vivien, M., Kulbicki, M., Lavergne, S., Lavorel, S., Mouquet, N., Paine, C.E.T., Renaud, J., Thuiller, W., 2013. Rare species support vulnerable functions in high-diversity ecosystems. PLoS Biol. 11, e1001569.

Mouquet, N., Moore, J.L., Loreau, M., 2002. Plant species richness and community productivity: why the mechanism that promotes coexistence matters. Ecol. Lett. 5, 56–65.

Mousseau, T.A., Roff, D.A., 1987. Natural selection and the heritability of fitness components. Heredity 59, 181–197.

Mraja, A., Unsicker, S.B., Reichelt, M., Gershenzon, J., Roscher, C., 2011. Plant community diversity influences allocation to direct chemical defence in *Plantago lanceolata*. PLoS One 6, e28055.

Mulder, C.P.H., Koricheva, J., Huss-Danell, K., Högberg, P., Joshi, J., 1999. Insects affect relationships between plant species richness and ecosystem processes. Ecol. Lett. 2, 237–246.

Murphy, G.E.P., Romanuk, T.N., 2014. A meta-analysis of declines in local species richness from human disturbances. Ecol. Evol. 4, 91–103.

Naeem, S., Li, S., 1997. Biodiversity enhances ecosystem reliability. Nature 390, 507–509.

Naeem, S., Thompson, L.J., Lawler, S.P., Lawton, J.H., Woodfin, R.M., 1994. Declining biodiversity can alter the performance of ecosystems. Nature 368, 734–737.

Naeem, S., Duffy, J.E., Zavaleta, E., 2012. The functions of biological diversity in an age of extinction. Science 336, 1401–1406.

Narum, S.R., Buerkle, C.A., Davey, J.W., Miller, M.R., Hohenlohe, P.A., 2013. Genotyping-by-sequencing in ecological and conservation genomics. Mol. Ecol. 22, 2841–2847.

Newbold, T., Hudson, L.N., Hill, S.L.L., Contu, S., Lysenko, I., Senior, R.A., Börger, L., Bennett, D.J., Choimes, A., Collen, B., Day, J., De Palma, A., et al., 2015. Global effects of land use on local terrestrial biodiversity. Nature 520, 45–50.

Niklaus, P.A., Leadley, P.W., Schmid, B., Körner, C., 2001. A long-term field study on biodiversity X elevated $CO_2$ interactions in grassland. Eco. Inform. 71, 341–356.

Niklaus, P.A., Le Roux, X., Poly, F., Buchmann, N., Scherer-Lorenzen, M., Weigelt, A., Barnard, R.L., 2016. Plant species diversity affects soil–atmosphere fluxes of methane and nitrous oxide. Oecologia 181, 919–930.

Norberg, J., Swaney, D.P., Dushoff, J., Lin, J., Casagrandi, R., Levin, S.A., 2001. Phenotypic diversity and ecosystem functioning in changing environments: a theoretical framework. Proc. Natl. Acad. Sci. U. S. A. 98, 11376–11381.

O'Connor, M.I., Gonzalez, A., Byrnes, J.E.K., Cardinale, B.J., Duffy, J.E., Gamfeldt, L., Griffin, J.N., Hooper, D., Hungate, B.A., Paquette, A., Thompson, P.L., Dee, L.E., Dolan, K.L., 2017. A general biodiversity–function relationship is mediated by trophic level. Oikos 126, 18–31.

Oehri, J., Schmid, B., Schaepman-Strub, G., Niklaus, P.A., 2017. Biodiversity promotes primary productivity and growing season lengthening at the landscape scale. Proc. Natl. Acad. Sci. U. S. A. 114, 10160–10165.

Paaby, A.B., Rockman, M.V., 2014. Cryptic genetic variation, evolution's hidden substrate. Nat. Rev. Genet. 15, 247–258.

Paine, C.E.T., Amissah, L., Auge, H., Baraloto, C., Baruffol, M., Bourland, N., Bruelheide, H., Daïnou, K., de Gouvenain, R.C., Doucet, J.L., Doust, S., Fine, P.V.A., et al., 2015. Globally, functional traits are weak predictors of juvenile tree growth, and we do not know why. J. Ecol. 103, 978–989.

Paquette, A., Hector, A., Vanhellemont, M., Koricheva, J., Scherer-Lorenzen, M., Verheyen, K., Abdala-Roberts, L., Auge, H., Barsoum, N., Bauhus, J., Baum, C., Bruelheide, H., et al., 2018. A million and more trees for science. Nat. Ecol. Evol. 2, 763–766.

Pasari, J.R., Levi, T., Zavaleta, E.S., Tilman, D., 2013. Several scales of biodiversity affect ecosystem multifunctionality. Proc. Natl. Acad. Sci. U. S. A. 110 (25), 10219–10222.

Pelletier, F., Garant, D., Hendry, A.P., 2009. Eco-evolutionary dynamics. Philos. Trans. R. Soc. B 364, 1483–1489.

Pennekamp, F., Pontarp, M., Tabi, A., Altermatt, F., Alther, R., Choffat, Y., Fronhofer, E.A., Ganesanandamoorthy, P., Garnier, A., Griffiths, J.I., Greene, S., Horgan, K., et al., 2018. Biodiversity increases and decreases ecosystem stability. Nature 563, 109–112.

Pérès, G., Cluzeau, D., Menasseri, S., Soussana, J.F., Bessler, H., Engels, C., Habekost, M., Gleixner, G., Weigelt, A., Weisser, W.W., Scheu, S., Eisenhauer, N., 2013. Mechanisms linking plant community properties to soil aggregate stability in an experimental grassland plant diversity gradient. Plant Soil 373, 285–299.

Petermann, J.S., Fergus, A.J.F., Turnbull, L.A., Schmid, B., 2008. Janzen-Connell effects are widespread and strong enough to maintain diversity in grasslands. Ecology 89, 2399–2406.

Petermann, J.S., Fergus, A.J.F., Roscher, C., Turnbull, L.A., Weigelt, A., Schmid, B., 2010. Biology, chance, or history? The predictable reassembly of temperate grassland communities. Ecology 91, 408–421.

Pfaff, C.T., Eichenberg, D., Liebergesell, M., König-Ries, B., Wirth, C., 2017. Essential annotation schema for ecology (EASE)—a framework supporting the efficient data annotation and faceted navigation in ecology. PLoS One 12, 1–13.

Poisot, T., Mouquet, N., Gravel, D., 2013. Trophic complementarity drives the biodiversity-ecosystem functioning relationship in food webs. Ecol. Lett. 16, 853–861.

Potthast, T., 2014. The values of biodiversity. In: Lanzerath, D., Minou, F. (Eds.), Concepts and Values in Biodiversity. Routledge, London, pp. 131–146.

Pretty, J., 2018. Intensification for redesigned and sustainable agricultural systems. Science 362, eaav0294.

Ratcliffe, S., Wirth, C., Jucker, T., van der Plas, F., Scherer-Lorenzen, M., Verheyen, K., Allan, E., Benavides, R., Bruelheide, H., Ohse, B., Paquette, A., Ampoorter, E., et al., 2017. Biodiversity and ecosystem functioning relations in European forests depend on environmental context. Ecol. Lett. 20, 1414–1426.

Reich, P.B., Knops, J., Tilman, D., Craine, J.M., Ellsworth, D., Tjoelker, M., Lee, T., Wedin, D., Naeem, S., Bahauddin, D., Hendrey, G., Jose, S., et al., 2001. Plant diversity enhances ecosystem responses to elevated $CO_2$ and nitrogen deposition. Nature 410, 809–812.

Reich, P.B., Tilman, D., Isbell, F., Mueller, K., Hobbie, S.E., Flynn, D.F.B., Eisenhauer, N., 2012. Impacts of biodiversity loss escalate through time as redundancy fades. Science 336, 589–592.

Ricketts, T.H., Watson, K.B., Koh, I., Ellis, A.M., Nicholson, C.C., Posner, S., Richardson, L.L., Sonter, L.J., 2016. Disaggregating the evidence linking biodiversity and ecosystem services. Nat. Commun. 7, 13106.

Rodrigo, G.P., Henderson, M., Weber, G.H., Ophus, C., Antypas, K., Ramakrishnan, L., 2018. Science search: enabling search through automatic metadata generation. In: 2018 IEEE 14th International Conference on E-Science. IEEE, pp. 93–104.

Roscher, C., Schumacher, J., Baade, J., Wilcke, W., Gleixner, G., Weisser, W.W., Schmid, B., Schulze, E.D., 2004. The role of biodiversity for element cycling and trophic interactions: an experimental approach in a grassland community. Basic Appl. Ecol. 5, 107–121.

Roscher, C., Temperton, V.M., Scherer-Lorenzen, M., Schmitz, M., Schumacher, J., Schmid, B., Buchmann, N., Weisser, W.W., Schulze, E.D., 2005. Overyielding in experimental grassland communities—irrespective of species pool or spatial scale. Ecol. Lett. 8, 419–429.

Roscher, C., Schumacher, J., Foitzik, O., Schulze, E.D., 2007. Resistance to rust fungi in Lolium perenne depends on within-species variation and performance of the host species in grasslands of different plant diversity. Oecologia 153, 173–183.

Roscher, C., Temperton, V.M., Buchmann, N., Schulze, E.D., 2009. Community assembly and biomass production in regularly and never weeded experimental grasslands. Acta Oecol. 35, 206–217.

Roslin, T., Hardwick, B., Novotny, V., Petry, W.K., Andrew, N.R., Asmus, A., Barrio, I.C., Basset, Y., Boesing, A.L., Bonebrake, T.C., Cameron, E.K., Dáttilo, W., et al., 2017. Higher predation risk for insect prey at low latitudes and elevations. Science 356, 742–744.

Rottstock, T., Joshi, J., Kummer, V., Fischer, M., 2014. Higher plant diversity promotes higher diversity of fungal pathogens, while it decreases pathogen infection per plant. Ecology 95, 1907–1917.

Ruiz-González, C., Archambault, E., Laforest-Lapointe, I., del Giorgio, P.A., Kembel, S.W., Messier, C., Nock, C.A., Beisner, B.E., 2018. Soils associated to different tree communities do not elicit predictable responses in lake bacterial community structure and function. FEMS Microbiol. Ecol. 94, fiy115.

Ryo, M., Rillig, M.C., 2017. Statistically reinforced machine learning for nonlinear patterns and variable interactions. Ecosphere 8, e01976.

Savolainen, O., Lascoux, M., Merilä, J., 2013. Ecological genomics of local adaptation. Nat. Rev. Genet. 14, 807–820.

Scherber, C., 2015. Insect responses to interacting global change drivers in managed ecosystems. Curr. Opin. Insect Sci. 11, 56–62.

Scherber, C., Eisenhauer, N., Weisser, W.W., Schmid, B., Voigt, W., Fischer, M., Schulze, E.D., Roscher, C., Weigelt, A., Allan, E., Beler, H., Bonkowski, M., et al., 2010. Bottom-up effects of plant diversity on multitrophic interactions in a biodiversity experiment. Nature 468, 553–556.

Scherer-Lorenzen, M., 2014. The functional role of biodiversity in the context of global change. In: Burslem, D., Coomes, D., Simonson, W. (Eds.), Forests and Global Change. Cambridge University Press, Cambridge, pp. 195–238.

Scherer-lorenzen, A.M., Palmborg, C., Prinz, A., Schulze, E.D., 2003. The role of plant diversity and composition for nitrate leaching in grasslands the role of plant diversity and composition for nitrate leaching in grasslands. Ecology 84, 1539–1552.

Scherer-Lorenzen, M., Schulze, E.D., Don, A., Schumacher, J., Weller, E., 2007. Exploring the functional significance of forest diversity: a new long-term experiment with temperate tree species (BIOTREE). Perspect. Plant Ecol. Evol. Syst. 9, 53–70.

Schielzeth, H., Husby, A., 2014. Challenges and prospects in genome-wide quantitative trait loci mapping of standing genetic variation in natural populations. Ann. N. Y. Acad. Sci. 1320, 35–57.

Schläpfer, F., Pfisterer, A.B., Schmid, B., 2005. Non-random species extinction and plant production: implications for ecosystem functioning. J. Appl. Ecol. 42, 13–24.

Schmid, B., Hector, A., 2004. The value of biodiversity experiments. Basic Appl. Ecol. 5, 535–542.

Schmid, B., Hector, A., Saha, P., Loreau, M., 2008. Biodiversity effects and transgressive overyielding. J. Plant Ecol. 1, 95–102.

Schmid, M.W., Heichinger, C., Coman Schmid, D., Guthörl, D., Gagliardini, V., Bruggmann, R., Aluri, S., Aquino, C., Schmid, B., Turnbull, L.A., Grossniklaus, U., 2018. Contribution of epigenetic variation to adaptation in Arabidopsis. Nat. Commun. 9, 4446.

Schneider, F.D., Scheu, S., Brose, U., 2012. Body mass constraints on feeding rates determine the consequences of predator loss. Ecol. Lett. 15, 436–443.

Schneider, F.D., Brose, U., Rall, B.C., Guill, C., 2016. Animal diversity and ecosystem functioning in dynamic food webs. Nat. Commun. 7, 12718.

Schnitzer, S.A., Klironomos, J.N., Hillerislambers, J., Kinkel, L.L., Reich, P.B., Xiao, K., Rillig, M.C., Sikes, B.A., Callaway, R.M., Scott, A., van Nes, E.H., Scheffer, M., et al., 2011. Soil microbes drive the classic plant-productivity diversity pattern. Ecology 92, 296–303.

Schöb, C., Brooker, R.W., Zuppinger-Dingley, D., 2018. Evolution of facilitation requires diverse communities. Nat. Ecol. Evol. 2, 1381–1385.

Schuldt, A., Fornoff, F., Bruelheide, H., Klein, A.M., Staab, M., 2017a. Tree species richness attenuates the positive relationship between mutualistic ant-hemipteran interactions and leaf chewer herbivory. Proc. R. Soc. B Biol. Sci. 284, 20171489.

Schuldt, A., Hönig, L., Li, Y., Fichtner, A., Härdtle, W., von Oheimb, G., Welk, E., Bruelheide, H., 2017b. Herbivore and pathogen effects on tree growth are additive, but mediated by tree diversity and plant traits. Ecol. Evol. 7, 7462–7474.

Schuldt, A., Assmann, T., Brezzi, M., Buscot, F., Eichenberg, D., Gutknecht, J., Härdtle, W., He, J.S., Klein, A.M., Kühn, P., Liu, X., Ma, K., et al., 2018. Biodiversity across trophic levels drives multifunctionality in highly diverse forests. Nat. Commun. 9, 2989.

Schuldt, A., Ebeling, A., Kunz, M., Staab, M., Guimarães-Steinicke, C., Bachmann, D., Buchmann, N., Durka, W., Fichtner, A., Fornoff, F., Härdtle, W., Hertzog, L., et al., 2019. Multiple plant diversity components drive consumer communities across ecosystems. Nat. Commun. 10, 1460.

Schulze, E.D., Mooney, H.A., 1994. Biodiversity and Ecosystem Function. Springer, Berlin, Heidelberg.

Schwarz, B., Barnes, A.D., Thakur, M.P., Brose, U., Ciobanu, M., Reich, P.B., Rich, R.L., Rosenbaum, B., Stefanski, A., Eisenhauer, N., 2017. Warming alters energetic structure and function but not resilience of soil food webs. Nat. Clim. Change 7, 895–900.

Seabloom, E.W., Kinkel, L., Borer, E.T., Hautier, Y., Montgomery, R.A., Tilman, D., 2017. Food webs obscure the strength of plant diversity effects on primary productivity. Ecol. Lett. 20, 505–512.

Seibold, S., Brandl, R., Buse, J., Hothorn, T., Schmidl, J., Thorn, S., Müller, J., 2015. Association of extinction risk of saproxylic beetles with ecological degradation of forests in Europe. Conserv. Biol. 29, 382–390.

Seibold, S., Cadotte, M.W., MacIvor, J.S., Thorn, S., Müller, J., 2018. The necessity of multitrophic approaches in community ecology. Trends Ecol. Evol. 33, 754–764.

Sobral, M., Silvius, K.M., Overman, H., Oliveira, L.F.B., Raab, T.K., Fragoso, J.M.V., 2017. Mammal diversity influences the carbon cycle through trophic interactions in the Amazon. Nat. Ecol. Evol. 1, 1670–1676.

Soliveres, S., van der Plas, F., Manning, P., Prati, D., Gossner, M.M., Renner, S.C., Alt, F., Arndt, H., Baumgartner, V., Binkenstein, J., Birkhofer, K., Blaser, S., et al., 2016a. Biodiversity at multiple trophic levels is needed for ecosystem multifunctionality. Nature 536, 456–459.

Soliveres, S., Manning, P., Prati, D., Gossner, M.M., Alt, F., Arndt, H., Baumgartner, V., Binkenstein, J., Birkhofer, K., Blaser, S., Blüthgen, N., Boch, S., et al., 2016b. Locally rare species influence grassland ecosystem multifunctionality. Philos. Trans. R. Soc. B 371, 20150269.

Spiesman, B.J., Stapper, A.P., Inouye, B.D., 2018. Patch size, isolation, and matrix effects on biodiversity and ecosystem functioning in a landscape microcosm. Ecosphere 9, e02173.

Stachowicz, J.J., Bruno, J.F., Duffy, J.E., 2007. Understanding the effects of marine biodiversity on communities and ecosystems. Annu. Rev. Ecol. Evol. Syst. 38, 739–766.

Stachowicz, J.J., Best, R.J., Bracken, M.E.S., Graham, M.H., 2008a. Complementarity in marine biodiversity manipulations: reconciling divergent evidence from field and mesocosm experiments. Proc. Natl. Acad. Sci. U. S. A. 105, 18842–18847.

Stachowicz, J.J., Graham, M., Bracken, M.E.S., Szoboszlai, A.I., 2008b. Diversity enhances cover and stability of seaweed assemblages: the role of heterogeneity and time. Ecology 89, 3008–3019.

Stapley, J., Reger, J., Feulner, P.G.D., Smadja, C., Galindo, J., Ekblom, R., Bennison, C., Ball, A.D., Beckerman, A.P., Slate, J., 2010. Adaptation genomics: the next generation. Trends Ecol. Evol. 25, 705–712.

Steinauer, K., Jensen, B., Strecker, T., Luca, E., Scheu, S., Eisenhauer, N., 2016. Convergence of soil microbial properties after plant colonization of an experimental plant diversity gradient. BMC Ecol. 16, 19.

Steudel, B., Hector, A., Friedl, T., Löfke, C., Lorenz, M., Wesche, M., Kessler, M., 2012. Biodiversity effects on ecosystem functioning change along environmental stress gradients. Ecol. Lett. 15, 1397–1405.

Stocker, R., Korner, C., Schmid, B., Niklaus, P.A., Leadley, P.W., 1999. A field study of the effects of elevated $CO_2$ and plant species diversity on ecosystem-level gas exchange in a planted calcareous grassland. Glob. Chang. Biol. 5, 95–105.

Suding, K.N., Collins, S.L., Gough, L., Clark, C., Cleland, E.E., Gross, K.L., Milchunas, D.G., Pennings, S., 2005. Functional- and abundance-based mechanisms explain diversity loss due to N fertilization. Proc. Natl. Acad. Sci. U. S. A. 102, 4387–4392.

Swift, M.J., Izac, A.M.N., van Noordwijk, M., 2004. Biodiversity and ecosystem services in agricultural landscapes—are we asking the right questions? Agric. Ecosyst. Environ. 104, 113–134.

Thakur, M.P., Milcu, A., Manning, P., Niklaus, P.A., Roscher, C., Power, S., Reich, P.B., Scheu, S., Tilman, D., Ai, F., Guo, H., Ji, R., et al., 2015. Plant diversity drives soil microbial biomass carbon in grasslands irrespective of global environmental change factors. Glob. Chang. Biol. 21, 4076–4085.

Thakur, M.P., Tilman, D., Purschke, O., Ciobanu, M., Cowles, J., Isbell, F., Wragg, P.D., Eisenhauer, N., 2017. Climate warming promotes species diversity, but with greater taxonomic redundancy, in complex environments. Sci. Adv. 3 e1700866.

Thebault, E., Loreau, M., 2003. Food-web constraints on biodiversity–ecosystem functioning relationships. Proc. Natl. Acad. Sci. U. S. A. 100, 14949–14954.

Thompson, P.L., Gonzalez, A., 2016. Ecosystem multifunctionality in metacommunities. Ecology 97, 2867–2879.

Thompson, R.M., Brose, U., Dunne, J.A., Hall, R.O., Hladyz, S., Kitching, R.L., Martinez, N.D., Rantala, H., Romanuk, T.N., Stouffer, D.B., Tylianakis, J.M., 2012. Food webs: reconciling the structure and function of biodiversity. Trends Ecol. Evol. 27, 689–697.

Thompson, P.L., Isbell, F., Loreau, M., O'connor, M.I., Gonzalez, A., 2018. The strength of the biodiversity-ecosystem function relationship depends on spatial scale. Proc. R. Soc. B Biol. Sci. 285, 20180038.

Thorn, S., Bässler, C., Bernhardt-Römermann, M., Cadotte, M., Heibl, C., Schäfer, H., Seibold, S., Müller, J., 2016. Changes in the dominant assembly mechanism drive species loss caused by declining resources. Ecol. Lett. 19, 163–170.

Tiede, J., Wemheuer, B., Traugott, M., Daniel, R., Tscharntke, T., Ebeling, A., Scherber, C., 2016. Trophic and non-trophic interactions in a biodiversity experiment assessed by next- generation sequencing. PLoS One 11 e0148781.

Tilman, D., Downing, J.A., 1994. Biodiversity and stability in grasslands. Nature 367, 363–365.

Tilman, D., Snell-Rood, E.C., 2014. Diversity breeds complementarity. Nature 515, 44.

Tilman, D., Lehman, C.L., Thompson, K.T., 1997a. Plant diversity and ecosystem productivity: theoretical considerations. Proc. Natl. Acad. Sci. U. S. A. 94, 1857–1861.

Tilman, D., Knops, J., Wedin, D., Reich, P., Ritchie, M., Siemann, E., 1997b. The influence of functional diversity and composition on ecosystem processes. Science 277, 1300–1302.

Tilman, D., Hill, J., Lehman, C., 2006. Carbon-negative biofuels from low-input high-diversity grassland biomass. Science 314, 1598–1600.

Tilman, D., Isbell, F., Cowles, J.M., 2014. Biodiversity and ecosystem function. Annu. Rev. Ecol. Evol. Syst. 45, 471–493.

Trenbath, B.R., 1974. Biomass productivity of mixtures. Adv. Agron. 26, 177–210.

Trogisch, S., Schuldt, A., Bauhus, J., Blum, J.A., Both, S., Buscot, F., Castro-Izaguirre, N., Chesters, D., Durka, W., Eichenberg, D., Erfmeier, A., Fischer, M., et al., 2017. Toward a methodical framework for comprehensively assessing forest multifunctionality. Ecol. Evol. 7, 10652–10674.

Tscharntke, T., Tylianakis, J.M., Rand, T.A., Didham, R.K., Fahrig, L., Batáry, P., Bengtsson, J., Clough, Y., Crist, T.O., Dormann, C.F., Ewers, R.M., Fründ, J., et al., 2012. Landscape moderation of biodiversity patterns and processes—eight hypotheses. Biol. Rev. 87, 661–685.

Turnbull, L.A., Levine, J.M., Loreau, M., Hector, A., 2013. Coexistence, niches and biodiversity effects on ecosystem functioning. Ecol. Lett. 16, 116–127.

Turnbull, L.A., Isbell, F., Purves, D.W., Loreau, M., Hector, A., 2016. Understanding the value of plant diversity for ecosystem functioning through niche theory. Proc. R. Soc. B Biol. Sci. 283, 20160536.

Tylianakis, J.M., Didham, R.K., Bascompte, J., Wardle, D.A., 2008. Global change and species interactions in terrestrial ecosystems. Ecol. Lett. 11, 1351–1363.

Urban, M.C., Bocedi, G., Hendry, A.P., Mihoub, J.B., Pe'er, G., Singer, A., Bridle, J.R., Crozier, L.G., De Meester, L., Godsoe, W., Gonzalez, A., Hellmann, J.J., et al., 2016. Improving the forecast for biodiversity under climate change. Science 353, aad8466.

Valverde-Barrantes, O.J., Freschet, G.T., Roumet, C., Blackwood, C.B., 2017. A worldview of root traits: the influence of ancestry, growth form, climate and mycorrhizal association on the functional trait variation of fine-root tissues in seed plants. New Phytol. 215, 1562–1573.

van der Heijden, M.G.A., Klironomos, J.N., Ursic, M., Moutoglis, P., Streitwolf-Engel, R., Boller, T., Wiemken, A., Sanders, I.R., 1998. Mycorrhizal fungal diversity determines plant biodiversity, ecosystem variability and productivity. Nature 74, 69–72.

van der Plas, F., 2019. Biodiversity and ecosystem functioning in naturally assembled communities. Biol. Rev. 94, 1220–1245.

van der Plas, F., Manning, P., Allan, E., Scherer-Lorenzen, M., Verheyen, K., Wirth, C., Zavala, M.A., et al., 2016a. Jack-of-all-trades effects drive biodiversity-ecosystem multifunctionality relationships in European forests. Nat. Commun. 7, 11109.

van der Plas, F., Manning, P., Soliveres, S., Allan, E., Scherer-Lorenzen, M., Verheyen, K., Wirth, C., Zavala, M.A., Ampoorter, E., Baeten, L., Barbaro, L., Bauhus, J., et al., 2016b. Biotic homogenization can decrease landscape-scale forest multifunctionality. Proc. Natl. Acad. Sci. U. S. A. 113, 3557–3562.

van der Plas, F., Ratcliffe, S., Ruiz-Benito, P., Scherer-Lorenzen, M., Verheyen, K., Wirth, C., Zavala, M.A., Ampoorter, E., Baeten, L., Barbaro, L., Bastias, C.C., Bauhus, J., et al., 2018. Continental mapping of forest ecosystem functions reveals a high but unrealised potential for forest multifunctionality. Ecol. Lett. 21, 31–42.

Van Kleunen, M., Dawson, W., Essl, F., Pergl, J., Winter, M., Weber, E., Kreft, H., Weigelt, P., Kartesz, J., Nishino, M., Antonova, L.A., Barcelona, J.F., et al., 2015. Global exchange and accumulation of non-native plants. Nature 525, 100.

van Moorsel, S.J., Hahl, T., Wagg, C., De Deyn, G.B., Flynn, D.F.B., Zuppinger-Dingley, D., Schmid, B., 2018. Community evolution increases plant productivity at low diversity. Ecol. Lett. 21, 128–137.

van Ruijven, J., Berendse, F., 2003. Positive effects of plant species diversity on productivity in the absence of legumes. Ecol. Lett. 6, 170–175.

Vandermeer, J.H., 1981. The interference production principle: an ecological theory for agriculture. Bioscience 31, 361–364.

Vandermeer, J.H., 1990. Intercropping. In: Carroll, C.R., Vandermeer, J.H., Rosset, P. (Eds.), Agroecology. McGraw-Hill, New York, pp. 481–516.

Veiga, A.K., Saraiva, A.M., Chapman, A.D., Morris, P.J., Gendreau, C., Schigel, D., Robertson, T.J., 2017. A conceptual framework for quality assessment and management of biodiversity data. PLoS One 12, e0178731.

Vellend, M., Geber, M.A., 2005. Connections between species diversity and genetic diversity. Ecol. Lett. 8, 767–781.

Verheyen, K., Vanhellemont, M., Auge, H., Baeten, L., Baraloto, C., Barsoum, N., Bilodeau-Gauthier, S., Bruelheide, H., Castagneyrol, B., Godbold, D., Haase, J., Hector, A., et al., 2016. Contributions of a global network of tree diversity experiments to sustainable forest plantations. Ambio 45, 29–41.

Vitousek, P.M., D'Antonio, C.M., Loope, L.L., Rejmanek, M., Westbrooks, R., 1997. Introduced species: A significant component of human-caused global change. N. Z. J. Ecol. 21, 1–16.

Vogel, A., Scherer-Lorenzen, M., Weigelt, A., 2012. Grassland resistance and resilience after drought depends on management intensity and species richness. PLoS One 7, e36992.

Vogel, A., Manning, P., Cadotte, M.W., Cowles, J., Isbell, F., Jousset, A.C., Kimmel, K., Meyer, S.T., et al., 2019a. Lost in trait space: species-poor communities are inflexible in properties that drive ecosystem functioning. Adv. Ecol. Res. 61, 91–131.

Vogel, A., Ebeling, A., Gleixner, G., Roscher, C., Scheu, S., Ciobanu, M., Koller-France, E., Lange, M., et al., 2019b. A new experimental approach to test why biodiversity effects strengthen as ecosystems age. Adv. Ecol. Res. 61, 221–264.

Voigt, W., Perner, J., Davis, A.J., Eggers, T., Schumacher, J., Bährmann, R., Fabian, B., Heinrich, W., Köhler, G., Lichter, D., Marstaller, R., Sander, F.W., 2003. trophic levels are differentially sensitive to climate. Ecology 84, 2444–2453.

Wacker, L., Baudois, O., Eichenberger-Glinz, S., Schmid, B., 2008. Environmental heterogeneity increases complementarity in experimental grassland communities. Basic Appl. Ecol. 9, 467–474.

Wall, D.H., Nielsen, U.N., Six, J., 2015. Soil biodiversity and human health. Nature 528, 69–76.

Wang, S., Brose, U., 2018. Biodiversity and ecosystem functioning in food webs: the vertical diversity hypothesis. Ecol. Lett. 21, 9–20.

Wang, L., Delgado-Baquerizo, M., Wang, D., Isbell, F., Liu, J., Zhu, H., Zhong, Z., Liu, J., Yuan, X., Feng, C., Chang, Q., 2019. Diversifying livestock promotes multidiversity and multifunctionality in managed grasslands. Proc. Natl. Acad. Sci. U. S. A. 116 (13), 6187–6192.

Wardle, D.A., 2016. Do experiments exploring plant diversity-ecosystem functioning relationships inform how biodiversity loss impacts natural ecosystems? J. Veg. Sci. 27, 646–653.

Wardle, D.A., Zackrisson, O., 2005. Effects of species and functional group loss on island ecosystem properties. Nature 435, 806–810.

Wardle, D.A., Bardgett, R.D., Callaway, R.M., van der Putten, W.H., 2011. Terrestrial ecosystem responses to species gains and losses. Science 332, 1273–1277.

Weigelt, A., Weisser, W.W., Buchmann, N., Scherer-Lorenzen, M., 2009. Biodiversity for multifunctional grasslands: equal productivity in high-diversity low-input and low-diversity high-input systems. Biogeosciences 6, 1695–1706.

Weisser, W.W., Roscher, C., Meyer, S.T., Ebeling, A., Luo, G., Allan, E., Beßler, H., Barnard, R.L., Buchmann, N., Buscot, F., Engels, C., Fischer, C., et al., 2017. Biodiversity effects on ecosystem functioning in a 15-year grassland experiment: patterns, mechanisms, and open questions. Basic Appl. Ecol. 23, 1–73.

Wilkinson, M.D., Dumontier, M., Aalbersberg, I.J., Appleton, G., Axton, M., Baak, A., Blomberg, N., Boiten, J.W., da Silva Santos, L.B., Bourne, P.E., Bouwman, J., Brookes, A.J., 2016. The fair guiding principles for scientific data management and stewardship. Sci. Data 3, 160018.

Wilsey, B.J., Polley, H.W., 2004. Realistically low species evenness does not alter grassland species-richness–productivity relationships. Ecology 85, 2693–2700.

Winfree, R., Reilly, J.R., Bartomeus, I., Cariveau, D.P., Williams, N.M., Gibbs, J., 2018. Species turnover promotes the importance of bee diversity for crop pollination at regional scales. Science 359, 791–793.

Worm, B., Duffy, J.E., 2003. Biodiversity, productivity and stability in real food webs. Trends Ecol. Evol. 18, 628–632.

Wright, A.J., Schnitzer, S.A., Reich, P.B., 2014. Size matters when living close to your neighbors—the importance of both competition and facilitation in plant communities. Ecology 95, 2213–2223.

Wright, A.J., Ebeling, A., De Kroon, H., Roscher, C., Weigelt, A., Buchmann, N., Buchmann, T., Fischer, C., Hacker, N., Hildebrandt, A., Leimer, S., et al., 2015. Flooding disturbances increase resource availability and productivity but reduce stability in diverse plant communities. Nat. Commun. 6, 6092.

Wright, A.J., Wardle, D.A., Callaway, R., Gaxiola, A., 2017. The overlooked role of facilitation in biodiversity experiments. Trends Ecol. Evol. 32, 383–390.

Wuest, S.E., Niklaus, P.A., 2018. A plant biodiversity effect resolved to a single chromosomal region. Nat. Ecol. Evol. 2, 1933–1939.

Yachi, S., Loreau, M., 1999. Biodiversity and ecosystem productivity in a fluctuating environment: the insurance hypothesis. Proc. Natl. Acad. Sci. U. S. A. 96, 1463–1468.

Yang, J., Cao, M., Swenson, N.G., 2018. Why functional traits do not predict tree demographic rates. Trends Ecol. Evol. 33, 326–336.

Zavaleta, E.S., Pasari, J.R., Hulvey, K.B., Tilman, G.D., 2010. Sustaining multiple ecosystem functions in grassland communities requires higher biodiversity. Proc. Natl. Acad. Sci. U. S. A. 107, 1443–1446.

Zeng, X., Durka, W., Fischer, M., 2017. Species-specific effects of genetic diversity and species diversity of experimental communities on early tree performance. J. Plant Ecol. 10, 252–258.

Zhang, Q.G., Zhang, D.Y., 2006. Resource availability and biodiversity effects on the productivity, temporal variability and resistance of experimental algal communities. Oikos 114, 385–396.

Zuppinger-Dingley, D., Schmid, B., Petermann, J.S., Yadav, V., De Deyn, G.B., Flynn, D.F.B., 2014. Selection for niche differentiation in plant communities increases biodiversity effects. Nature 515, 108–111.

Zuppinger-Dingley, D., Flynn, D.F.B., De Deyn, G.B., Petermann, J.S., Schmid, B., 2016. Plant selection and soil legacy enhance long-term biodiversity effects. Ecology 97, 918–928.

CHAPTER TWO

# Above- and belowground overyielding are related at the community and species level in a grassland biodiversity experiment

**Kathryn E. Barry**[a,b,†], **Alexandra Weigelt**[a,b,†], **Jasper van Ruijven**[c], **Hans de Kroon**[d], **Anne Ebeling**[e], **Nico Eisenhauer**[a,b], **Arthur Gessler**[f,g,h], **Janneke M. Ravenek**[d], **Michael Scherer-Lorenzen**[i], **Natalie J. Oram**[c], **Anja Vogel**[a,b,e], **Cameron Wagg**[j], **Liesje Mommer**[c,*]

[a]Institute of Biology, Leipzig University, Leipzig, Germany
[b]German Centre for Integrative Biodiversity Research (iDiv) Halle-Jena-Leipzig, Leipzig, Germany
[c]Plant Ecology and Nature Conservation Group, Wageningen University, Wageningen, The Netherlands
[d]Department of Experimental Plant Ecology, Institute for Water and Wetland Research, Radboud University, Nijmegen, The Netherlands
[e]Institute of Ecology and Evolution, Friedrich Schiller University Jena, Jena, Germany
[f]Institute for Landscape Biochemistry, Leibniz Centre for Agricultural Landscape Research (ZALF), Müncheberg, Germany
[g]Swiss Federal Research Institute for Forest, Snow and Landscape Research WSL, Birmensdorf, Switzerland
[h]Berlin-Brandenburg Institute of Advanced Biodiversity Research (BBIB), Berlin, Germany
[i]Faculty of Biology, Geobotany, University of Freiburg, Freiburg, Germany
[j]Fredericton Research and Development Centre, Agriculture and Agri-Food Canada, Fredericton, NB, Canada
*Corresponding author: e-mail address: liesje.mommer@wur.nl

## Contents

1. Introduction 56
2. Methods 60
   2.1 Site description 60
   2.2 Biomass sampling 61
   2.3 Estimating species root biomass using molecular methods 62
   2.4 Data analysis 63
3. Results 67
   3.1 Hypothesis 1: At the community level, above- and belowground overyielding are correlated 67
   3.2 Hypothesis 2: At the pool level, species from the 'spatial' pool overyield more aboveground, whereas the species from the 'temporal' pool overyield more belowground 69
   3.3 Hypothesis 3: At the species level, some species exhibit trade-offs between above- and belowground overyielding 71

[†] Joint first authorship.

4. Discussion                                                                                      77
   4.1 Are above- and belowground overyielding correlated at the
       community level?                                                                            79
   4.2 Above- and belowground overyielding relationships differ between
       species pools                                                                               80
5. Conclusions                                                                                     83
Authorship statement                                                                               83
Acknowledgements                                                                                   83
References                                                                                         84

## Abstract

Plant species richness positively affects plant productivity both above- and belowground. While this suggests that they are related at the community level, few studies have calculated above- and belowground overyielding simultaneously. It thus remains unknown whether above- and belowground overyielding are correlated. Moreover, it is unknown how belowground community level overyielding translates to the species level.

We investigated above- and belowground overyielding in the Jena Trait-Based Biodiversity Experiment, at both the community and species level and across two 8-species pools. We found that above- and belowground overyielding were positively correlated at the community level and at the species level—for seven out of the 13 investigated species. Some plant species performed better in mixtures compared to monocultures and others performed worse, but the majority did so simultaneously above- and belowground. However, plants invested more in aboveground overyielding than belowground. Based on this disproportional investment in overyielding aboveground, we conclude that light was more limiting than belowground resources in the present study, which requires individual species to compete more for light than for belowground resources.

## 1. Introduction

Plant species richness positively affects ecosystem functioning (Cardinale et al., 2012; Hooper et al., 2005). In particular, plant species mixtures on average produce more biomass aboveground than their respective monocultures (Cardinale et al., 2012; Hector et al., 1999; Marquard et al., 2009; Tilman et al., 2001; van Ruijven and Berendse, 2005), a phenomenon known as overyielding. While aboveground overyielding is well established, more recently, a number of long-term studies have shown that belowground biomass also increases with diversity (Fornara and Tilman, 2009; Meyer et al., 2016; Ravenek et al., 2014; Tilman et al., 2001). At the biodiversity experiment in Jena, aboveground biomass increased with plant species richness from the beginning of the experiment (Roscher et al., 2005) while belowground biomass did not (Bessler et al., 2009). However, both above- and belowground biomass increased with species richness once the community was established (Ravenek et al., 2014). In the long-term biodiversity experiment at Cedar Creek (Tilman et al., 2001), above- and belowground biomass were positively related (Fornara and Tilman, 2009) and both increased strongly with plant species richness (Fornara and Tilman, 2009; Mueller et al., 2013). Similarly, in the Wageningen biodiversity experiment (van Ruijven and Berendse, 2003), root biomass increased with plant species richness, when measured after 12 years (Cong et al., 2014), as did aboveground biomass (van Ruijven and Berendse, 2009). While the majority of these studies were unable to calculate belowground 'overyielding' per se due to the lack of species-specific data belowground, these results suggest that at the level of the plant community, aboveground and belowground biomass production are positively related.

However, if and how the relationship between above- and belowground community level overyielding translates to the species level is far less clear. In principle, community overyielding might be caused by at least two different patterns of biomass allocation at the species level. First, all species may show a simultaneous and proportional increase in above- and belowground overyielding and thus a positive correlation. Second, sets of species may exhibit trade-offs for different resources. That is, some species may invest more in competing for belowground resources and therefore in belowground overyielding, while others invest more in competing for aboveground resources and therefore in aboveground overyielding. At the community level, these different trade-offs likely result in simultaneous total community overyielding above- and belowground even as individual species experience trade-offs.

Mechanistically, simultaneous above- and belowground overyielding may occur if species experience competitive release in more diverse mixtures. This competitive release may occur because plants partition resources in space or time. In more diverse mixtures, plants may occupy different resource partitions in space and/or time (reviewed by Barry et al. 2019). The occupation of these different resource partitions may alleviate competition for limiting resources and decrease the extent to which plants compete for limiting resources. Evidence that resource partitioning in either space or time drives enhanced ecosystem functioning in biodiversity experiments is, however, limited. In the Jena Main Experiment, Jesch et al. (2018) found no evidence for resource partitioning in space, time, or across different forms of proxies for potassium and water (see also Bachmann et al., 2015). Similarly, Oram et al. (2018) found in the Jena Trait Based Experiment that belowground overyielding was associated with deeper rooting communities rather than communities that appeared to partition resources by depth. However, rather than allowing for enhanced ecosystem functioning directly, resource partitioning may release plants from competition for specific resources and allow simultaneous above- and belowground overyielding.

Alternatively, we may expect a trade-off between overyielding above- and belowground because allocation to one compartment (e.g. aboveground biomass) affects allocation to the others (i.e. belowground biomass; Poorter et al., 2012). The 'functional equilibrium' model predicts that a plant must sacrifice investment in roots to invest in leaves (Brouwer, 1962). According to the functional equilibrium model of plant biomass allocation, plants will allocate relatively more biomass towards competing for the most limiting resource (Brouwer, 1962; see also Gedroc, 1996; Shipley and Meziane, 2002). That is, if light is the most limiting resource for a plant it will invest disproportionately in aboveground biomass to increase its ability to compete for light (reviewed by Poorter and Nagel, 2000; Poorter et al., 2012; Poorter et al., 2019). Alternatively, if nutrients and water are most limiting then the plant will invest disproportionately in belowground biomass (Poorter et al., 2012). The result of this differential allocation may be a trade-off in investment between above- and belowground biomass at the species level depending on the most limiting resource. Several studies show that individual species differ strongly in their contribution to aboveground overyielding, ranging from clearly positive to negative effects on community overyielding (Hector et al., 2002; HilleRisLambers et al., 2004; Roscher et al., 2007; van Ruijven and Berendse, 2005). In addition, species overyielding may differ between functional groups. In a biodiversity experiment with nine potentially dominant species (five grasses, two forbs and two legumes), most grasses and one legume species overyielded, whereas the forbs and the other legume underyielded (i.e. negative overyielding or better growth in monocultures compared to mixtures; Roscher et al., 2007). Similarly, HilleRisLambers et al. (2004) found that C4 grasses and legumes overyielded, while most forb species underyielded. However, in an experiment without legumes, no clear differences in overyielding between grasses and forbs were found (van Ruijven and Berendse, 2003). Instead, in both groups, some species overyielded, while others underyielded (Roscher et al., 2011a; van Ruijven and Berendse, 2005).

Current approaches to understanding overyielding above- and belowground are limited by the lack of species-specific data belowground. While species-specific data is relatively easy to obtain aboveground using traditional morphological methods, roots of different species are, generally, morphologically indistinguishable. This lack of morphological differences restricts species identification belowground (but see: Genney et al., 2002; Hendriks et al., 2013; Janecek et al., 2004; Mommer et al., 2011; Padilla et al., 2013). To overcome this limitation, DNA-based techniques have been developed to identify plant roots at the species level in species-rich plant communities (Dumbrell et al., 2011; Frank et al., 2015; Hendriks et al., 2015; Jackson et al., 1999; Jones et al., 2011; Kesanakurti et al., 2011; Linder et al., 2000; Mommer et al., 2010). These techniques make it possible to quantify species-specific root biomass and determine species overyielding belowground (e.g. Mommer et al., 2010) and investigate relationships between aboveground and belowground overyielding at the species level.

Here, we explicitly test the correlation between aboveground and belowground overyielding at the community- and the species-level in the Jena Trait-Based Experiment (Ebeling et al., 2014). This experiment is ideal for examining the potential causes of overyielding because it consists of two species pools which were deliberately chosen to vary in specific functional traits related to their resource acquisition (Wagg et al., 2017). The first pool, referred to as the *'spatial'* pool, contains species that differ in traits reflecting spatial resource use (e.g. plant height, rooting depth, root length density, and leaf size). The 'spatial pool' spans a gradient from small statured, short, dense rooted species to larger statured species with long, coarse roots. This combination of species, hypothetically, maximizes their potential to enhance spatial resource partitioning in mixtures, especially belowground. The second species pool, the *'temporal'* pool, contains species that differ in their temporal resource use (e.g. starting date of growth or flowering). This pool stretches a gradient from species with early onset of growth and flowering to species with late onset of growth and flowering. This 'temporal' species pool, theoretically, maximizes temporal resource partitioning in mixtures, in particular aboveground. The type of traits and thus the resulting resource partitioning (spatial vs. temporal) may have consequences for the relationships between aboveground and belowground overyielding. For example, spatial resource partitioning belowground, as maximized in the *'spatial'* pool, may release belowground competition, which will lead to more intense competition for light (Harpole et al., 2016; Hautier et al., 2009), which in turn leads to increased biomass allocation and overyielding aboveground, (Aerts et al., 1991; Anten and Hirose, 1999; Roscher et al., 2007), but not belowground (Cahill, 2003; Lamb et al., 2009). In contrast, temporal resource partitioning during the growing season, as maximized in the 'temporal' pool, may primarily release competition for light due to differences in the onset of growth. Temporal partitioning should reduce the extent to which species experience light limitation in high diverse communities and therefore release aboveground competition, which may increase belowground competition and therefore biomass allocation.

We compared biomass in mixture to expected biomass based on monoculture performance at both the community and species level. Specifically, we tested the following hypotheses:

1. At the community level, above- and belowground overyielding are correlated.
2. At the pool level, species from the 'spatial' pool overyield more aboveground, whereas the species from the 'temporal' pool overyield more belowground.
3. At the species level, some species exhibit trade-offs between above- and belowground overyielding (i.e. above- and belowground overyielding will be negatively correlated).

## 2. Methods
### 2.1 Site description

We conducted this study at the Trait-Based Experiment (TBE) at the field site of the Jena Experiment (Thuringia, Germany, 50°55′ N, 11°35′ E, 130m above sea level; Roscher et al., 2004; Weisser et al., 2017). The TBE was established in 2010 and manipulates trait diversity of plant communities together with plant species richness (Ebeling et al., 2014; Wagg et al., 2017). The 48 non-legume species of the original Jena Experiment pool of 60 species were ordinated in a PCA based on six traits relevant for resource acquisition in space and time (plant height, leaf size, rooting depth, root length density, the onset of flowering, the start of the growing period). The first ordination axis was positively correlated with leaf size, plant height, and rooting depth and negatively to root length density. Thus, species along this axis vary from short plants with small leaves and with shallow and dense roots to taller plants with large leaves and deeper roots. The second axis spans a phenological gradient, which was negatively correlated with species with earlier flowering and start of growth and positively correlated with species with late flowering and onset of growth (Ebeling et al., 2014). PCA axes were separated into four sectors, and two species from each sector along both axes were selected to build the experimental species pool of eight species (the *'spatial'* pool along axis 1 related to spatial resource uptake; the *'temporal'* pool along axis 2 spanning a phenological gradient). The *'spatial'* pool consisted of four forbs (*Centaurea jacea, Knautia arvensis, Leucanthemum vulgare, Plantago lanceolata*) and four grasses (*Festuca rubra, Helictotrichon pubescens, Phleum pratense, Poa pratensis*). The *'temporal'* pool also consisted of four forbs (*Geranium pratense, Leucanthemum vulgare, Plantago lanceolata, Ranunculus acris*) and four grasses (*Anthoxanthum odoratum, Dactylis glomerata, Holcus lanatus, Phleum pratense*). Please note that *L. vulgare, P. lanceolata* and *P. pratense* were present in both pools.

We used monocultures for each species in the species pool, where the three species that are in both pools also have two monocultures ($n=16$), and further combinations of two species ($n=32$), three species ($n=24$), and four species ($n=18$). Together this is a total of 90 plots which were $3.5 \times 3.5$ m. Plots were sown in autumn 2010 and arranged in three blocks to account for spatial variation in soil texture. Plots were mowed twice a year in June and September to mimic traditional grassland management in the region. Finally, plots were weeded three times per year (April, July, and October) to maintain the target plant community.

## 2.2 Biomass sampling

We sampled aboveground plant community biomass directly before mowing in May and August 2012, 2014, and 2016 by clipping the vegetation of two 0.2 m × 0.5 m rectangles per plot 3 cm above the ground. The position of the rectangles was randomly changed on all sampling dates to avoid clipping on the same positions. We then sorted the samples into the species that were sown and unsown weeds and dried the samples at 70 °C for at least 72 h to constant weight. We then weighed the dried samples. Finally, we averaged the two samples per date and summed these two sampling dates (May and August) to estimate annual aboveground biomass production per area.

To determine root biomass, we took eight root cores (diameter 4 cm) up to 40 cm depth in each plot in August 2012, 2014, and 2016. We pooled all of the cores in a given plot and stored these soil cores at 4 °C for <32 h. We processed these samples as quickly as possible by washing the root cores over a 0.5 mm sieve with tap water (see Oram et al., 2018; Ravenek et al., 2014). We then separated these roots into two size classes: coarse (>2 mm diameter) and fine (<2mm diameter). We homogenized the fine root samples and then subsampled (50 mg) them for molecular analyses of species-specific root biomass. Molecular samples were stored at −80 °C. We dried all of the remaining fine root biomass at 65 °C for at least 48 h to a constant weight and weighed these samples. Importantly, our procedure for calculating species-specific root biomass is only calibrated for fine root biomass and thus we were unable to use the coarse root biomass that was collected.

## 2.3 Estimating species root biomass using molecular methods

We used real-time quantitative polymerase chain reactions (RT-PCR) to estimate the relative proportion of the eight species in each mixed root sample (as in Hendriks et al., 2015; Mommer et al., 2008). To extract all root DNA, we used the DNEasy 96 plant mini kit and followed the included protocol (Qiagen, Venlo, The Netherlands). For each sample, we separately amplified each species using species-specific primer pairs. Mommer et al. (2008) provides complete methods for primer development for *A. odoratum*, *F. rubra*, and *L. vulgare*. Similarly, Oram et al. (2018) used the same procedure to develop primer pairs for *C. jacea*, *D. glomerata*, *G. pratense*, *H. lanatus*, *H. pubescens*, *K. arvensis*, *P. lanceolata*, *P. pratense*, *P. pratensis*, and *R. acris*.

We performed all RT-PCR reactions with HOT FIREPol Eva Green qPCR Mix Plus (Solis BioDyne, Tartu, Estonia) and added 0.94 µM $MgCl_2$. We used 60 nM primer for *A. odoratum* and *C. jacea*. For all other species, we used 120 nM primer. We then added 1 ng of genomic DNA (for all species except *P. lanceolata*, which required 4 ng of genomic DNA) to a total reaction volume of 20 µL. In a CFX Touch Real-Time PCR Detection System (Bio-rad Laboratories, Hercules, CA, USA), all PCR used the following cycle: 15 min at 95 °C, then 41 65-s cycles: 95 °C for 20 s, 62 °C for 30 s, and 72 °C for 15 s. The cycle ended with a 5 s melting curve analysis from 70 °C to 91 °C with an increment of 0.5 °C per cycle (see also Oram et al., 2018).

We used the 16 monoculture plots to create 'standard samples' that contained equal proportions of all plant species and used these standardized samples to validate the experimental samples. We created a further six reference samples with known proportions of species from 0% to 50%. We then regressed the measured (RT-PCR) proportions against the known proportions (see Oram et al., 2018). We used the five samples with the smallest summed discrepancy between the measured and actual abundance as reference standards for all RT-PCR. We calculated species-specific root biomass using the RT-PCR proportion and the measured dry mass of the fine root sample. Thus, fine root biomass is equal to the proportion of the RT-PCR result represented by a given species multiplied by the dry mass of a given sample (Mommer et al., 2010; Oram et al., 2018).

## 2.4 Data analysis

We asked the question of whether plants have the same biomass allocation strategy above- and belowground from two perspectives: at the community (plot, in this case) level and at the species level. At the community level, we calculated total biomass above- or belowground as the sum of the biomass collected for all species in a plot for each year. We also calculated community-level overyielding in terms of the community log-response ratio as:

$$\log \text{response ratio}_{community} = \log \left( Biomass_{community} \right) - \log \left( Expected\ biomass_{community} \right)$$

where the community expected biomass was:

$$Expected\ biomass_{community} = \frac{\sum Monoculture\ biomass\ component\ species}{Plot\ species\ richness}$$

The community log-response ratio is a standardized version of the net biodiversity effect (sensu Loreau, 1998) and tells us how much a community over- or underperforms relative to the monocultures of its component species. If the community log-response ratio is positive, the community produced more biomass than expected based on the monocultures of the species that comprise it. If the community log-response ratio is negative, the community produced less biomass than expected based on the monocultures of the species that comprise it.

At the species level, we determined species biomass for each year and calculated the species overyielding as the species log-response ratio:

$$\log \text{response ratio}_{species} = \log \left( Biomass_{species} \right) - \log \left( Expected\ biomass_{species} \right)$$

where the species expected biomass was:

$$Expected\ biomass_{species} = \frac{Monoculture\ biomass_{species}}{Plot\ species\ richness}$$

The species log-response ratio tells us how much a species over- or underperforms when it is in mixture relative to when it is in monoculture. If the species log-response ratio is positive, it indicates that a species produced more biomass than expected based on its monoculture biomass. If the species log-response ratio is negative, it indicates that a species produced less biomass than expected based on its monoculture biomass. We chose to use the log-response ratio over other methods of calculated overyielding because it provides a more balanced view of overyielding and underyielding. In general, overyielding can increase to infinity because biomass production can always exceed the initial amount of biomass, while underyielding is limited by the amount of biomass initially present. This difference in limits on over- vs. underyielding often biases studies of overyielding and underyielding towards extremely high overyielding values and may underestimate the contribution of underyielding. The log-response ratio (in contrast to other methods such as relative yield total or relative net effect) decreases the emphasis on these extremely high values and likely provides a more conservative measure of overyielding. However, the overall outcome of our study did not change with the use of relative net effects as demonstrated with Figs S1–S3 in Supplementary Material in the online version at https://doi.org/10.1016/bs.aecr.2019.05.001.

Once community and species biomasses and log-response ratios were calculated (Table 1), we used the R package 'lme4' (Bates et al., 2015) to perform mixed effect linear models and 'lmerTest' (Kuznetsova et al., 2015) to calculate Satterthwaite approximations of $P$-values. Because we included many fixed and random factors in our models and allowed for interactions between all of them, we used a backwards stepwise Akaike Information Criterion (AIC) analysis to eliminate factors and avoid overfitting our models using the 'step' function as implemented by the 'lmerTest' package. This procedure first eliminates random effects using likelihood ratio tests. Once the random effect structure is optimized, fixed effects are eliminated. However, because model selection procedures can exaggerate differences in model fit ($R^2$), we avoid comparing fit between our final models and instead refer only to the individual factors that remained in the model. We, therefore, report both the initial model and the final model (Table 1).

We then performed a Tukey Honest Significant Difference post-hoc test on the fixed factors included in the final model in the package 'lsmeans' with a Bonferroni-Holm correction to correct for multiple comparisons (Lenth and Herv, 2016). We include contrasts from this analysis with the results

**Table 1** Model parameters.

| Nr | Dependent variable | Initial model | Final model |
|---|---|---|---|
| 1 | Community total biomass | Species richness × Pool (1\|Above.Below) + (1\|Year) + (1\|Block/Plot) | Species richness + (1\|Above. Below) + (1\|Year) + (1\|Block/Plot) |
| 2 | Community aboveground biomass | Species richness × Pool + (1\|Year) + (1\|Block/Plot) | 1 + (1\|Year) + (1\|Block/Plot) |
| 3 | Community belowground biomass | Species richness × Pool + (1\|Year) + (1\|Block/Plot) | Species richness + Pool + (1\|Year) + (1\|Block/Plot) |
| 4 | Community log-response ratio | Species richness × Pool +(1\|Above.Below) + (1\|Year) + (1\|Block/Plot) | 1 + (1\|Plot × Block) + (1\|Block) |
| 5 | Community aboveground log-response ratio | Species richness × Pool + (1\|Year) + (1\|Block/Plot) | 1 + (1\|Block) |
| 6 | Community belowground log-response ratio | Species richness × Pool + (1\|Year) + (1\|Block/Plot) | Pool + (1\|Block/Plot) |
| 7 | Community belowground log-response ratio | Aboveground community log-response ratio × Species richness × Pool + (1\|Year) + (1\|Block/Plot) | Aboveground community log-response ratio + Pool + (1\|Block × Year) |
| 8 | Species log-response ratio | Species × Species richness × Pool + (1\|Above.Below) + (1\|Year) + (1\|Block/Plot) | Species + Species richness + Pool + Species × Species richness + Species × Pool + (1\|Plot × Block) |
| 9 | Species aboveground log-response ratio | Species × Species richness × Pool + (1\|Year) + (1\|Block/Plot) | Species + Pool + Species × Pool + (1\|Year) |
| 10 | Species belowground log-response ratio | Species × Species richness × Pool + (1\|Year) + (1\|Block/Plot) | Species + Pool + Species × Pool + (1\|Plot × Block) + (1\|Block) |

*Continued*

**Table 1** Model parameters.—cont'd

| Nr | Dependent variable | Initial model | Final model |
|---|---|---|---|
| 11 | Species belowground log-response ratio | Aboveground species log-response ratio × Species × Species richness × Pool + (1 | Year) + (1 | Block/Plot) | Aboveground species log-response ratio + Species + Pool + Aboveground species log-response ratio × Species + Species × Pool + (1 | Plot × Block) |

Given are fixed and random effects of the initial linear mixed effects models for all dependent variables as well as the remaining final model parameters after backward stepwise selection to remove factors from our models that contributed to overfitting. The term 'Species' refers to the test of species identity effects in the models. Interactions are represented with x. Nested random effects are indicated with /. Results from final models with more than one remaining fixed effect and/or specifications of contrasts for the effect of 'Pool' are given in Tables 2–7 as indicated by bold model numbers.

of the full model where relevant (see e.g., Table 3). In addition to these general community and species-level analyses, we also report the results of individual species models. However, because we have relatively low power for individual species models, we did not conduct a stepwise AIC and do not report Tukey comparisons with the Bonferroni-Holm correction for the individual species models. Our data were collected over 4 years (three sampling dates) and the design of the experiment is a randomized block design with two pools. All plots were sampled three times. To account for this nestedness and the potential that variation over time may contribute to some of the patterns reported here, we include an independent random factor of (1 | Year) and a nested random factor of (1 | Block/Pool/Plot) in all models with random effects except when pool is a fixed effect in which case (1 | Block/Plot) was used. Table 1 provides an overview over all of the 11 initial and final models tested in this study. We always tested the response variable in three ways: 1. The total community (above- and belowground), 2. aboveground only, and 3. belowground only to differentiate between the effects on aboveground vs. belowground biomass, respectively. Detailed effects of the final models with more than one remaining fixed effect are given in tables two through nine. To demonstrate that our findings are robust, we also used plot as a random factor, and year and above/belowground as fixed factors with model selection on only fixed effects first with lme4. We then compared these model runs in nlme with model selection second where possible (see Supplementary Material in the online version at https://doi.org/10.1016/bs.aecr.2019.05.001 in Tables S1–S10 for these results). The results and conclusions did not change with this alternative approach which is why we present the original model results only in the main text.

## 3. Results

At the community level, we found that communities with higher species richness produced more total biomass (above- + belowground biomass, $F_{1,86} = 5.682$, $P = 0.019$, Table 1, model 1; Fig. 1). However, this effect was primarily driven via a significant effect of species richness on community root biomass (Table 1, model 3 and 2; Fig. 1). The effect of species richness on community aboveground biomass was not significant (Table 1, model 2; Fig. 1). In contrast, there was no general effect of species richness on the total community log-response ratio, or the above- or belowground community log-response ratio (Table 1, models 4–6, respectively; Fig. 1). This lack of a significant relationship indicates that an increase in species richness from two to four species did not further increase mixture performance relative to what would have been expected based on monocultures. Importantly, however, the total community (above- and belowground biomass combined) log-response ratio was significantly larger than zero ($t = 3.06$, $P = 0.05$, Table 1, model 4). This significantly positive intercept indicates an overall positive effect of plant species growing in mixture compared to growing in their respective monocultures. That is, mixtures perform better than what would have been expected from monocultures of the constituent species though this does not increase with increasing species richness.

### 3.1 Hypothesis 1: At the community level, above- and belowground overyielding are correlated

As predicted, at the community level, the aboveground community log-response ratio and the belowground community log-response ratio were positively correlated (Tables 1, model 7 and 4; Fig. 2). That is, the more individual communities overyielded aboveground, the more they overyielded belowground. Alternatively, the more communities underyielded aboveground, the more they also underyielded belowground. This correlation, however, did not depend on the species richness of the plant community. In general, we expect that if the proportional increase in aboveground biomass in mixtures relative to monocultures matches the proportional increase in belowground biomass in mixtures, the correlation between above- and belowground overyielding will approach the 1:1 line (Fig. 2).

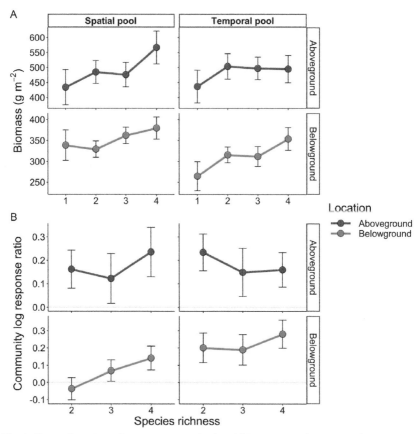

**Fig. 1** Does diversity enhance community level biomass production and community overyielding above and belowground? Given are mean community biomass production (A) and plot log-response ratio (B) over 3 years (± SE) along the species richness gradient separated by compartment and pool. Lines are simple graphical connections and do not indicate significant relationships.

**Table 2** Results of linear mixed effect model on community belowground biomass (model 3, Table 1) asking: Does community belowground biomass change along a species richness gradient or differ between pools of spatial vs. temporal resource partitioning?

|  | Sum Sq | Mean Sq | NumDF | DenDF | F value | P (>F) |
|---|---|---|---|---|---|---|
| **Species richness** | **74,057.07** | **74,057.07** | **1** | **85.023** | **4.586** | **0.035** |
| *Pool* | *57,779.74* | *57,779.74* | *1* | *85.000* | *3.578* | *0.062* |

Significant results ($p < 0.05$) presented in bold. Marginally significant results ($P < 0.10$) are presented in italics. The final formula for this model was: Species richness + Pool + (1 | Year) + (1 | Block/Plot).

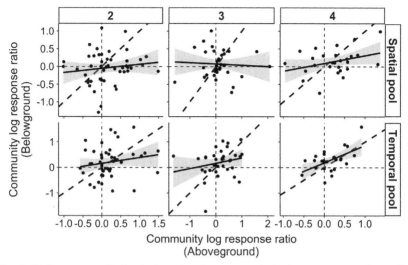

**Fig. 2** At the community level, does the aboveground plot log-response ratio predict belowground community response ratio? Given are plot log-response ratios for different diversity levels (2, 3 and 4) and pools (spatial and temporal) separately. Solid lines indicate significant regressions, dashed line represents the 1:1 line. Light grey lines represent $y=0$ and $x=0$. Points that fall to the left of $x=0$ represent aboveground underyielding. Points that fall below $y=0$ represent belowground underyielding.

Therefore, a community that invests proportionally more in overyielding aboveground will lie below the 1:1 line and a community that invests proportionally more in overyielding belowground will lie above the 1:1 line. At the community level, the slope of the positive correlation between above- and belowground log-response ratio was less than one, indicating that communities overyielded proportionally more above- compared to belowground (Fig. 2).

## 3.2 Hypothesis 2: At the pool level, species from the 'spatial' pool overyield more aboveground, whereas the species from the 'temporal' pool overyield more belowground

The experimental design of the Trait Based Experiment provided two pools of species differing in functional traits that maximized the potential for either spatial or temporal resource partitioning. Overall, there was no difference in community biomass production between the 'spatial' and 'temporal' pools in both total and aboveground biomass production. However, community root biomass was marginally higher in the 'spatial' compared to the 'temporal' pool (Tables 1, model 3 and 2; Fig. 1), indicating that species that are separated to enhance spatial resource partitioning produce more biomass belowground. These differences in species pools are mirrored in community overyielding, where again we detected no differences between species pools in total and aboveground log-response ratio (Table 1, model 4 and 5; Fig. 1).

Community belowground overyielding differed between the 'spatial' and 'temporal' pools of species in that 'temporal' species overyielded more than 'spatial' species (Tables 1, model 6 and 3; Fig. 1). Importantly, the relationship between above- and belowground overyielding also differed between species pools: the slope was steeper in the 'temporal' compared to the 'spatial' pool (Tables 1, model 7 and 4; Fig. 2). Overall, this indicated

**Table 3** Results of linear mixed effect model on community belowground log-response ratio (model 6, Table 1) asking: Does community belowground community log-response ratio change along a species richness gradient or differ between pools of spatial vs. temporal resource partitioning?

|  | Sum Sq. | Mean Sq. | NumDF | DenDF | F value | P |
|---|---|---|---|---|---|---|
| Pool | 1.648 | 1.648 | 1 | 218.166 | 7.513 | **0.007** |

| Contrast | Estimate of difference | SE | DF | T ratio | P |
|---|---|---|---|---|---|
| Spatial pool—Temporal pool | −0.172 | 0.063 | 218.166 | −2.741 | **0.007** |

Significant results ($P<0.05$) presented in bold. Contrasts from Tukey post hoc test of pool with Bonferroni-Holm correction provided below general model results.

**Table 4** Results of linear mixed effect model on plot belowground community log-response ratio (model 7, Table 1) asking: Does community aboveground log-response ratio predict community belowground log-response ratio and does this relationship change along a species richness gradient or differ between pools of spatial vs. temporal resource partitioning?

|  | Sum Sq. | Mean Sq. | NumDF | DenDF | F value | P |
|---|---|---|---|---|---|---|
| Aboveground community log-response ratio | 1.740 | 1.740 | 1 | 218.136 | 8.391 | **0.004** |
| Pool | 1.657 | 1.657 | 1 | 211.275 | 7.990 | **0.005** |

| Contrast | Estimate of difference | SE | DF | T ratio | P |
|---|---|---|---|---|---|
| Spatial pool—Temporal pool | −0.173 | 0.061 | 211.275 | −2.827 | **0.005** |

Significant results ($P<0.05$) presented in bold. Contrasts from Tukey post hoc test of pool with Bonferroni-Holm correction provided below general model results.

that species with the potential to partition resources in space produce more root biomass, but the slope of root biomass increase along the aboveground biomass gradient was steeper for species differing in temporal resource partitioning. In addition, 'temporal' species invested more simultaneously in above- and belowground overyielding (slope closer to one) compared to the 'spatial' species pool, where species invested more in aboveground overyielding.

### 3.3 Hypothesis 3: At the species level, some species exhibit trade-offs between above- and belowground overyielding

Table 5 summarizes the number of samples and means ($\pm$ SE) of species-specific biomass production and species log-response ratio. The table shows that sample sizes ranged from 21 to 45 per species indicating substantial differences in individual sample numbers. This difference was due to the fact that species were not equally abundant in the plots and the rarer ones were, therefore, not always sampled in the biomass subsamples of the plot, both, above- or belowground. This was most obvious for the grass species *A. odoratum* and *P. pratensis,* where we were able to retrieve only 21 and 22 of 45 possible samples, respectively. We studied 13 different grassland species, seven of which are grasses and six of which forbs, with four species of each functional group in each of the two species pools (three species overlapped between pools). All of our models testing species level differences in log-response ratios revealed significant effects of species identity.

At the species level, we found that species differed in the extent to which they overyielded above- and belowground (Table 6; Fig. 3). Seven out of these 13 species underyielded both above- and belowground indicating they grew worse in mixture compared to their monocultures and may suffer less from intraspecific competition compared to interspecific competition (*C. jacea, F. rubra, H. pubescens, P. pratensis, A. odoratum, G. pratense, R. acris,* Table 5; Fig. 3). In contrast, only *K. arvensis* appeared to do significantly better above- and belowground, while three other species overyielded belowground and not significantly aboveground (*L. vulgare* in the temporal pool, *P. lanceolata, D. glomerata*). *Holcus lanatus* was the only species that significantly overyielded at two diversity levels aboveground and significantly underyielded at two diversity levels belowground, showing a clear trade-off (Fig. 3). Some species were more likely to change log-response ratios in higher-diversity mixtures than others, both positive (*L. vulgare*) or negative (*F. rubra*) (Species identity × Species richness: Table 6; Fig. 3). Three species were present in both species pools: *L. vulgare, P. lanceolata,* and

**Table 5** Summary table for species biomass and species log-response ratio separated by pool.

| Pool | Sampling location | Species | Biomass (g m$^{-2}$) | | | Species log-response ratio | | |
|---|---|---|---|---|---|---|---|---|
| | | | N | Mean | SE | N | Mean | SE |
| Spatial | Aboveground | H. pubescens | 39 | 136.527 | 21.900 | 36 | −0.854 | 0.278 |
| | | C. jacea | 38 | 171.049 | 35.177 | 35 | −0.312 | 0.236 |
| | | F. rubra | 33 | 147.004 | 37.092 | 30 | −1.421 | 0.311 |
| | | K. arvensis | 42 | 386.827 | 45.093 | 39 | 0.531 | 0.156 |
| | | L. vulgare | 38 | 264.299 | 36.802 | 35 | −0.597 | 0.220 |
| | | P. pratense | 41 | 325.686 | 37.663 | 38 | 0.332 | 0.091 |
| | | P. lanceolata | 40 | 90.879 | 13.588 | 37 | −0.133 | 0.183 |
| | | P. pratensis | 36 | 106.686 | 23.819 | 22 | −1.404 | 0.312 |
| | Belowground | H. pubescens | 42 | 156.586 | 22.490 | 39 | −0.336 | 0.211 |
| | | C. jacea | 39 | 182.228 | 25.880 | 36 | −0.606 | 0.173 |
| | | F. rubra | 39 | 89.594 | 21.087 | 36 | −2.046 | 0.322 |
| | | K. arvensis | 42 | 204.903 | 16.071 | 39 | 0.355 | 0.100 |
| | | L. vulgare | 45 | 94.675 | 12.112 | 42 | −0.103 | 0.265 |
| | | P. pratense | 45 | 141.170 | 21.501 | 42 | −0.101 | 0.199 |
| | | P. lanceolata | 45 | 118.001 | 13.158 | 42 | 0.580 | 0.130 |
| | | P. pratensis | 39 | 141.708 | 21.996 | 36 | −1.057 | 0.306 |
| Temporal | Aboveground | A. odoratum | 34 | 56.045 | 18.448 | 21 | −1.537 | 0.222 |
| | | D. glomerata | 37 | 368.964 | 48.130 | 34 | −0.015 | 0.283 |
| | | G. pratense | 33 | 142.621 | 28.926 | 30 | −0.880 | 0.189 |
| | | H. lanatus | 32 | 287.482 | 39.062 | 29 | 0.377 | 0.146 |
| | | L. vulgare | 37 | 322.498 | 34.581 | 34 | −0.009 | 0.138 |
| | | P. pratense | 38 | 309.509 | 42.080 | 35 | −0.204 | 0.172 |
| | | P. lanceolata | 33 | 92.844 | 14.275 | 30 | −0.179 | 0.183 |
| | | R. acris | 38 | 46.168 | 8.954 | 35 | −0.669 | 0.215 |
| | Belowground | A. odoratum | 39 | 39.511 | 7.298 | 36 | −0.991 | 0.246 |
| | | D. glomerata | 45 | 126.211 | 13.506 | 42 | 0.556 | 0.134 |

**Table 5** Summary table for species biomass and species log-response ratio separated by pool.—cont'd

| Pool | Sampling location | Species | Biomass (g m$^{-2}$) | | | Species log-response ratio | | |
|---|---|---|---|---|---|---|---|---|
| | | | N | Mean | SE | N | Mean | SE |
| | | G. pratense | 39 | 110.589 | 21.385 | 36 | −1.091 | 0.180 |
| | | H. lanatus | 39 | 116.419 | 20.201 | 36 | −0.497 | 0.289 |
| | | L. vulgare | 45 | 239.799 | 18.778 | 42 | 0.792 | 0.091 |
| | | P. pratense | 42 | 109.209 | 18.172 | 39 | −0.826 | 0.208 |
| | | P. lanceolata | 45 | 123.134 | 13.221 | 42 | 0.666 | 0.185 |
| | | R. acris | 42 | 126.263 | 21.817 | 39 | −0.276 | 0.200 |

Given are numbers of measured plots (N) and mean (± SE) of the measures for all species. Note that L. vulgare, P. pratense, and P. lanceolata are present in both pools. Number of measured instances (N) differ between biomass and species log-response ratio because the species log-response ratio does not include monocultures. This difference is usually by an N of 3 (1 monoculture measurement for each of 3 years). However, for P. pratensis and A. odoratum aboveground biomass was unavailable for 1 year and therefore we could only calculate log-response ratios for two of the 3 years.

**Table 6** Results of linear mixed effect model on species log-response ratio (model 8, Table 1) asking: Do species differ in their contribution to overyielding/underyielding?

| | Sum Sq | Mean Sq | NumDF | DenDF | F value | P (>F) |
|---|---|---|---|---|---|---|
| **Species identity** | **425.577** | **35.465** | **12** | **823.890** | **23.360** | **0.000** |
| Species richness | 2.989 | 1.494 | 2 | 54.315 | 0.984 | 0.380 |
| Pool | 0.251 | 0.251 | 1 | 111.184 | 0.166 | 0.685 |
| **Species identity × Species richness** | **74.852** | **3.119** | **24** | **686.494** | **2.054** | **0.002** |
| **Species identity × Pool** | **30.843** | **15.421** | **2** | **907.170** | **10.158** | **0.000** |

Significant results ($P < 0.05$) presented in bold. Interactions are represented with ×.

*P. pratense*. These three species behaved differently depending on which pool they were in (Species identity × Pool: Table 6; Fig. 3). *Leucanthemum vulgare* neither underyielded nor overyielded in the 'spatial' pool above- or belowground but significantly overyielded belowground in the 'temporal' pool. *Plantago lanceolata* overyielded belowground in both the 'spatial' and

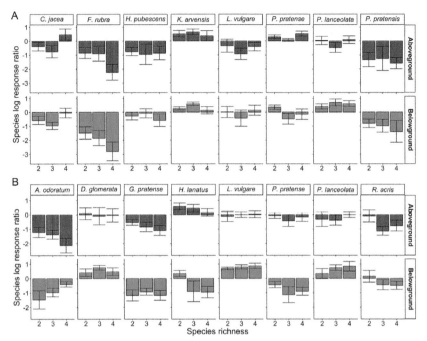

**Fig. 3** Which species contribute to biodiversity effects? Given are mean species-specific log-response ratios across 4 years (three sampling times, ±SE) along the species richness gradient separated by compartment for the spatial pool (A) and the temporal pool (B). Note the three species occurring in both pools (*L. vulgare, P. pratense* and *P. lanceolata*). Species in the spatial pool are arranged from the shortest statured with smallest leaves and the shallowest roots on the left to the largest statured with largest leaves and the longest roots on the right. Species in the temporal pool are arranged from the earliest growth and flowering times on the left to the species with the latest growth and flowering times on the right.

'temporal' pools. Alternatively, *P. pratense* neither overyielded nor underyielded in the 'spatial' pool both above- and belowground and significantly underyielded in the 'temporal' pool. This interaction was not isolated to either aboveground (Table 7) or belowground (Table 8).

We expected that at the species level some species would exhibit trade-offs between investing in overyielding above- vs. belowground and that the result would be a negative correlation between above- and belowground overyielding. Instead, at the species level, high aboveground overyielding was strongly associated with high belowground overyielding (Table 9; Fig. 4). However, if there was a correlation between above- and belowground overyielding, it depended upon species identity (Table 9; Fig. 5). In addition, the extent to which aboveground overyielding was associated

**Table 7** Results of linear mixed effect model on species aboveground log-response ratio (model 9, Table 1) asking: Do species differ in their contribution to biodiversity effects aboveground?

|  | Sum Sq | Mean Sq | NumDF | DenDF | F value | P (>F) |
|---|---|---|---|---|---|---|
| **Species identity** | **173.120** | **14.427** | **12** | **502.471** | **10.024** | **0.000** |
| Pool | 0.003 | 0.003 | 1 | 502.066 | 0.002 | 0.963 |
| **Species identity × Pool** | **11.126** | **5.563** | **2** | **502.163** | **3.865** | **0.022** |

Significant results ($P < 0.05$) presented in bold. Interactions are represented with ×.

**Table 8** Results of linear mixed effect model on species belowground log-response ratio (model 10, Table 1) asking: Do species differ in their contribution to biodiversity effects belowground?

|  | Sum Sq | Mean Sq | NumDF | DenDF | F value | P (>F) |
|---|---|---|---|---|---|---|
| **Species identity** | **335.536** | **27.961** | **12** | **515.612** | **17.368** | **0.000** |
| Pool | 0.223 | 0.223 | 1 | 147.550 | 0.138 | 0.711 |
| **Species identity × Pool** | **25.333** | **12.666** | **2** | **554.638** | **7.867** | **0.000** |

Significant results ($P < 0.05$) presented in bold. Interactions are represented with ×.

**Table 9** Results of linear mixed effect model on species belowground log-response ratio (model 11, Table 1) asking: Does the species aboveground log-response ratio predict the species belowground log-response ratio?

|  | Sum Sq | Mean Sq | NumDF | DenDF | F value | P (>F) |
|---|---|---|---|---|---|---|
| **Aboveground species log-response ratio** | **87.298** | **87.298** | **1** | **585.114** | **62.299** | **0.000** |
| **Species identity** | **191.216** | **15.935** | **12** | **551.664** | **11.371** | **0.000** |
| Pool | 0.416 | 0.416 | 1 | 157.286 | 0.297 | 0.587 |
| **Aboveground species log-response ratio × Species identity** | **40.699** | **3.392** | **12** | **587.870** | **2.420** | **0.005** |
| **Species identity × Pool** | **13.334** | **6.667** | **2** | **564.565** | **4.758** | **0.009** |

Significant results ($P < 0.05$) presented in bold. Interactions represented with ×.

with belowground overyielding also depended upon the species identity (Species aboveground log-response ratio × Species identity: Table 9; Fig. 5). Further, the extent to which species identity determined belowground overyielding depended upon the pool in which those species were found (Species identity × Pool: Table 9; Fig. 5).

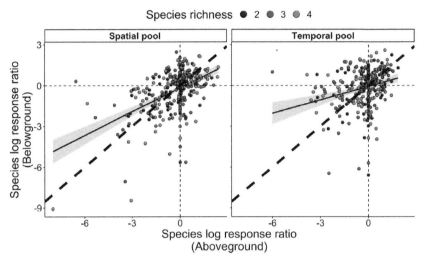

**Fig. 4** At the overall species level, are above- and belowground overyielding correlated? Given are the correlations (solid lines) between above- and belowground species-specific log-response ratios in the spatial pool (left) and temporal pool (right). Dashed lines represent the 1:1 lines. Light grey lines represent $y=0$ and $x=0$. Points that fall to the left of $x=0$ represent aboveground underyielding. Points that fall below $y=0$ represent belowground underyielding.

We found that for seven out of 16 individual species models, the belowground log-response ratio increased with increasing aboveground log-response ratio (Table 10; Fig. 5). Further, for five out of these seven species the slope of the correlation was significantly less than one indicating higher allocation above- than belowground (Table 10; Fig. 5). Furthermore, the belowground log-response ratio never decreased with increasing aboveground log-response ratio.

Species appeared to drive strong positive relationships between their aboveground and belowground overyielding in two separate ways: simultaneous underyielding above- and belowground and simultaneous overyielding above- and belowground. Simultaneous underyielding appeared to drive a strong positive relationship between above- and belowground species log-response ratios for *C. jacea* and *F. rubra*, in the 'spatial' pool and *P. pratense* and *R. acris* in the 'temporal' pool. Simultaneous overyielding appeared to drive a strong positive relationship between above- and belowground log-response ratios for *K. arvensis* and *P. lanceolata* in the 'spatial' pool and no species significantly in the 'temporal' pool. Only one species (*H. lanatus*) appeared to invest in both strategies

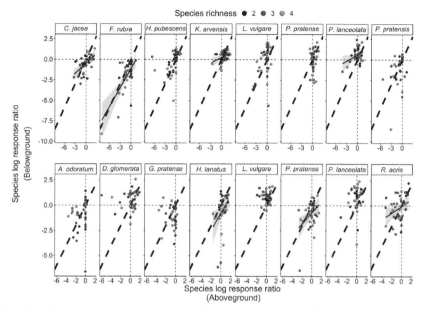

**Fig. 5** At the species level, are above- and belowground overyielding correlated? Given are significant correlations (solid lines) between above- and belowground species-specific log-response ratios in spatial pool (top) and the temporal pool (bottom). Dashed line represent the 1:1 line. Light grey lines represent y = 0 and x = 0. Points that fall to the left of x = 0 represent aboveground underyielding. Points that fall below y = 0 represent belowground underyielding. Species in the spatial pool are arranged from the shortest statured with the smallest leaves and the shallowest roots on the left to the largest statured with the largest leaves and the longest roots on the right. Species in the temporal pool are arranged from the earliest growth and flowering times on the left to the species with the latest growth and flowering times on the right.

(simultaneous overyielding and simultaneous underyielding) equally. Finally, while no species demonstrated a strict trade-off in terms of a negative relationship between above- and belowground overyielding, in the 'spatial' pool *C. jacea*, *F. rubra*, *K. arvensis*, and *P. lanceolata* invested significantly more in overyielding aboveground relative to belowground. Alternatively, in the 'temporal' pool, only *R. acris* invested significantly more in aboveground overyielding (Table 10). *P. pratense* and *H. lanatus* in the temporal pool invested equally in above- and belowground overyielding.

## 4. Discussion

In this paper, we test – for the first time – if and how above- and belowground overyielding are related at both the community and the

**Table 10** Results of individual species-level mixed effect models (Species belowground log-response ratio ~ Species aboveground log-response ratio + (1 | Year/Block).

| Pool | Species | Slope | SE | DF | T | p | Deviation 1:1 line |
|---|---|---|---|---|---|---|---|
| Spatial pool | **C. jacea** | **0.336** | **0.082** | **26.110** | **4.103** | **0.000** | ↓ |
| | **F. rubra** | **0.707** | **0.132** | **31.965** | **5.357** | **0.000** | ↓ |
| | H. pubescens | 0.033 | 0.100 | 28.356 | 0.325 | 0.748 | |
| | **K. arvensis** | **0.254** | **0.096** | **32.975** | **2.653** | **0.012** | ↓ |
| | L. vulgare | 0.280 | 0.169 | 34.709 | 1.654 | 0.107 | |
| | P. pratense | 0.427 | 0.290 | 35.746 | 1.474 | 0.149 | |
| | **P. lanceolata** | **0.229** | **0.093** | **36.358** | **2.460** | **0.019** | ↓ |
| | P. pratensis | 0.320 | 0.209 | 17.065 | 1.530 | 0.144 | |
| Temporal pool | A. odoratum | 0.317 | 0.191 | 28.406 | 1.663 | 0.107 | |
| | D. glomerata | 0.080 | 0.085 | 24.832 | 0.940 | 0.356 | |
| | G. pratense | −0.008 | 0.192 | 25.303 | −0.044 | 0.966 | |
| | **H. lanatus** | **0.800** | **0.388** | **33.644** | **2.062** | **0.047** | |
| | L. vulgare | 0.100 | 0.125 | 23.967 | 0.795 | 0.434 | |
| | **P. pratense** | **0.781** | **0.233** | **29.035** | **3.348** | **0.002** | |
| | P. lanceolata | 0.307 | 0.223 | 34.755 | 1.377 | 0.177 | |
| | **R. acris** | **0.333** | **0.110** | **35.006** | **3.028** | **0.005** | ↓ |

Significant results are indicated in bold. The relationship of the regression line to the identity line is presented as the direction with a downwards arrow indicating that the slope of the line is significantly less than one

species level in a field-based biodiversity experiment. We show that the positive effect of plant species richness on total biomass production is based on simultaneous overyielding of both above- and belowground biomass at the community level. Thus, plots that perform better aboveground also produce more biomass belowground in mixture relative to monoculture and vice versa. Interestingly, the positive correlation between above- and belowground overyielding was biased towards investing in aboveground allocation for species differing in spatial resource uptake, while species separated along a phenological gradient invested in a more balanced way in above- and belowground overyielding.

Above- and belowground overyielding were positively related for seven species out of 13 species. However, for five out of these seven species, species invested significantly more in overyielding aboveground than belowground.

## 4.1 Are above- and belowground overyielding correlated at the community level?

Grassland communities produce more biomass in mixtures compared to monoculture both aboveground and belowground (Dimitrakopoulos and Schmid, 2004; Fornara and Tilman, 2009; Mueller et al., 2013; Ravenek et al., 2014; Tilman et al., 2001). In our experiment, we also observed overyielding at the overall community level. However, we found no gradual increase of above- and belowground overyielding with plant species richness (also see Wagg et al., 2017 for aboveground and Oram et al., 2018 for belowground biomass). This finding is in contrast with observations in other biodiversity experiments, aboveground (Cardinale et al., 2007; Marquard et al., 2009; van Ruijven and Berendse, 2009) and belowground (Cong et al., 2014; Fornara and Tilman, 2009; Mueller et al., 2013; Ravenek et al., 2014; Reich et al., 2004).

A frequently invoked explanation for the fact that overyielding is not always affected by plant species richness (as seen in model 4) is that overyielding establishes over time. Several studies have shown that diversity effects tend to increase over time (e.g. Eisenhauer et al., 2012; van Ruijven and Berendse, 2009). However, at the Jena Main Experiment (Roscher et al., 2004), which started ten years before the Trait Based Experiment reported here, it took only one year for positive effects of plant species richness on biomass production to establish aboveground and four years belowground (Ravenek et al., 2014). The current study uses biomass sampled up to seven years after community establishment, and the community should thus be mature enough to reveal an effect of species richness on overyielding. We believe a far more relevant argument to explain why our results differ from others reported in similar experiments is that we used only a gradient from 1 to 4 species, which is a much narrower range than in most other biodiversity studies (Cardinale et al., 2007; Isbell et al., 2011). In the Trait Based Experiment, there is an eight species plot available in both pools, but eight species mixtures are not replicated and are thus prone to error (but see Oram et al., 2018). We believe that the lack of a progressive overyielding with plant species richness in our experiment may be due to the limited species richness range used (see also Roscher et al., 2016; Siebenkäs et al., 2016).

Despite this limitation, our results reveal a positive correlation between above- and belowground overyielding at the community level and support the first hypothesis: above- and belowground overyielding occur simultaneously in one plant community. In general, plants invested slightly but significantly more in overyielding aboveground than they did in overyielding belowground at the community level. This increased investment towards overyielding aboveground may indicate that light is more limiting than belowground resources at this site. Light limitation may be common in grassland biodiversity experiments in general (Hautier et al., 2009), and also in the Jena Main Experiment (Roscher et al., 2011b). In addition, evidence for belowground resource partitioning has not been found in the Jena Main Experiment (Bachmann et al., 2015; Jesch et al., 2018). Two other studies of grassland community root: shoot ratios across diversity gradients found similarly that plants invested more in aboveground biomass than belowground biomass in diverse mixtures (via the root: shoot ratio: Bessler et al., 2009; Dimitrakopoulos and Schmid, 2004), yet others found no effect (Siebenkäs and Roscher, 2016).

## 4.2 Above- and belowground overyielding relationships differ between species pools

At the pool level, we predicted that the selection of species along a spatial resource partitioning gradient in the 'spatial pool' would release the community from belowground competition and, therefore, increase investment in aboveground overyielding. Similarly, we predicted that the selection of species along a temporal resource partitioning gradient in the 'temporal pool' would release the community from competition aboveground and therefore increase investment in belowground overyielding. We observed a significant effect of species pool on belowground overyielding, but not on aboveground overyielding. As we predicted, belowground overyielding was greater in the 'temporal' species pool than in the 'spatial' pool. In fact, the most diverse communities in the 'temporal' species pool invested equally in above- and belowground overyielding. This 1:1 relationship would indicate that neither below- nor aboveground resources were more limiting.

Alternatively, in the spatial pool, the slope of the above−/belowground overyielding relationship was always less than one, indicating that plants in the spatial pool always invested proportionally more in aboveground overyielding than belowground overyielding. This proportionally higher investment in aboveground overyielding may indicate that light is more limiting than belowground resources as we hypothesized.

Competitive release is not the only potential explanation why resource allocation above- and belowground may be correlated rather than showing a trade-off. Resource availability is highly variable in time and space, and thus resource limitation also varies across time and space (Anten and Hirose, 2001; Březina et al., 2019; Farley and Fitter, 1999; Lepik et al., 2005; Schmitt and Wulff, 1993). Given this resource variability, investing disproportionately in biomass allocation towards competing for an individual limiting resource may be inefficient for plants (Fransen et al., 1998; Jansen et al., 2006).

Contrary to what we predicted based on the functional equilibrium hypothesis (Brouwer, 1962), none of the 13 species demonstrated a significant trade-off (i.e. negative relationship) between overyielding above- vs. belowground. Rather, for the 16 different species relationships we investigated (three species occurred in both species pools), we found seven significant positive relationships between above- and belowground overyielding and no significant negative relationships between above- and belowground overyielding. These findings indicate that plant species simultaneously overyield or simultaneously underyield above- and belowground. If this general trend holds true in other biodiversity studies, aboveground overyielding may be a good proxy for belowground overyielding. If aboveground overyielding is indeed a good proxy for belowground overyielding, this may decrease costs associated with intense manpower necessary for sampling and washing roots in biodiversity experiments. Further, aboveground overyielding may also be a reasonable proxy for inputs of roots into belowground ecosystem functions such as carbon sequestration, nutrient cycling and water storage (Bardgett et al., 2014; Fischer et al., 2015; Fischer et al., 2018; Gould et al., 2016; Hacker et al., 2015; Lange et al., 2015).

However, individual species revealed strong differences in biomass allocation patterns and overyielding in our species pool. These differences may partially reflect generic differences in grasses versus forbs: five of the seven species that showed a significant relationship between above- and belowground overyielding were forbs. Further, all forbs invested significantly more into aboveground overyielding, while only one grass species (*F. rubra*) invested significantly more in aboveground overyielding. For tall growing forbs such as *K. arvensis, C. jacea, L. vulgare* and *R. acris*, much of the aboveground biomass is allocated to the stem. This stem investment may enable these species to use aboveground space more efficiently and potentially dominate the community via asymmetric competition. In recent decades, there has been intense discussion about the asymmetry of competition above- vs. belowground (Cahill and Casper, 2000; DeMalach et al., 2016; Goldberg and Novoplansky, 1997; Lamb et al., 2009; Rajaniemi, 2003; Schwinning and Weiner, 1998). Because asymmetric competition drives investment into increasing the height of biomass, the asymmetry of aboveground competition compared to belowground competition may partially explain why there is more investment in aboveground overyielding than in belowground overyielding for five of the seven significant relationships between above- and belowground overyielding.

Alternatively, the lack of a trade-off between above- and belowground overyielding at the species level found here may suggest that species do not experience a single limiting resource in our biodiversity experiment; rather, they may experience multiple resources as limiting. Although light may be the most limiting resource in diverse grassland communities (Hautier et al., 2009), plant species can be simultaneously limited by many factors (Fay et al., 2015; Harpole et al., 2016). Navigating in this multiple limitation landscape may require that plant species invest heavily in competing for both above- and belowground resources. Indeed, many studies theorize that co-limitation by multiple limiting resources should be the status quo for plant communities (Bloom et al., 1985; Chapin et al., 1987; Gleeson and Tilman, 1992). Further, empirical evidence from global networks of nutrient addition studies suggests that co-limitation is common (Ågren et al. 2012; Bracken et al. 2015; Craine et al., 2008; Craine & Jackson 2010; Elser et al. 2009; Harpole et al. 2011). For example, Fay et al. (2015) found that grasslands previously assumed to be largely nitrogen limited often produced more aboveground biomass after the addition of nitrogen, potassium, and phosphorus than when they received nitrogen alone. Further, Harpole et al. (2016) found that the addition of multiple belowground resources (i.e. reducing belowground resource limitation) in a grassland vegetation, increased investment in aboveground biomass indicating that upon the addition of belowground resources, light availability is an important limiting factor. Further, many studies have shown that plant species richness leads to enhanced carbon and nutrient stocks in the soil (Cong et al., 2014; Cong et al., 2015; Fornara et al., 2009; Hacker et al., 2015; Weisser et al., 2017), leading to enhanced decomposition (Chen et al., 2017) and nutrient availability (Hacker et al., 2015), potentially shifting resource limitation from below- to aboveground. The potential for multiple limiting resources is also suggested at the Jena Main experiment where (Roscher et al., 2011a) found that traits related to light and nutrient acquisition best explain species performance.

## 5. Conclusions

We showed a positive correlation between above- and belowground overyielding – at the community level. For seven out of the 13 species tested here we observed also a positive relationship between above- and belowground overyielding. However, for five out of these seven species, species invested significantly more in overyielding aboveground than belowground, suggesting that individual species are competing more strongly for light than they are for belowground resources. This finding is in contrast to what we have been focussing on as a research community in the last few years: finding evidence for belowground resource partitioning among plant species, which is one of the most commonly invoked potential drivers of positive biodiversity–ecosystem functioning relationships in grasslands.

## Authorship statement

KB, AW, JvR, LM designed the study; KB, AW, JvR, HdK, AE, KB, JMR, NJO, NE, AV, CW and LM collected the shoot and root samples; together with technicians; JMR, NJO, and LM performed the molecular analyses of the root samples with help of technicians; KB, AW, JvR and LM analysed the data and wrote the first draft of the manuscript. All co-authors discussed the results, contributed substantially to the drafts and gave final approval for publication. The authors declare no conflicts of interest.

## Acknowledgements

We thank the gardener team of the Jena Experiment and many field assistants and student helpers for maintaining the field. We are grateful for the support of Hannie de Caluwe, Hongmei Chen, Peter Cruijsen, Anja Kahl, Frans Möller, Roman Patzak, Jan-Willem van der Paauw, Annemiek Smit-Tiekstra and Jan van Walsem during field and laboratory campaigns in Leipzig and Wageningen. The Jena Experiment is funded by the German Research Foundation (DFG, FOR 1451). Further support came from the German Centre for Integrative Biodiversity Research (iDiv) Halle-Jena-Leipzig, funded by the German Research Foundation (FZT 118). LM is supported by a VIDI-NWO grant (864.14.006). KB is funded by the iDiv Flexible Pool (grant #43600900).

## References

Aerts, R., Boot, R.G.A., Vanderaart, P.J.M., 1991. The relation between aboveground and belowground biomass allocation patterns and competitive ability. Oecologia 87 (4), 551–559.

Ågren, G.I., Wetterstedt, J.Å., Billberger, M.F., 2012. Nutrient limitation on terrestrial plant growth – modeling the interaction between nitrogen and phosphorus. New Phytol. 953–960. https://doi.org/10.1111/j.1469-8137.2012.04116,x@10.1002/(ISSN)1469-8137(CAT)VirtualIssues(VI)Ecologicalstoichiometryglobalchange.

Anten, N.P.R., Hirose, T., 1999. Interspecific differences in above-ground growth patterns result in spatial and temporal partitioning of light among species in a tall-grass meadow. J. Ecol. 87 (4), 583–597.

Anten, N.P.R., Hirose, T., 2001. Limitations on photosynthesis of competing individuals in stands and the consequences for canopy structure. Oecologia 129 (2), 186–196.

Bachmann, D., Gockele, A., Ravenek, J.M., Roscher, C., Strecker, T., Weigelt, A., Buchmann, N., 2015. No evidence of complementary water use along a plant species richness gradient in temperate experimental grasslands. PLoS One 10 (1), e0116367.

Bardgett, R.D., Mommer, L., De Vries, F.T., 2014. Going underground: root traits as drivers of ecosystem processes. Trends Ecol. Evol. 29 (12), 692–699.

Barry, K.E., Mommer, L., van Ruijven, J., Wirth, C., Wright, A.J., Bai, Y.F., Connolly, J., De Deyn, G.B., de Kroon, H., Isbell, F., et al., 2019. The future of complementarity: disentangling causes from consequences. Trends Ecol. Evol. 34 (2), 167–180.

Bates, D., Maechler, M., Bolker, B., Walker, S., 2015. Fitting linear mixed-effects models using lme4. J. Stat. Softw. 67 (1), 1–48.

Bessler, H., Temperton, V.M., Roscher, C., Buchmann, N., Schmid, B., Schulze, E.D., Weisser, W.W., Engels, C., 2009. Aboveground overyielding in grassland mixtures is associated with reduced biomass partitioning to belowground organs. Ecology 90 (6), 1520–1530.

Bloom, A.J., Chapin III, F.S., Mooney, H.A., 1985. Resource limitation in plants-an economic analogy. Annu. Rev. Ecol. Syst. 16 (1), 363–392.

Bracken, M.E.S., Hillebrand, H., Borer, E.T., Seabloom, E.W., Cebrian, J., Cleland, E.E., Elser, J.J., et al., 2015. Signatures of nutrient limitation and co-limitation: responses of autotroph internal nutrient concentrations to nitrogen and phosphorus additions. Oikos 124 (2), 113–121. https://doi.org/10.1111/oik.01215.

Březina, S., Jandová, K., Pecháčková, S., Hadincová, V., Skálová, H., Krahulec, F., Herben, T., 2019. Nutrient patches are transient and unpredictable in an unproductive mountain grassland. Plant Ecol. 220 (1), 111–123.

Brouwer, R., 1962. Distribution of dry matter in the plant. Research Report. Instituut voor biologisch en scheikundig onderzoek van landbouwgewassen no. 203.

Cahill, J.F., 2003. Lack of relationship between below-ground competition and allocation to roots in 10 grassland species. J. Ecol. 91 (4), 532–540.

Cahill, J.F., Casper, B.B., 2000. Investigating the relationship between neighbor root biomass and belowground competition: field evidence for symmetric competition belowground. Oikos 90 (2), 311–320.

Cardinale, B.J., Wrigh, J.P., Cadotte, M.W., Carroll, I.T., Hector, A., Srivastava, D.S., Loreau, M., Weis, J.J., 2007. Impacts of plant diversity on biomass production increase through time because of species complementarity. Proc. Natl. Acad. Sci. U. S. A. 104, 18123–18128.

Cardinale, B.J., Duffy, J.E., Gonzalez, A., Hooper, D.U., Perrings, C., Venail, P., Narwani, A., Mace, G.M., Tilman, D., Wardle, D.A., 2012. Biodiversity loss and its impact on humanity. Nature 486 (7401), 59–67.

Chapin, F.S., Bloom, A.J., Field, C.B., Waring, R.H., 1987. Plant responses to multiple environmental factors. Bioscience 37, 49–57.
Chen, H.M., Oram, N.J., Barry, K.E., Mommer, L., van Ruijven, J., de Kroon, H., Ebeling, A., Eisenhauer, N., Fischer, C., Gleixner, G., Gessler, A., Mace, O.G., Hacker, N., Hildebrandt, A., Lange, M., Scherer-Lorenzen, M., Scheu, S., Oelmann, Y., Wagg, C., Wilcke, W., Wirth, C., Weigelt, A., 2017. Root chemistry and soil fauna, but not soil abiotic conditions explain the effects of plant diversity on root decomposition. Oecologia 185 (3), 499–511.
Cong, W., Ruijven, J.v., Mommer, L., Deyn, G.B.d., Berendse, F., Hoffland, E., 2014. Plant species richness promotes soil carbon and nitrogen stocks in grasslands without legumes. J. Ecol. 102 (5), 1163–1170.
Cong, W.F., Hoffland, E., Li, L., Six, J., Sun, J.H., Bao, X.G., Zhang, F.S., van der Werf, W., 2015. Intercropping enhances soil carbon and nitrogen. Glob. Chang. Biol. 21 (4), 1715–1726.
Craine, J.M., Jackson, R.D., 2010. Plant nitrogen and phosphorus limitation in 98 North American grassland soils. Plant and Soil 334 (1), 73–84. https://doi.org/10.1007/s11104-009-0237-1.
Craine, J.M., Morrow, C., Stock, W.D., 2008. Nutrient concentration ratios and co-limitation in South African Grasslands. New Phytol. 179 (3), 829–836. https://doi.org/10.1111/j.1469-8137.2008.02513.x.
DeMalach, N., Zaady, E., Weiner, J., Kadmon, R., 2016. Size asymmetry of resource competition and the structure of plant communities. J. Ecol. 104 (4), 899–910.
Dimitrakopoulos, P.G., Schmid, B., 2004. Biodiversity effects increase linearly with biotope space. Ecol. Lett. 7 (7), 574–583.
Dumbrell, A.J., Ashton, P.D., Aziz, N., Feng, G., Nelson, M., Dytham, C., Fitter, A.H., Helgason, T., 2011. Distinct seasonal assemblages of arbuscular mycorrhizal fungi revealed by massively parallel pyrosequencing. New Phytol. 190 (3), 794–804.
Ebeling, A., Pompe, S., Baade, J., Eisenhauer, N., Hillebrand, H., Proulx, R., Roscher, C., Schmid, B., Wirth, C., Weisser, W.W., 2014. A trait-based experimental approach to understand the mechanisms underlying biodiversity-ecosystem functioning relationships. Basic Appl. Ecol. 15 (3), 229–240.
Eisenhauer, N., Reich, P.B., Scheu, S., 2012. Increasing plant diversity effects on productivity with time due to delayed soil biota effects on plants. Basic Appl. Ecol. 13 (7), 571–578.
Elser, J.J., Andersen, T., Baron, J.S., Bergström, A.-K., Jansson, M., Kyle, M., Nydick, K.R., Steger, L., Hessen, D.O., 2009. Shifts in lake N:P stoichiometry and nutrient limitation driven by atmospheric nitrogen deposition. Science 326 (5954), 835–837. https://doi.org/10.1126/science.1176199.
Farley, R.A., Fitter, A.H., 1999. Temporal and spatial variation in soil resources in a deciduous woodland. J. Ecol. 87 (4), 688–696.
Fay, P.A., Prober, S.M., Harpole, W.S., Knops, J.M., Bakker, J.D., Borer, E.T., Lind, E.M., MacDougall, A.S., Seabloom, E.W., Wragg, P.D., 2015. Grassland productivity limited by multiple nutrients. Nat. Plants 1 (7), 15080.
Fischer, C., Tischer, J., Roscher, C., Eisenhauer, N., Ravenek, J., Gleixner, G., Attinger, S., Jensen, B., de Kroon, H., Mommer, L., Scheu, S., Hildebrandt, A., 2015. Plant species diversity affects infiltration capacity in an experimental grassland through changes in soil properties. Plant and Soil 397 (1-2), 1–16.
Fischer, C., Leimer, S., Roscher, C., Ravenek, J., de Kroon, H., Kreutziger, Y., Baade, J., Beßler, H., Eisenhauer, N., Weigelt, A., 2018. Plant species richness and functional groups have different effects on soil water content in a decade-long grassland experiment. J. Ecol. 107 (1), 127–141.

Fornara, D.A., Tilman, D., 2009. Ecological mechanisms associated with the positive diversity-productivity relationship in an N-limited grassland. Ecology 90 (2), 408–418.

Fornara, D.A., Tilman, D., Hobbie, S.E., 2009. Linkages between plant functional composition, fine root processes and potential soil N mineralization rates. J. Ecol. 97 (1), 48–56.

Frank, D.A., Pontes, A.W., Maine, E.M., Fridley, J.D., 2015. Fine-scale belowground species associations in temperate grassland. Mol. Ecol. 24 (12), 3206–3216.

Fransen, B., de Kroon, H., Berendse, F., 1998. Root morphological plasticity and nutrient acquisition of perennial grass species from habitats of different nutrient availability. Oecologia 115 (3), 351–358.

Gedroc, J., 1996. Plasticity in root/shoot partitioning: optimal, ontogenetic, or both. Funct. Ecol. 10, 44–50.

Genney, D.R., Alexander, I.J., Hartley, S.E., 2002. Soil organic matter distribution and below-ground competition between Calluna vulgaris and Nardus stricta. Funct. Ecol. 16 (5), 664–670.

Gleeson, S.K., Tilman, D., 1992. Plant allocation and the multiple limitation hypothesis. Am. Nat. 139 (6), 1322–1343.

Goldberg, D., Novoplansky, A., 1997. On the relative importance of competition in unproductive environments. J. Ecol. 85 (4), 409–418.

Gould, I.J., Quinton, J.N., Weigelt, A., De Deyn, G.B., Bardgett, R.D., 2016. Plant diversity and root traits benefit physical properties key to soil function in grasslands. Ecol. Lett. 19 (9), 1140–1149.

Hacker, N., Ebeling, A., Gessler, A., Gleixner, G., Mace, O.G., de Kroon, H., Lange, M., Mommer, L., Eisenhauer, N., Ravenek, J., Scheu, S., Weigelt, A., Wagg, C., Wilcke, W., Oelmann, Y., 2015. Plant diversity shapes microbe-rhizosphere effects on P mobilisation from organic matter in soil. Ecol. Lett. 18 (12), 1356–1365.

Harpole, W.S., Ngai, J.T., Cleland, E.E., Seabloom, E.W., Borer, E.T., Bracken, M.E.S., Elser, J.J., et al., 2011. Nutrient co-limitation of primary producer communities. Ecol. Lett. 14 (9), 852–862. https://doi.org/10.1111/j.1461-0248.2011.01651.x.

Harpole, W.S., Sullivan, L.L., Lind, E.M., Firn, J., Adler, P.B., Borer, E.T., Chase, J., Fay, P.A., Hautier, Y., Hillebrand, H., 2016. Addition of multiple limiting resources reduces grassland diversity. Nature 537 (7618), 93.

Hautier, Y., Niklaus, P.A., Hector, A., 2009. Competition for light causes plant biodiversity loss after eutrophication. Science 324 (5927), 636–638.

Hector, A., Schmid, B., Beierkuhnlein, C., Caldeira, M.C., Diemer, M., Dimitrakopoulos, P.G., Finn, J.A., Freitas, H., Giller, P.S., Good, J., Harris, R., Hogberg, P., Huss-Danell, K., Joshi, J., Jumpponen, A., Korner, C., Leadley, P.W., Loreau, M., Minns, A., Mulder, C.P.H., O'Donovan, G., Otway, S.J., Pereira, J.S., Prinz, A., Read, D.J., Scherer-Lorenzen, M., Schulze, E.D., Siamantziouras, A.S.D., Spehn, E.M., Terry, A.C., Troumbis, A.Y., Woodward, F.I., Yachi, S., Lawton, J.H., 1999. Plant diversity and productivity experiments in European grasslands. Science 286 (5442), 1123–1127.

Hector, A., Bazeley-White, E., Loreau, M., Otway, S., Schmid, B., 2002. Overyielding in grassland communities: testing the sampling effect hypothesis with replicated biodiversity experiments. Ecol. Lett. 5 (4), 502–511.

Hendriks, M., Mommer, L., de Caluwe, H., Smit-Tiekstra, A.E., van der Putten, W.H., de Kroon, H., 2013. Independent variations of plant and soil mixtures reveal soil feedback effects on plant community overyielding. J. Ecol. 101 (2), 287–297.

Hendriks, M., Ravenek, J.M., Smit-Tiekstra, A.E., van der Paauw, J.W., de Caluwe, H., van der Putten, W.H., de Kroon, H., Mommer, L., 2015. Spatial heterogeneity of plant-soil feedback affects root interactions and interspecific competition. New Phytol. 207 (3), 830–840.

HilleRisLambers, J., Harpole, W.S., Tilman, D., Knops, J., Reich, P.B., 2004. Mechanisms responsible for the positive diversity–productivity relationship in Minnesota grasslands. Ecol. Lett. 7 (8), 661–668.
Hooper, D.U., Chapin, F.S., Ewel, J.J., Hector, A., Inchausti, P., Lavorel, S., Lawton, J.H., Lodge, D.M., Loreau, M., Naeem, S., Schmid, B., Setala, H., Symstad, A.J., Vandermeer, J., Wardle, D.A., 2005. Effects of biodiversity on ecosystem functioning: A consensus of current knowledge. Ecol. Monogr. 75 (1), 3–35.
Isbell, F., Calcagno, V., Hector, A., Connolly, J., Harpole, W.S., Reich, P.B., Scherer-Lorenzen, M., Schmid, B., Tilman, D., Van Ruijven, J., 2011. High plant diversity is needed to maintain ecosystem services. Nature 477 (7363), 199.
Jackson, R.B., Moore, L.A., Hoffmann, W.A., Pockman, W.T., Linder, C.R., 1999. Ecosystem rooting depth determined with caves and DNA. Proc. Natl. Acad. Sci. U. S. A. 96 (20), 11387–11392.
Janecek, S., Janeckova, P., Leps, J., 2004. Influence of soil heterogeneity and competition on growth features of three meadow species. Flora 199 (1), 3–11.
Jansen, C., van Kempen, M., Bögemann, G.M., Bouma, T.J., de Kroon, H., 2006. Limited costs of wrong root placement in Rumex palustris in heterogeneous soils. New Phytol. 171 (1), 117–126.
Jesch, A., Barry, K.E., Ravenek, J.M., Bachmann, D., Strecker, T., Weigelt, A., Buchmann, N., Kroon, H., Gessler, A., Mommer, L., Roscher, C., Scherer-Lorenzen, M., 2018. Below-ground resource partitioning alone cannot explain the biodiversity–ecosystem function relationship: a field test using multiple tracers. J. Ecol. 106 (5), 2002–2018.
Jones, F.A., Erickson, D.L., Bernal, M.A., Bermingham, E., Kress, W.J., Allen Herre, E., Muller-Landau, H.C., Turner, B.L., 2011. The roots of diversity: belowground richness and rooting distributions in a tropical forest revealed by DNA barcodes and inverse modelling. PLoS One 6.9 (2011), e24506.
Kesanakurti, P.R., Fazekas, A.J., Burgess, K.S., Percy, D.M., Newmaster, S.G., Graham, S.W., Barrett, S.C.H., Hajibabaei, M., Husband, B.C., 2011. Spatial patterns of plant diversity below-ground as revealed by DNA barcoding. Mol. Ecol. 20 (6), 1289–1302.
Kuznetsova, A., Brockhoff, P.B., Christensen, R.H.B., 2015. lmerTest: tests in linear mixed effects models (version R package version 2.0-29). R Core Team.
Lamb, E.G., Kembel, S.W., Cahill, J.F., 2009. Shoot, but not root, competition reduces community diversity in experimental mesocosms. J. Ecol. 97 (1), 155–163.
Lange, M., Eisenhauer, N., Sierra, C.A., Bessler, H., Engels, C., Griffiths, R.I., Mellado-Vazquez, P.G., Malik, A.A., Roy, J., Scheu, S., Steinbeiss, S., Thomson, B.C., Trumbore, S.E., Gleixner, G., 2015. Plant diversity increases soil microbial activity and soil carbon storage. Nat. Commun. 6, 6707.
Lenth, R., Herv, M., 2016. Least-squares means: the package. J. Stat. Softw. 69 (1), 1–33.
Lepik, M., Liira, J., Zobel, K., 2005. High shoot plasticity favours plant coexistence in herbaceous vegetation. Oecologia 145 (3), 465–474.
Linder, C.R., Moore, L.A., Jackson, R.B., 2000. A universal molecular method for identifying underground plant parts to species. Mol. Ecol. 9 (10), 1549–1559.
Loreau, M., 1998. Separating sampling and other effects in biodiversity experiments. Oikos 82 (3), 600–602.
Marquard, E., Weigelt, A., Roscher, C., Gubsch, M., Lipowsky, A., Schmid, B., 2009. Positive biodiversity–productivity relationship due to increased plant density. J. Ecol. 97 (4), 696–704.
Meyer, S.T., Ebeling, A., Eisenhauer, N., Hertzog, L., Hillebrand, H., Milcu, A., Pompe, S., Abbas, M., Bessler, H., Buchmann, N., De Luca, E., Engels, C., Fischer, M., Gleixner, G., Hudewenz, A., Klein, A.M., de Kroon, H., Leimer, S., Loranger, H.,

Mommer, L., Oelmann, Y., Ravenek, J.M., Roscher, C., Rottstock, T., Scherber, C., Scherer-Lorenzen, M., Scheu, S., Schmid, B., Schulze, E.D., Staudler, A., Strecker, T., Temperton, V., Tscharntke, T., Vogel, A., Voigt, W., Weigelt, A., Wilcke, W., Weisser, W.W., 2016. Effects of biodiversity strengthen over time as ecosystem functioning declines at low and increases at high biodiversity. Ecosphere 7 (12).

Mommer, L., Wagemaker, N., de Kroon, H., Ouborg, N.J., 2008. Unravelling belowground plant distributions: a real time PCR method for quantifying species proportions in mixed root samples. Mol. Ecol. Notes 8, 947–953.

Mommer, L., van Ruijven, J., de Caluwe, H., Smit-Tiekstra, A.E., Wagemaker, C.A.M., Ouborg, N.J., Bogemann, G.M., van der Weerden, G.M., Berendse, F., de Kroon, H., 2010. Unveiling below-ground species abundance in a biodiversity experiment: a test of vertical niche differentiation among grassland species. J. Ecol. 98 (5), 1117–1127.

Mommer, L., Van Ruijven, J., Jansen, C., Van de Steeg, H.M., De Kroon, H., 2011. Interactive effects of nutrient heterogeneity and competition: implications for root foraging theory. Funct. Ecol. 26 (1), 66–73.

Mueller, K.E., Tilman, D., Fornara, D.A., Hobbie, S.E., 2013. Root depth distribution and the diversity-productivity relationship in a long-term grassland experiment. Ecology 94 (4), 787–793.

Oram, N.J., Ravenek, J.M., Barry, K.E., Weigelt, A., Chen, H.M., Gessler, A., Gockele, A., de Kroon, H., van der Paauw, J.W., Scherer-Lorenzen, M., Smit-Tiekstra, A., van Ruijven, J., Mommer, L., 2018. Below-ground complementarity effects in a grassland biodiversity experiment are related to deep-rooting species. J. Ecol. 106 (1), 265–277.

Padilla, F.M., Mommer, L., Caluwe, H.d., Smit-Tiekstra, A.E., Wagemaker, C.A.M., Ouborg, N.J., Kroon, H.d., 2013. Early root overproduction not triggered by nutrients decisive for competitive success belowground. PLoS One 8 (1), 9.

Poorter, H., Nagel, O., 2000. The role of biomass allocation in the growth response of plants to different levels of light CO2, nutrients and water: a quantitative review. Aust. J. Plant Physiol. 27, 595–607.

Poorter, H., Niklas, K.J., Reich, P.B., Oleksyn, J., Poot, P., Mommer, L., 2012. Biomass allocation to leaves, stems and roots: meta-analyses of interspecific variation and environmental control. New Phytol. 193 (1), 30–50.

Poorter, H., Niinemets, Ü., Ntagkas, N., Siebenkäs, A., Mäenpää, M., Matsubara, S., Pons, T., 2019. A meta-analysis of plant responses to light intensity for 70 traits ranging from molecules to whole plant performance. New Phytol. https://doi.org/10.1111/nph.15754.

Rajaniemi, T.K., 2003. Evidence for size asymmetry of belowground competition. Basic Appl. Ecol. 4 (3), 239–247.

Ravenek, J.M., Bessler, H., Engels, C., Scherer-Lorenzen, M., Gessler, A., Gockele, A., De Luca, E., Temperton, V.M., Ebeling, A., Roscher, C., Schmid, B., Weisser, W.W., Wirth, C., de Kroon, H., Weigelt, A., Mommer, L., 2014. Long-term study of root biomass in a biodiversity experiment reveals shifts in diversity effects over time. Oikos 123 (12), 1528–1536.

Reich, P.B., Tilman, D., Naeem, S., Ellsworth, D.S., Knops, J., Craine, J., Wedin, D., Trost, J., 2004. Species and functional group diversity independently influence biomass accumulation and its response to $CO_2$ and N. Proc. Natl. Acad. Sci. U. S. A. 101 (27), 10101–10106.

Roscher, C., Schumacher, J., Baade, J., Wilcke, W., Gleixner, G., Weisser, W.W., Schmid, B., Schulze, E.D., 2004. The role of biodiversity for element cycling and trophic interactions: an experimental approach in a grassland community. Basic Appl. Ecol. 5 (2), 107–121.

Roscher, C., Temperton, V.M., Scherer-Lorenzen, M., Schmitz, M., Schumacher, J., Schmid, B., Buchmann, N., Weisser, W.W., Schulze, E.D., 2005. Overyielding in experimental grassland communities–irrespective of species pool or spatial scale. Ecol. Lett. 8 (4), 419–429.

Roscher, C., Schumacher, J., Weisser, W.W., Schmid, B., Schulze, E.-D., 2007. Detecting the role of individual species for overyielding in experimental grassland communities composed of potentially dominant species. Oecologia 154 (3), 535–549.

Roscher, C., Scherer-Lorenzen, M., Schumacher, J., Temperton, V.M., Buchmann, N., Schulze, E.D., 2011a. Plant resource-use characteristics as predictors for species contribution to community biomass in experimental grasslands. Perspect. Plant Ecol. Syst. 13 (1), 1–13.

Roscher, C., Kutsch, W.L., Kolle, O., Ziegler, W., Schulze, E.-D., 2011b. Adjustment to the light environment in small-statured forbs as a strategy for complementary resource use in mixtures of grassland species. Ann. Bot. 107 (6), 965–979.

Roscher, C., Schmid, B., Kolle, O., Schulze, E.D., 2016. Complementarity among four highly productive grassland species depends on resource availability. Oecologia 181 (2), 571–582.

Schmitt, J., Wulff, R.D., 1993. Light spectral quality, phytochrome and plant competition. Tree 8 (2), 47–51.

Schwinning, S., Weiner, J., 1998. Mechanisms determining the degree of size asymmetry in competition among plants. Oecologia 113 (4), 447–455.

Shipley, B., Meziane, D., 2002. The balanced-growth hypothesis and the allometry of leaf and root biomass allocation. Funct. Ecol. 16 (3), 326–331.

Siebenkäs, A., Roscher, C., 2016. Functional composition rather than species richness determines root characteristics of experimental grasslands grown at different light and nutrient availability. Plant and Soil 404 (1), 399–412.

Siebenkas, A., Schumacher, J., Roscher, C., 2016. Resource availability alters biodiversity effects in experimental Grass-Forb mixtures. PLoS One 11 (6), e0158110.

Tilman, D., Reich, P.B., Knops, J., Wedin, D., Mielke, T., Lehman, C., 2001. Diversity and productivity in a long-term grassland experiment. Science 294 (5543), 843–845.

van Ruijven, J., Berendse, F., 2003. Positive effects of plant species diversity on productivity in the absence of legumes. Ecol. Lett. 6 (3), 170–175.

van Ruijven, J., Berendse, F., 2005. Diversity-productivity relationships: initial effects, long-term patterns, and underlying mechanisms. Proc. Natl. Acad. Sci. U. S. A. 102 (3), 695–700.

van Ruijven, J., Berendse, F., 2009. Long-term persistence of a positive plant diversity-productivity relationship in the absence of legumes. Oikos 118 (1), 101–106.

Wagg, C., Ebeling, A., Roscher, C., Ravenek, J., Bachmann, D., Eisenhauer, N., Mommer, L., Buchmann, N., Hillebrand, H., Schmid, B., Weisser, W.W., 2017. Functional trait dissimilarity drives both species complementarity and competitive disparity. Funct. Ecol. 31 (12), 2320–2329.

Weisser, W.W., Roscher, C., Meyer, S.T., Ebeling, A., Luo, G.J., Allan, E., Besser, H., Barnard, R.L., Buchmann, N., Buscot, F., Engels, C., Fischer, C., Fischer, M., Gessler, A., Gleixner, G., Halle, S., Hildebrandt, A., Hillebrand, H., de Kroon, H., Lange, M., Leimer, S., Le Roux, X., Milcu, A., Mommer, L., Niklaus, P.A., Oelmann, Y., Proulx, R., Roy, J., Scherber, C., Scherer-Lorenzen, M., Scheu, S., Tscharntke, T., Wachendorf, M., Wagg, C., Weigelt, A., Wilcke, W., Wirth, C., Schulze, E.D., Schmid, B., Eisenhauer, N., 2017. Biodiversity effects on ecosystem functioning in a 15-year grassland experiment: Patterns, mechanisms, and open questions. Basic Appl. Ecol. 23, 1–73.

CHAPTER THREE

# Lost in trait space: species-poor communities are inflexible in properties that drive ecosystem functioning

Anja Vogel[a,b,c,*,†], Peter Manning[d,†], Marc W. Cadotte[e,†], Jane Cowles[f,†], Forest Isbell[f,†], Alexandre L.C. Jousset[g,†], Kaitlin Kimmel[f,†], Sebastian T. Meyer[h,†], Peter B. Reich[i,†], Christiane Roscher[a,j,†], Michael Scherer-Lorenzen[k,†], David Tilman[f,†], Alexandra Weigelt[a,b,†], Alexandra J. Wright[l,†], Nico Eisenhauer[a,b,†], Cameron Wagg[m,n,†]

[a]German Centre for Integrative Biodiversity Research (iDiv) Halle-Jena-Leipzig, Leipzig, Germany
[b]Institute of Biology, Leipzig University, Leipzig, Germany
[c]Institute of Ecology and Evolution, Friedrich Schiller University Jena, Jena, Germany
[d]Senckenberg Biodiversity and Climate Research Centre (SBiK-F), Frankfurt am Main, Germany
[e]Deptartment of Biological Sciences, University of Toronto Scarborough, Toronto, ON, Canada
[f]Department of Ecology, Evolution, and Behavior, University of Minnesota, St. Paul, MN, United States
[g]Institute for Environmental Biology, Utrecht University, Utrecht, The Netherlands
[h]Department of Ecology and Ecosystem Management, Terrestrial Ecology Research Group, School of Life Sciences Weihenstephan, Technical University of Munich, Freising, Germany
[i]Department of Forest Resources, University of Minnesota, St. Paul, MN, United States
[j]UFZ, Helmholtz Centre for Environmental Research, Physiological Diversity, Leipzig, Germany
[k]Institute of Biology/Geobotany, University of Freiburg, Freiburg, Germany
[l]Department of Biological Sciences, California State University Los Angeles, Los Angeles, CA, United States
[m]Department of Evolutionary Biology and Environmental Studies, University of Zürich, Zürich, Switzerland
[n]Fredericton Research and Development Centre, Agriculture and Agri-Food Canada, Fredericton, NB, Canada
*Corresponding author: e-mail address: anja.vogel@uni-jena.de

## Contents

| | |
|---|---|
| 1. Introduction | 92 |
| 2. Methods | 97 |
|    2.1 Study sites and experimental designs | 97 |
|    2.2 Data | 99 |
|    2.3 Calculation of functional and phylogenetic indices and statistical analyses | 102 |

[†] This study was motivated during a workshop on community assembly with JC, MWC, NE, FI, ALCJ, KK, PM, STM, AW, AJW, CW, AV, CR, contributing to discussions. NE and FI developed the general research idea; MSL and AW established and maintained the management and drought experiment together with AV; JC, FI, AW, AV, CR, PBR, MSL, DT contributed data; CW, AV, MWC analysed the data; AV and PM wrote the first draft of the manuscript with equal contributions. All authors contributed to revisions of the manuscript.

3. Results 104
   3.1 Temporal shift of the functional and phylogenetic diversity of plant
       communities 104
   3.2 Temporal shifts in community trait space 106
4. Discussion 116
   4.1 Temporal shift of the functional and phylogenetic diversity of plant
       communities 117
   4.2 Temporal shift of the trait space of plant communities 118
5. Conclusions 121
Acknowledgements 122
References 122

## Abstract

It is now well established that biodiversity plays an important role in determining ecosystem functioning and its stability over time. A possible mechanism for this positive effect of biodiversity is that more diverse plant communities have a greater capacity to respond to environmental changes through shifts in species dominance and composition. In our study, we utilized data from five long-term grassland biodiversity experiments located in North America (three studies) and Central Europe (two studies), in which plant species richness and global change drivers were manipulated simultaneously. The global change drivers included warming, drought, elevated atmospheric $CO_2$ concentrations, elevated N inputs, or intensive management. Across drivers, functional change over time was significantly greater for communities of high plant diversity than that of low diversity because of a higher functional and phylogenetic richness and mostly associated with a dominance by species with a 'slow and tall' strategy. Community functional shifts in response to global change drivers were, however, relatively weak and mostly not influenced by diversity. The exception to this was warming, where diverse communities showed stronger shifts than species-poor communities. Our results confirm the hypothesis that diverse communities have a greater capacity for functional change than species-poor communities, particularly in their successional dynamics, but also potentially in their responses to environmental change. This capacity could underlie the positive biodiversity-stability relationship and buffer ecosystem responses to environmental change.

## 1. Introduction

There is now significant evidence that more diverse plant communities exhibit greater productivity and stability compared with low-diversity plant communities (Cardinale et al., 2012; Craven et al., 2018; Eisenhauer et al., 2019; Hautier et al., 2015; Isbell et al., 2017). One hypothesis to explain the positive relationship between biodiversity and stability is the insurance hypothesis (Folke et al., 1996; Naeem and Li, 1997; Yachi and Loreau, 1999), which postulates that species rich communities contain a wider range of functional strategies that can maintain ecosystem function over time in variable environments because they are more likely to contain species with traits suited to the new environmental conditions (Allan et al., 2011; Cadotte et al., 2012; Craven et al., 2018; de Mazancourt et al., 2013; Diaz and Cabido, 2001; Isbell et al., 2011, 2015; Lepš et al., 2018; Wagg et al., 2017). This greater flexibility of diverse communities should also be visible in the capacity of plant communities change their functional composition over time, resulting in greater successional shifts with greater diversity (Roscher et al., 2013). The ability for diverse plant communities to shift in species composition through time may also be apparent in response to drivers of global environmental change (e.g. land use intensification and drought) compared to species-poor communities.

To our knowledge, however, this hypothesis has not been tested, and therefore, we investigated, whether diversity begets greater flexibility of successional dynamics under global environmental change. A higher flexibility of diverse communities could be manifested in various ways. For instance, greater species richness may have a higher probability of directional shifts of community properties through time and a wider range of options for changes in different directions. This expectation is based on the concept that species within a community are distributed in a multidimensional space of their functional traits, and assumes that the volume of the trait space occupied by the community (i.e. functional or phylogenetic richness) increases with increasing number of species (Cadotte et al., 2008, 2009; Diaz and Cabido, 2001; Flynn et al., 2011; Roscher et al., 2012).

In natural communities, shifts in community trait space under environmental change will result from a combination of phenotypic plasticity and genetic differentiation (Cornwell and Ackerly, 2009; Lepš et al., 2011) of species present in communities, changes in species abundance, and species immigrations and losses. Such changes will, in turn, affect the functional capacity of the ecosystem and the rates of ecological processes that it regulates (Boeddinghaus et al., 2019). Over time and under global change drivers, the abundance distribution of species will be structured due to biotic and abiotic filters and limiting similarity, i.e., species with traits that fit to specific environmental conditions do coexist if they are not overlapping in their niches (Chesson, 2000). The functional and phylogenetic diversity of traits in a community, can therefore inform us, whether communities converge or diverge during succession (Allan et al., 2013; Cadotte et al., 2012;

Gerhold et al., 2015; Webb, 2000; Webb et al., 2002) and in response to global change drivers (Valladares et al., 2015). Taking the phylogenetic relatedness among species into account can give information on the importance of phylogenetically conserved traits, for which trait data are not available (Kraft and Ackerly, 2010) and therefore complement functional trait approaches, which usually use predefined ecological traits (Chao et al., 2014; Srivastava et al., 2012). In contrast to natural communities, experimental communities do not allow for species immigration, making them an ideal set to study the role of species diversity for community dynamics over time.

We hypothesize that, in the absence of immigration, higher initial functional and phylogenetic diversity of diverse communities gives them a wider range of options through which abundance changes within these communities shift their trait distribution towards functional strategies that are well-suited to new environmental conditions (Fig. 1). This means, we predict greater convergence in their functional and phylogenetic diversity during succession, because they have a greater initial variation for filtering than species-poor communities. In contrast, low diversity communities are less likely to contain species with functional traits well suited to the new conditions, making them effectively 'lost in trait space' as they are unlikely to shift their trait distribution to match new environmental conditions (Fig. 1).

The direction of a temporal shift within the trait space of the communities can be hypothesized from processes in real-world communities. In secondary succession, disturbance creates an initially fertile and high light environment that favours acquisitive fast-growing species (Roscher et al., 2015; Vile et al., 2006). Such species are characterized by high specific leaf area (SLA), high tissue nitrogen (N) and phosphorus (P) concentrations, low leaf dry matter content (LDMC) ('fast' traits of the leaf economic spectrum; Díaz et al., 2016, Reich, 2014), and low plant height and low seed mass ('short' traits of the height spectrum). Over time, nutrients become sequestered in plant biomass and soil organic matter, and light availability declines due to the development of a competitive plant canopy. This favours plants with a 'slow' strategy that is more conservative in resource use (low SLA, leaf N and P and high LDMC) (Purschke et al., 2013; Török et al., 2018) and those with tall stature and large seeds, as they are able to recruit in shade and outcompete other plants for light (Douma et al., 2012; Kahmen and Poschlod, 2004; Roscher et al., 2015). We hypothesize that in diverse experimental plant communities there is a higher probability to make this transition over time, as they are more likely to possess species with relevant traits. In contrast, the probability, that species-poor communities by chance comprise

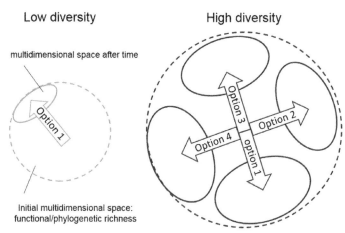

**Fig. 1** Hypothetical framework of the temporal effects of plant diversity and global environmental change drivers on experimental plant communities. We assume that species rich plant communities will typically occupy a larger volume of a multi-dimensional trait space (functional richness) and similarly higher phylogenetic richness than low-diversity plant communities (dashed circle). In the absence of invasion by new species over time, this functional/phylogenetic space should decrease due to species extinctions and abundance shifts, and the functional and phylogenetic diversity of the communities should converge (measured as mean pairwise and mean nearest neighbour distance, MPD and MNTD, as well as imbalance of abundances IAC). We expect that the larger trait/phylogenetic space of high-diversity communities beget more options to move in trait space through time compared to low-diversity communities. Depending on the global change driver and site conditions, this community-level flexibility will also allow for stronger directional shifts. In consequence of this directional shift, the community position within trait space will change, which changes community mean trait values and trait dispersion. We aimed to measure these changes with traits related to the leaf-height-seed trait spectrum.

species with a 'slow and tall strategy', is lower and therefore this transition would depend more on species immigration.

Global environmental change drivers (such as climate change, nutrient addition, land use change) are known to act as filters on the functional composition of communities (Cadotte and Tucker, 2017; Kraft et al., 2015). For example, grassland communities subjected to nutrient enrichment typically shift their trait distribution towards tall plants that can monopolize light and establish under shade (reflected by increased plant height and seed size), and which have a fast growth and tissue turnover (reflected by high leaf N concentration, specific leaf area, and low leaf dry matter content; Allan et al., 2015, Firn et al., 2019, Isbell et al., 2013, Manning et al., 2006). The optimal trait space therefore moves towards the 'fast and tall' corner of the global spectrum of plant form and function (Díaz et al., 2016; Reich, 2014). In contrast to species-poor communities, high-diversity communities are more likely to contain these types of species that outperform other species, thus causing a filtering effect that removes smaller and less acquisitive species from the community trait space. The remaining species may then perform well under the new conditions, and as with successional change, this may result in a loss of functional diversity. In contrast to the case of nutrient enrichment, there is little consensus on how other global change drivers, such as elevated $CO_2$, warming, and drought, affect functional trait distributions within plant communities (Cornwell and Ackerly, 2009; Crous et al., 2010; Liu et al., 2018). The greater flexibility in the trait distribution of diverse communities may underlie the greater resistance of diverse communities to extreme climate events (Isbell et al., 2015), and may be key to buffering ecosystem responses to an increasingly variable environment (IPCC, 2014).

Overall, we predict that for any given set of environmental conditions, coexisting plant species occupy a certain trait space as a community and that initial community diversity will determine how much succession and global change drivers will alter the centroid position and volume occupied by the community in trait space (Lin et al., 2011). Experiments that combine the manipulation of species richness with global change drivers, provide a unique opportunity to investigate these hypotheses, as they contain significant and controlled variation in both initial functional composition and functional diversity. Weeding of experimental communities manipulates regional species-loss and limits immigration of new species that are potentially better adapted to novel conditions. Furthermore, experimental communities are usually randomly selected from a local species pool according to a specific experimental design (e.g. by considering pre-defined functional groups), and therefore neither successional dynamics, nor global change drivers have filtered their initial composition.

Here, we tested the concepts presented above by analysing a unique dataset comprising long-term experiments, in which plant diversity and one global change driver (warming, drought, elevated atmospheric $CO_2$ concentrations, N addition, or land use intensification) were manipulated simultaneously in an orthogonal design. We first analysed (1) whether high plant diversity indeed covers a larger functional and phylogenetic space and whether this allows initially different high species richness compositions to converge over time and in response to global change drivers. Second, we assessed (2) to which position of the multivariate trait space the plant communities shifted over time and in response to global change drivers and whether high-diversity communities made a stronger shift over time compared to low-diversity communities. We assessed community change in multivariate trait space by measuring change the community weighted mean of traits related to global strategies of plant form and function (Díaz et al., 2016; Reich, 2014).

## 2. Methods
### 2.1 Study sites and experimental designs

We used long-term (>3 years; Craven et al., 2018, Guerrero-Ramírez et al., 2017) data on plant species-specific abundances within biodiversity experiments, which also manipulated a GEC driver in an orthogonal design. Four experiments met these criteria. Two were located in the Cedar Creek Ecosystem Science Reserve in Minnesota/USA (latitude 45.40, longitude −93.18), namely the BAC (Cowles et al., 2016) and BioCON (with two studies nested in this experiment; Reich et al., 2001) experiment, and two were part of the Jena Experiment in Thuringia, Germany (latitude 50.95, longitude 11.62; Weisser et al., 2017), namely the Jena drought experiment (Vogel et al., 2012) and Jena management experiment (Weigelt et al., 2009). Both research platforms are grasslands but differ in their climate, soils, and species composition.

Cedar Creek comprises native prairie grasslands on a glacial outwash sandplain with nutrient-poor soils. Climate is continental with cold winters (mean January temperature: −10.0°C) and hot summers (mean July temperature: 22.2°C), and mean annual precipitation of approximately 660 mm (Reich et al., 2001). The Jena Experiment represents semi-natural grasslands typical for Central Europe, where mowing and/or grazing prevent the development of woody species. It is located in an alluvial plain with soil categorized as nutrient-rich Eutric Fluvisol. The climate of the region is temperate oceanic with a mean annual temperature of 9.9°C, and mean annual precipitation of 610 mm (1980–2010; Hoffmann et al., 2014).

At both research sites, the vegetation was removed prior to experimental establishment, and experimental plots were sown with varying numbers of randomly chosen perennial plant species, selected from a local species pool. Colonization of unsown species was prevented by regular hand-weeding at both sites. Further management includes mowing twice a year at the Jena Experiment and regular burning at the Cedar Creek research site.

## 2.1.1 Cedar Creek

The BioCON experiment was established in 1997 and simulates elevated atmospheric $CO_2$ concentrations and N enrichment orthogonally crossed with an experimental plant species richness gradient (Reich et al., 2001). The sown species richness levels comprise monocultures and mixtures of 4, 9, and 16 species from a pool of 16 perennial grassland species, including legumes, herbs, C4- and C3-grasses. The grassland plots of $2 \times 2$ m are distributed within six rings. Three rings of this experiment are equipped with a free air $CO_2$ enrichment (FACE) that has enhanced atmospheric $CO_2$ concentration by 180 ppm since 1998 (Table 1). The remaining three rings are dummy controls with ambient $CO_2$ concentrations. Half of the plots within each ring are fertilized with $4 \, g \, N \, m^{-2} \, year^{-1}$. For our analysis, we used data from the plots sown with four species, representing low species diversity (15 replicates per treatment, replicates comprised of different species combinations) and those with 16 species representing high species diversity (12 replicates per treatment, replicates comprised of the same species composition). Moreover, we treated the $CO_2$ and N manipulations as separate experiments.

Atmospheric warming was manipulated within the BAC experiment (short for Biodiversity and Climate), which was established in 2009 on a subset of 32 plots of the BioDIV experiment at Cedar Creek (Cowles et al., 2016; Tilman et al., 2001, 2006). The grassland plots of this experiment were sown in 1994 and 1995. A species pool of 18 perennial grassland species (including legumes, non-leguminous forbs, C3- and C4-grasses) was used for the randomized plant communities. From this experiment, we used data

**Table 1** Experiments used in this study with details on the establishment of plots and the start of manipulation.

| Experiment name | GEC manipulation | Start of manipulation (year) | Establishment of plots (year) | Nr. of plots per treatment |
|---|---|---|---|---|
| BAC | Warming | 2009–ongoing | 1994 | 18 |
| BioCON | N enrichment | 1998–ongoing | 1998 | 27 |
| BioCON | Elevated $CO_2$ | 1998–ongoing | 1998 | 27 |
| Jena drought | Summer drought | 2008–2016 | 2002 | 30 |
| Jena management | Intensive mowing, fertilization | 2006–2009 | 2002 | 30 |

from nine replicates from both the low- (4 species) and high diversity (16 species) treatments. On each of the plots, two treatments were realized in areas of $3 \times 2.5$ m: a high warming treatment (1200 W) and a control receiving no extra warming (an additional low warming treatment was not included in this study). Warming was imposed from March throughout November each year by infrared heaters at a height of 1.8 m above the soil surface. The control treatments got the same installations without warming to account for potential confounding shade effects. For details on the infrared treatment, see (Kimball et al., 2008).

### *2.1.2 Jena*
The plots of the Jena Experiment were sown in 2002 from a species pool of 60 species comprising legumes, non-leguminous forbs, and C3-grasses (Roscher et al., 2004). From the different sown plant species richness levels, we used 16 replicates of a low diversity level (4 species) and 14 replicates of a high diversity level (16 species), to be as comparable as possible with the other studies.

The Jena management experiment was established on subplots of all plots of the Jena Experiment with two treatments per plot (control and intensive management; subplot size $1.6 \times 4$ m each) between 2006 and 2009. Control treatments were managed by two mowing events per year (at peak biomass and in late summer), while intensive management plots received four mowing events during the growing season and high fertilizer applications of $200.0 \, \text{kg} \, \text{N} \, \text{ha}^{-1} \, \text{a}^{-1}$, $87.2 \, \text{kg} \, \text{P} \, \text{ha}^{-1} \, \text{a}^{-1}$, $166.0 \, \text{kg} \, \text{K} \, \text{ha}^{-1} \, \text{a}^{-1}$. For details on the management experiment, see Weigelt et al. (2009).

The Jena drought experiment was established in 2008 on all plots of the Jena Experiment with two treatments. Two subplots ($1 \times 1$ m each) of a plot received either a summer drought or served as a control. Both treatments were covered by a transparent roof for around 6 weeks in summer (mid of July to end of August) each year until 2016. All rainfall was excluded from the drought treatment during this period, whereas the control treatment was watered with ambient amounts of rainwater after each rain event. This treatment resulted in a mean reduction of 42% of the annual precipitation. For further details on the rain shelter design see (Vogel et al., 2012, 2013b).

### 2.2 Data
For our analysis, we assembled plant species-specific abundance data from the first year of the manipulation of a global change driver onwards.

This corresponds to the time of plant community establishment in the BioCON experiment, whereas communities of BAC, the Jena management and the Jena drought experiment had been sown 15 years, 3 years, or 6 years earlier, respectively. We obtained species-specific cover estimates for the years 1998–2013 for BioCON and 2005–2009 for the Jena management experiment. From BAC, we used species-specific data on aboveground biomass in 2009–2014 obtained from a $0.1\,m^2$ strip of the subplots. For the Jena drought experiment, we used species-specific frequency data, obtained in the second (2009) and fourth year (2011) of the manipulation and the year after consecutive drought manipulations in 2017. Within each subplot, 30 $10 \times 10\,cm$ quadrats were arranged in three rows of 10 quadrats. Within each quadrat, the presence or absence of every sown plant species was recorded. The relative abundance of each species was calculated by dividing the sum of quadrats occupied by each species by the total sum of occupied quadrats of every sown species. Given these different assessment methods of plant community composition, we caution readers to not make direct comparisons of absolute values of functional trait composition across experiments, while temporal changes and effects of GEC drivers within studies can be explored.

We assembled trait data for all plant species representing a wide range of the global spectrum of plant form and function (Díaz et al., 2016): plant height (H), specific leaf area (SLA), leaf dry matter content (LDMC), leaf N concentration (leafN) and seed mass. Those traits reflect the capacity of light preemption (H), resource acquisition (SLA, LDMC, leafN), and generative reproduction (seed mass). We derived most of the trait data from the TRY database (Kattge et al., 2011). We completed missing values for one species (*Dalea villosa*) using our own measurements in the BioCON study. As we could not get sufficient information on plant height from TRY, we used site-specific measurements for all species in monocultures from the local databases.

We compiled the phylogenetic information of all species in our analysis from the megaphylogeny of Zanne et al. (2013). This phylogeny is based on GenBank sequences of seven gene regions of a wide range of angiosperms across the globe (Zanne et al., 2014) and comprised almost all species of the species pool in our analysis. The phyloGenerator package allowed us to bind three missing species on this phylogeny based on genetic information (Pearse and Purvis, 2013). The resulting phylogenetic tree is visible in Fig. 2, which includes also trait information.

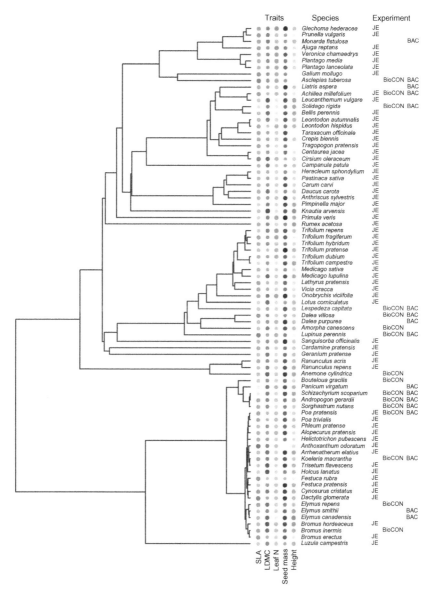

**Fig. 2** Plant species sown at the experimental sites including information on their phylogenetic relatedness, trait values and their presence in the species pool of the experimental platforms (JE = Jena Experiment, which includes the Jena drought and the Jena management study; BioCON includes the $CO_2$ as well as the N fertilization treatments and BAC the warming treatment). The size of symbols visualizes the relative value of traits. Abbreviations of the traits are given in the text.

## 2.3 Calculation of functional and phylogenetic indices and statistical analyses

To analyse, whether communities varied in their functional and phylogenetic diversity, in response to plant diversity, treatment, and time, we first calculated measures reflecting functional and phylogenetic space, which we defined as the volume of the overall multidimensional trait space or phylogenetic space that is covered by all existing target species within a plant community, irrespective of their abundances. Functional trait space is calculated as functional richness (Boersma et al., 2016; Villeger et al., 2008), while we used phylogenetic diversity (Faith, 1992) as an broadly equivalent measure of phylogenetic richness, and thus phylogenetic space (Tucker et al., 2016). As there could be significant changes in the abundances of species within trait space, even in the absence of species extinction (species introductions are not allowed) we additionally calculated functional and phylogenetic divergence and evenness (Mouchet et al., 2010; Tucker et al., 2016; Villeger et al., 2008). We generated a dendrogram based on a trait-distance matrix using a hierarchical cluster analysis (package *stats* in R; R Development Core Team, 2011), to have a comparable mathematical logic for both functional and phylogenetic diversity indices. For divergence measures, we calculated mean pairwise distances (MPD) and mean nearest neighbour distances (MNTD) of the species per community per year. As evenness measure, we calculated the IAC (=imbalance of abundances at the clade level; Cadotte et al., 2010). Distances were obtained from the phylogenetic tree and the trait dendrogram, respectively (Fig. 2). For all indices requiring a distance matrix, we created one single matrix across all experiments and did not use separate matrices for experiments based on different species pools (Jena, Cedar Creek).

To get an understanding of the direction of change in community trait space, we calculated community-weighted mean trait values (CWM), and functional dispersion (FDis), both weighted by species' relative abundances (Laliberte and Legendre, 2010). In cases where all sown species went extinct, we set CWM=0 and FDis=0. If only one species was present, we set FDis=0 and assigned the trait value of this species as CWM. This allows us to keep important information for our analysis on community assembly over time, as directional changes could also include extinctions. For the FDis calculation, LDMC, seed mass, and height were log-transformed to achieve a normal distribution. We standardized all traits for CWM calculations to make them independent of units.

Mixed effects models were performed separately for each of the five experiments in order to analyse the effects of year (as continuous variable testing for temporal trends followed by a factor testing for non-linear effects), GEC treatment (control vs. treatment), plant species diversity (4 species vs. 16 species), and the interactive effects of those factors on all diversity indices. We accounted for different experimental designs in the random terms of the models. For the Jena management experiment, Jena drought experiment, and BAC experiment, we used subplot nested in plot (= treatment) and plot as well as the interaction of year by plot, and year by plot by subplot as random terms. For BioCON, we used only plot and the plot by year interaction as random term. CWM values of seed mass and height were log-transformed and all FDis measures were square root transformed prior to analysis to meet assumptions of normality. We performed all mixed effects models using the function asreml of the package *asreml* in R 3.3.1 (R Core Team, 2018).

To assess potential temporal shifts of the trait space occupied by the communities, we performed a principal components analysis based on euclidian distances using standardized (zero mean and unit variance) CWM values of all experiments, years and treatments together. We applied varimax rotation to the PCA to enhance the association of each trait CWM to one of the two major axes in order to simplify the interpretation of the two major axes of the PCA (package *psych*; Revelle, 2018). To assess whether temporal shifts in community functional trait composition differed between plant diversity levels and treatments or their interaction, we calculated the vector length (Euclidian distance) of the plant community position of each plot in the multivariate space between the initial and final years. These vector lengths represent how far community functional trait composition has shifted over time. Since the experiment assessing N fertilization and $CO_2$ enrichment (BioCON) lasted 16 years, we avoided confounding effects of atypical years (e.g. due to weather events) by averaging the PCA coordinates of the first and last 3 years of the experiment as the initial and final compositions. The distances were then analysed using mixed models (package *nlme*; Pinheiro et al., 2018) with species richness, GEC driver and their interaction as fixed effects. We fitted the initial functional or phylogenetic richness of the communities as covariates before plant diversity in separate models to get an understanding, whether those properties explained the effect of plant diversity on the compositional shift over time. In experiments where treatments were nested (drought, warming and intensive management), plot identity was included as a random term. Means between diversity levels within a given treatment were then compared using the function 'lsmeans' (R package *lsmeans*; Lenth, 2016).

## 3. Results
### 3.1 Temporal shift of the functional and phylogenetic diversity of plant communities

At the start of the global change manipulation, functional and phylogenetic richness was higher at high plant diversity compared to communities with low diversity. Plant species richness declined over time in all experiments except the Jena drought experiment, especially when plant diversity was initially high (Fig. 3 and Table S1 in Supplementary Material in the online version at https://doi.org/10.1016/bs.aecr.2019.06.002). In parallel, functional and phylogenetic richness decreased in those experiments over time (Fig. 3 and Table S1 in Supplementary Material in the online version

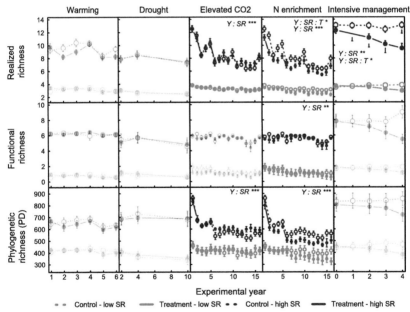

**Fig. 3** Changes in realized species numbers and functional trait as well as phylogenetic richness of species-poor (four species) and species-rich (16 species) plant communities over time under different global environmental change drivers. Circles and error bars represent the means and standard error, respectively, of all replicates in a given year (Y) by each treatment (T) × plant diversity (SR) combination. When SR did not significantly affect the temporal trend of diversity properties or the treatment effect (see model results in Table S1 in Supplementary Material in the online version at https://doi.org/10.1016/bs.aecr.2019.06.002), we faded out the error bars and lines. Significant interactions of Y × SR or Y × SR × T were obtained from the mixed effects models.

at https://doi.org/10.1016/bs.aecr.2019.06.002), indicating a general trend of decreasing overall trait and phylogenetic space within the experimental plant communities. The decrease of phylogenetic richness over time was stronger at high plant diversity in both BioCON experiments, but not in the other experiments. Here, the strong loss of species, and with it phylogenetic richness, was strongest in the first five years and levelled out afterwards. By contrast, functional richness did not show this strong decrease at high plant diversity beyond the general decreasing temporal trend. In the BioCON-N experiment, the decrease in functional richness was stronger at low plant diversity (Fig. 3 and Table S1 in Supplementary Material in the online version at https://doi.org/10.1016/bs.aecr.2019.06.002). This indicates that the strong decrease in the phylogenetic space in originally high-diversity communities in BioCON was not mirrored by an equally strong decrease of the functional trait space. Instead, the loss of functional trait space was stronger at low plant diversity.

Beyond a temporal trend of decreasing overall community trait and phylogenetic space due to species extinctions, we found evidence of community convergence caused by abundance shifts within communities (Table S1 and Fig. S1 in Supplementary Material in the online version at https://doi.org/10.1016/bs.aecr.2019.06.002). In the warming study, functional mean pairwise distance and mean nearest neighbour distance decreased over time in high-diversity communities, while phylogenetic mean pairwise distance and mean nearest neighbour distance did not change significantly (Table S1 in Supplementary Material in the online version at https://doi.org/10.1016/bs.aecr.2019.06.002). This indicates a convergence in functional diversity, but not in phylogenetic diversity, due to abundance shifts in high-diversity communities of this study. Species-poor communities in this experiment did not show any significant temporal change. In the BioCON experiment, high-diversity communities increased in functional and phylogenetic imbalance over time (higher IAC, Fig. S1 and Table S1 in Supplementary Material in the online version at https://doi.org/10.1016/bs.aecr.2019.06.002) and decreased in phylogenetic mean pairwise distance (Fig. S1 and Table S1 in Supplementary Material). This also indicates a functional and phylogenetic convergence over time in high-diversity communities.

In the Jena management experiment, phylogenetic MNTD increased over time in communities of high plant diversity (Fig. S1 and Table S1 in Supplementary Material in the online version at https://doi.org/10.1016/bs.aecr.2019.06.002), indicating the loss of phylogenetically similar species, while species-poor communities did not change.

The global change drivers rarely affected the overall community trait and phylogenetic space. Only intensive management caused species loss, which decreased the functional and phylogenetic richness over time, especially for communities with an initially high sown plant diversity (Fig. 3 and Table S1 in Supplementary Material in the online version at https://doi.org/10.1016/bs.aecr.2019.06.002). N enrichment in the BioCON experiment also caused species loss, but this did not lead to changes in functional or phylogenetic richness. All other global change drivers (warming, drought, elevated $CO_2$) did not cause species extinctions, nor changes in the overall trait and phylogenetic space occupied by the experimental communities.

For metrics related to abundance shifts within the communities, we found that warming decreased functional mean pairwise distance, independent of time and sown plant diversity. This indicates convergence of the communities due to warming. N enrichment led to higher functional mean nearest taxon distance and higher phylogenetic imbalance (higher IAC, Fig. S1 and Table S1 in Supplementary Material in the online version at https://doi.org/10.1016/bs.aecr.2019.06.002) in communities of high plant diversity, compared to control conditions. Thus, initially diverse communities were more uneven in their phylogenetic composition and lost functionally similar species under N enrichment compared to ambient conditions. These effects of N enrichment were strongest at the beginning of the experiment and diminished over time (Fig. S1). This means that N enrichment caused an initial divergence within functionally similar species, but during succession, the communities converged again. The phylogenetic mean pairwise distance decreased due to N enrichment, which strengthened over time (Fig. S1 and Table S1 in Supplementary Material in the online version at https://doi.org/10.1016/bs.aecr.2019.06.002), indicating overall phylogenetic convergence of the communities due to N enrichment.

Functional and phylogenetic MPD were reduced by intensive management, independent of sown plant diversity (Fig. S1 and Table S1 in Supplementary Material in the online version at https://doi.org/10.1016/bs.aecr.2019.06.002). Drought and elevated $CO_2$ did not have a significant effect on the functional and phylogenetic diversity metrics.

## 3.2 Temporal shifts in community trait space

The ordination of the communities based on CWM trait values revealed two main axes, which explained ca. 72% of the variance (Fig. 4). Community-weighted LDMC, leaf N, SLA, and height were strongly loaded on the first

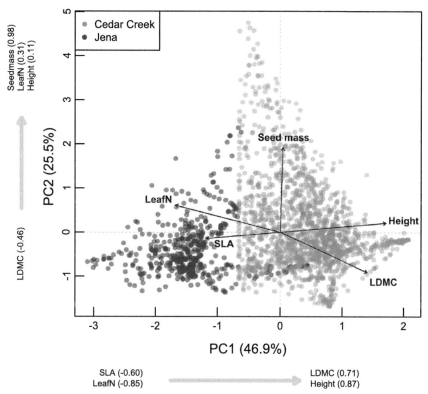

**Fig. 4** Principal components analysis of the community weighted mean traits (SLA, LDMC, leaf N, seed mass and height) of communities in this study. All levels of diversity, time points and global change treatments of the five experiments are presented. The loadings of each community mean trait on the axes are given in brackets. Colours separate the communities of the Jena Experiment (red) and Cedar Creek Ecosystem Science Reserve (blue), which each of them comprising several experiments

axis, whereas seed mass was strongly related to the second axis (Fig. 4). With the exception of height, this association of traits with the leading axes of the PCA represents the two main axes of the global spectrum of plant form and function, which shows the variation in the leaf economics spectrum of species versus the size spectrum, which at the global level includes both height and seed size. The communities were clearly separated along the first, leaf economics spectrum-related, axis according to their geographical origin. The Jena communities had higher mean SLA and mean leaf N, but lower mean LDMC and mean height compared to the Cedar Creek communities. This suggests that the Jena communities were in general dominated by 'fast' strategy species with small stature compared to the Cedar Creek communities with 'slow' strategies and a taller stature. This general separation of the two geographical regions along the first axis was independent of the global change driver, diversity treatment, and time.

Overall, the community position within the trait space shifted over time strongly along the first PCA axis in most communities of all experiments (Fig. 5). This temporal shift was stronger at high diversity compared to low diversity in the warming experiment, in the BioCON experiment with elevated $CO_2$ and in the drought experiment (Table 2, Figs. 5 and 6). The initially higher functional and/or phylogenetic richness of the communities were closely related to the plant diversity effect on the temporal shift of the communities. In case of the BioCON with N enrichment, it even explained more variance than plant diversity (Table 2). This means that if communities initially cover a larger multidimensional functional trait space, than they are indeed better able to move more through time compared to communities covering a smaller space. In most cases, this larger trait space was the consequence of higher initial plant diversity. Warming further increased the temporal shift in functional composition of diverse communities (Table 2, Figs. 5 and 6), while N enrichment and marginally also intensive management caused a stronger shift over time compared to control conditions, and this shift was independent of plant diversity.

Analysis of individual community weighted mean traits and functional dispersion measures revealed the mechanisms underlying general shifts in plant strategy in more detail. In the warming experiment, diverse communities shifted towards species with high LDMC values (Fig. 7, time × species richness interaction in Table S2 in Supplementary Material in the online version at https://doi.org/10.1016/bs.aecr.2019.06.002), high mean height (Fig. 7 and Table S2 in Supplementary Material in the online version at https://doi.org/10.1016/bs.aecr.2019.06.002), and lower leaf N (Fig. 7 and Table S2 in Supplementary Material in the online version at https://doi.org/10.1016/bs.aecr.2019.06.002), thus indicating a shift towards a 'slow and tall' strategy. Furthermore, diverse communities decreased in their mean seed mass over time, independent of treatment in this experiment (Fig. 7, time × species richness interaction in Table S2 in Supplementary Material in the online version at https://doi.org/10.1016/bs.aecr.2019.06.002). The temporal shifts in community mean traits at high species richness in the warming study were accompanied by decreases in the dispersion of community LDMC (Fig. 8, time × species richness interaction in Table S3), leaf N, and seed mass (Fig. 8 and Table S3 in Supplementary Material in the online version at https://doi.org/10.1016/bs.aecr.2019.06.002). This indicates that the shift of the plant communities to a 'slow

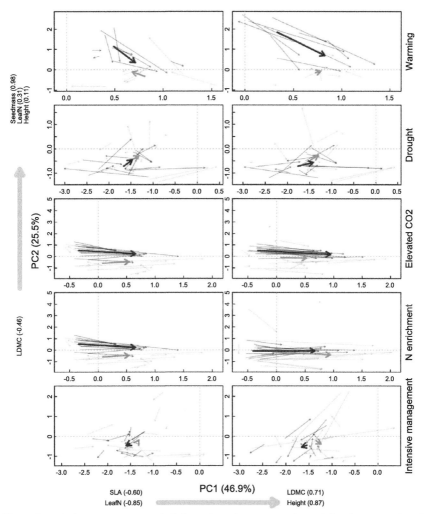

**Fig. 5** Principal components analysis based on the community weighted mean traits of SLA, LDMC, leaf N, seed mass and plant height used in Fig. 4 but separated for the five experiments and treatments. The left panel shows the communities under ambient control conditions, the right panel under conditions of the manipulated global environmental change driver. Yellow = low diversity (4 species), blue = high diversity (16 species). Arrows show the direction of the shifts over time (arrows in Fig. 1) for each community (faded lines) and the mean shift of all communities per treatment and control separated per plant diversity level and linearly connect the means of the hulls of the initial and final years.

and tall' strategy was accompanied by a filtering of those traits. Warming also generally favoured tall species (Fig. 7 and Table S2 in Supplementary Material in the online version at https://doi.org/10.1016/bs.aecr.2019.06.002).

**Table 2** ANOVA results testing for the effect of functional or phylogenetic richness, plant diversity (SR, 4 vs 16 sown plant species) and treatment (T, control vs. treatment) on the mean temporal change of trait composition in Fig. 5 from experimental start to end (SR=4 vs 16 species richness).

| | | Phylogenetic richness as covariate | | | | | Functional richness as covariate | | | |
|---|---|---|---|---|---|---|---|---|---|---|
| | | | numDF | denDF | F | P | | denDF | F | P |
| Warming | Intercept | | 1 | 16 | 48.19 | 0.000 | *** | Intercept | 16 | 43.59 | 0.000 | *** |
| | $PD$ | | 1 | 14 | 5.33 | 0.037 | * | $Fric$ | 14 | 10.88 | 0.005 | ** |
| | $SR_{Resid.}$ | | 1 | 16 | 8.51 | 0.010 | * | $SR_{Resid.}$ | 16 | 1.35 | 0.263 | |
| | **$SR_{Main}$** | | **1** | **16** | **12.38** | **0.003** | ** | **$SR_{Main}$** | **16** | **12.38** | **0.003** | ** |
| | $T$ | | 1 | 14 | 2.11 | 0.168 | | $T$ | 14 | 1.93 | 0.187 | |
| | $PD \times T$ | | 1 | 14 | 12.51 | 0.003 | ** | $Fric \times T$ | 14 | 9.48 | 0.008 | ** |
| | $SR_{Resid.} \times T$ | | 1 | 1 | 14 | 0.03 | | $SR_{Resid.} \times T$ | 14 | 0.11 | 0.741 | |
| | **$SR_{Main} \times T$** | | **1** | **16** | **11.10** | **0.004** | ** | **$SR_{Main} \times T$** | **16** | **11.10** | **0.004** | ** |
| Drought | Intercept | | 1 | 52 | 113.00 | 0.000 | *** | Intercept | 52 | 107.84 | 0.000 | *** |
| | $PD$ | | 1 | 52 | 0.30 | 0.589 | | $Fric$ | 52 | 3.93 | 0.053 | . |
| | $SR_{Resid.}$ | | 1 | 52 | 9.40 | 0.003 | ** | $SR_{Resid.}$ | 52 | 2.81 | 0.100 | . |
| | **$SR_{Main}$** | | **1** | **54** | **6.89** | **0.011** | * | **$SR_{Main}$** | **54** | **6.89** | **0.011** | * |
| | $T$ | | 1 | 52 | 0.01 | 0.937 | | $T$ | 52 | 0.00 | 0.952 | |
| | $PD \times T$ | | 1 | 52 | 0.20 | 0.656 | | $Fric \times T$ | 52 | 0.23 | 0.630 | |
| | $SR_{Resid.} \times T$ | | 1 | 52 | 0.21 | 0.647 | | $SR_{Resid.} \times T$ | 52 | 0.30 | 0.586 | |
| | **$SR_{Main} \times T$** | | **1** | **54** | **0.002** | **0.960** | | **$SR_{Main} \times T$** | **54** | **0.002** | **0.960** | |

| Treatment | Term | df | df2 | F | p | sig | Term | df | df2 | F | p | sig |
|---|---|---|---|---|---|---|---|---|---|---|---|---|
| Elevated CO2 | Intercept | 1 | 48 | 245.02 | 0.000 | *** | Intercept | 1 | 48 | 247.31 | 0.000 | *** |
| | PD | 1 | 48 | 20.73 | 0.000 | *** | Fric | 1 | 48 | 27.43 | 0.000 | *** |
| | $SR_{Resid.}$ | 1 | 48 | 6.51 | 0.014 | * | $SR_{Resid.}$ | 1 | 48 | 1.36 | 0.250 | |
| | **$SR_{Main}$** | **1** | **50** | **27.86** | **0.000** | **\*\*\*** | **$SR_{Main}$** | **1** | **50** | **27.86** | **0.000** | **\*\*\*** |
| | T | 1 | 48 | 0.27 | 0.604 | | T | 1 | 48 | 0.54 | 0.466 | |
| | $PD \times T$ | 1 | 48 | 1.03 | 0.315 | | $Fric \times T$ | 1 | 48 | 1.18 | 0.283 | |
| | $SR_{Resid.} \times T$ | 1 | 48 | 1.65 | 0.205 | | $SR_{Resid.} \times T$ | 1 | 48 | 0.42 | 0.521 | |
| | **$SR_{Main} \times T$** | **1** | **50** | **2.29** | **0.136** | | **$SR_{Main} \times T$** | **1** | **50** | **2.29** | **0.136** | |
| N enrichment | Intercept | 1 | 48 | 229.70 | 0.000 | *** | Intercept | 1 | 48 | 278.51 | 0.000 | *** |
| | PD | 1 | 48 | 1.28 | 0.263 | | Fric | 1 | 48 | 6.47 | 0.014 | * |
| | $SR_{Resid.}$ | 1 | 48 | 0.75 | 0.390 | | $SR_{Resid.}$ | 1 | 48 | 1.64 | 0.207 | |
| | **$SR_{Main}$** | **1** | **50** | **2.03** | **0.160** | | **$SR_{Main}$** | **1** | **50** | **2.03** | **0.160** | |
| | T | 1 | 48 | 6.39 | 0.015 | * | T | 1 | 48 | 9.48 | 0.003 | ** |
| | $PD \times T$ | 1 | 48 | 2.91 | 0.094 | . | $Fric \times T$ | 1 | 48 | 0.48 | 0.493 | |
| | $SR_{Resid.} \times T$ | 1 | 48 | 0.11 | 0.741 | | $SR_{Resid.} \times T$ | 1 | 48 | 6.01 | 0.018 | * |
| | **$SR_{Main} \times T$** | **1** | **50** | **2.99** | **0.090** | **.** | **$SR_{Main} \times T$** | **1** | **50** | **2.99** | **0.090** | **.** |

*Continued*

**Table 2** ANOVA results testing for the effect of functional or phylogenetic richness, plant diversity (SR, 4 vs 16 sown plant species) and treatment (T, control vs. treatment) on the mean temporal change of trait composition in Fig. 5 from experimental start to end (SR = 4 vs 16 species richness).—cont'd

| | | | Phylogenetic richness as covariate | | | | Functional richness as covariate | | | |
|---|---|---|---|---|---|---|---|---|---|---|
| | | numDF | | denDF | F | P | | denDF | F | P |
| Intensive management | Intercept | 1 | | 28 | 51.68 | 0.000 *** | Intercept | 28 | 52.29 | 0.000 *** |
| | *PD* | 1 | | 26 | 0.76 | 0.392 | *Fric* | 26 | 0.54 | 0.469 |
| | *SR*<sub>Resid.</sub> | 1 | | 28 | 0.60 | 0.447 | *SR*<sub>Resid.</sub> | 28 | 0.97 | 0.333 |
| | **SR**<sub>**Main**</sub> | 1 | | 28 | 1.40 | 0.246 | **SR**<sub>**Main**</sub> | 28 | 1.40 | 0.246 |
| | T | 1 | | 26 | 4.00 | 0.056 . | T | 26 | 4.02 | 0.056 . |
| | *PD* × *T* | 1 | | 26 | 0.82 | 0.374 | *Fric* × *T* | 26 | 1.35 | 0.255 |
| | *SR*<sub>Resid.</sub> × *T* | 1 | | 26 | 0.06 | 0.807 | *SR*<sub>Resid.</sub> × *T* | 26 | 0.06 | 0.813 |
| | **SR**<sub>**Main**</sub> × **T** | 1 | | 28 | 0.71 | 0.405 | **SR**<sub>**Main**</sub> × **T** | 28 | 0.71 | 0.405 |

The results are visualized in Fig. 6. Indented italicized terms are the covariates of the initial sown phylogenetic richness (PD) and Functional trait richness (Fric) and SR<sub>Resid</sub> is the residual effect of plant diversity after first explaining PD and Fric. Bold terms are the main effect of plant diversity (SR<sub>Main</sub>) when PD and Fric are not include in the model. Hence the difference in the effect of SR<sub>Main</sub> and SR<sub>Resid</sub> indicates how much the effect of plant diversity (SR<sub>Main</sub>) can be explained by greater sown richness having greater PD and Fric.

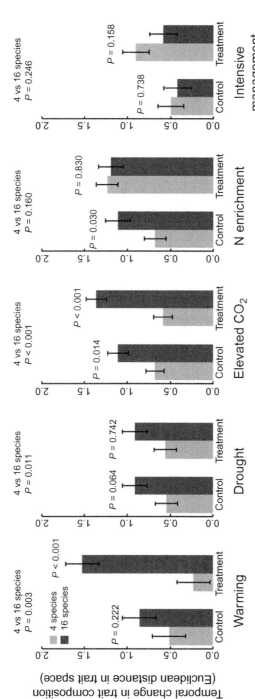

Fig. 6 Overall change in community level plant functional trait values (based on CWM) between the first and last years of the experiments (shifts in trait space are shown in Fig. 5). In general, the more diverse plant communities (16 species, dark shaded bars) shifted more in functional trait space compared to those of low diversity (4 species, light bars).

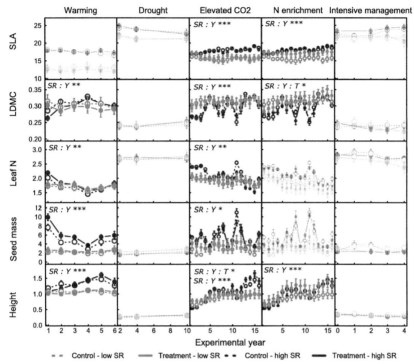

**Fig. 7** Community weighted mean trait (CWM) change over time under different global environmental change drivers. Error bars represent the means and standard error of all replicates in a given year (Y) by treatment (T) by diversity (SR) combination. We highlighted only significant interactions of T × Y × SR or SR × Y obtained from mixed effects model results (Table S2 in Supplementary Material in the online version at https://doi.org/10.1016/bs.aecr.2019.06.002). Non-significant results are shown in faded colours.

In the Jena drought experiment, community mean LDMC slightly increased over time under drought, but this trend was independent of plant diversity (time × treatment interaction in Table S2 in Supplementary Material in the online version at https://doi.org/10.1016/bs.aecr.2019.06.002). The functional dispersion in the single community traits did not change significantly in response to drought or plant diversity.

In the BioCON study, communities shifted towards a 'slow and tall' strategy over time, similar to the warming experiment, with species of high LDMC, tall height, and low leaf N (Fig. 7, time × species richness interaction in Table S2). Community mean SLA increased in both of the BioCON studies at high plant diversity (Fig. 7, time × species richness interaction in Table S2 in Supplementary Material in the online version at

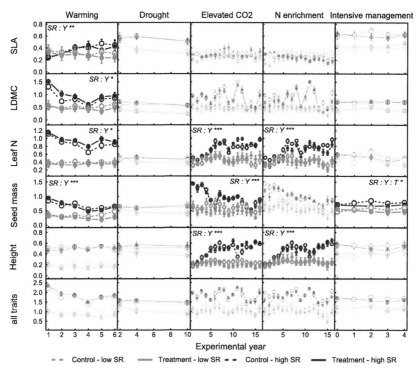

**Fig. 8** Changes in abundance weighted functional dispersion (FDis) measures of single traits or all five traits together under different global environmental change drivers. Error bars represent the means and standard error of all replicates in a given year (Y) by treatment (T) by diversity (SR) combination. We highlighted only significant interactions of T × Y × SR or SR × Y obtained from the mixed effects model results (Table S3 in Supplementary Material in the online version at https://doi.org/10.1016/bs.aecr.2019.06. 002). Non-significant results are shown in faded colours.

https://doi.org/10.1016/bs.aecr.2019.06.002). The dispersion of community leaf N and height (Fig. 8, time × species richness interaction in Table S3 in Supplementary Material in the online version at https://doi.org/10.1016/bs.aecr.2019.06.002) increased and the dispersion of community seed mass decreased over time in diverse communities (Fig. 8, time × species richness interaction in Table S3 in Supplementary Material in the online version at https://doi.org/10.1016/bs.aecr.2019.06.002). This indicates that the shift of diverse communities towards a 'slow and tall' strategy did not reduce their functional diversity. Under increased atmospheric $CO_2$ enrichment, there was a temporal trend towards tall species dominance in communities with high species richness (Fig. 7, time × species richness × treatment interaction in Table S2 in Supplementary Material in the online version at https://doi.org/10.1016/bs.aecr.2019.06.002). Under N enrichment, communities initially decreased in mean LDMC before increasing again later. This temporal trend was weaker in diverse communities (Fig. 7, time × species richness × treatment interaction in Table S2 in Supplementary Material in the online version at https://doi.org/10.1016/bs.aecr.2019.06.002). Furthermore, and independent of plant diversity, the relative abundance of species with high leaf N decreased over time under N enrichment (Fig. 7, time × treatment interaction in Table S2 in Supplementary Material in the online version at https://doi.org/10.1016/bs.aecr.2019.06.002).

In the Jena management experiment, mean LDMC and height decreased over time under intensive management (time × treatment interaction in Table S2 in Supplementary Material in the online version at https://doi.org/10.1016/bs.aecr.2019.06.002). The dispersion of all traits also decreased over time under intensive management (time × treatment interaction FDis$_{all\ traits}$ in Table S3 in Supplementary Material in the online version at https://doi.org/10.1016/bs.aecr.2019.06.002). Analysing the single community traits separately revealed that only the dispersion of seed mass decreased over time, but more so at low diversity (time × species richness × treatment interaction in Table S2 in Supplementary Material in the online version at https://doi.org/10.1016/bs.aecr.2019.06.002). This indicates a loss of diversity in seed mass under intensive management, when communities have low numbers of species.

## 4. Discussion

In our study, we examined how initial plant diversity affected successional trends in functional composition and diversity over time and under global change drivers. We found evidence for convergence in functional and phylogenetic diversity over time in some, but not all studies. This shift was associated with a compositional shift over time towards tall and slow-growing species, as is typical in secondary succession. As hypothesized, the functional composition of diverse communities shifted more across time than species-poor communities, in most experiments. Global change drivers, however, had little influence on these general trends. Our results suggest that while the temporal trends observed are generally consistent, the magnitude of change depends on contextual factors, such as community age, and duration of the study. In the following sections, we first discuss the general trends over time in response to plant diversity and second, the effects of global change drivers.

## 4.1 Temporal shift of the functional and phylogenetic diversity of plant communities

We found phylogenetic and functional convergence of communities over time (Fig. 3) in all experiments at Cedar Creek. This successional filtering was especially pronounced in diverse plant communities and was independent of the global change drivers, indicating that successional changes were a stronger force in structuring plant communities than the global change drivers. These shifts are likely to be associated with community assembly processes acting upon initially randomly assembled communities to generate assemblages that contain species which were better suited to, and more able to compete, under local site conditions. It has been shown that experimental plant communities converge in their trait composition and phylogenetic relatedness even more strongly when the natural colonization of unsown species is allowed (Allan et al., 2013; Cadotte and Strauss, 2011; Fukami et al., 2005; Roscher et al., 2014, 2016). This suggests that habitat filtering processes play a strong role in community assembly, at least during early stages of secondary succession (Li et al., 2015; Purschke et al., 2013). Successional change in response to soil development following grassland establishment involves changes in soil communities (Eisenhauer et al., 2012, 2011a, 2011b; Vogel et al., 2019) and changes to abiotic factors, such as soil carbon and N stocks (Fornara and Tilman, 2008; Lange et al., 2015, 2019; Steinbeiss et al., 2008). A combination of changes to vegetation properties and light availability, with soil-related changes, and their interaction with plant competition, could drive the observed plant community dynamics, as has been revealed in previous studies of succession (Bever, 1994; De Deyn et al., 2003; Tilman, 1988; Van der Putten et al., 1993).

We could not confirm higher community convergence over time at high plant diversity for both experiments in Jena. The increasing phylogenetic mean nearest neighbour distance over time in the high-diversity plant communities of the Jena management experiment indicates that limiting similarity, rather than habitat filtering, is the key process driving community assembly here. Traits other than those related to size and resource capture might have caused this pattern, since our measure of functional mean nearest neighbour distance, comprised of traits related to size and resource capture, did not show equivalent increases. In line with our observations, Allan et al. (2013) found that the communities of the Jena Experiment showed phylogenetic divergence across the first 7 years after establishment, which could be due to limiting similarity and complementarity among species. The Jena management experiment was carried out within this period (year 3–7), whereas the Jena drought experiment started later and had more experimental years (year 6–15 of the Jena Experiment). It is possible that, the divergence of communities after establishment of the Jena Experiment did not continue in the later years of the experiment that were studied here, indicating constraints to community change, or the arrival at a relative state of equilibrium. We also could not detect any successional trends in the community mean traits or trait dispersion in both experiments at Jena, unlike another long-term study carried out in the same system (Roscher et al., 2013). Together these results indicate that, the very first years after establishment of the Jena Experiment were very influential for community assembly, but that these were not covered in the drought and management sub experiments studied here. We suggest that the nutrient-rich soil at the Jena site accelerated community assembly processes compared to Cedar Creek, and therefore overall patterns were less likely to be observable in the periods investigated with this study.

In contrast to our hypothesis, there were few significant effects of global change drivers on community assembly over time. Only N enrichment and intensive management decreased community functional and phylogenetic diversity. This finding is in line with other studies, which reported decreased functional diversity after fertilization due to light competition (Chollet et al., 2018; Gerhold et al., 2013; Janeček et al., 2013; Pakeman et al., 2011). Notably, in the same experiment, the soil community composition was significantly affected by this treatment (Eisenhauer et al., 2012), and therefore changes to soil feedback effects may partly explain our findings, as these are known to be affected by nitrogen enrichment (Bever, 1994; De Deyn et al., 2003; Johnson et al., 1991; Manning et al., 2008). However, the soil community was not investigated before the 13th year of this experiment, limiting causal explanations here.

## 4.2 Temporal shift of the trait space of plant communities

The trait space occupied by the communities in our analysis was fundamentally different between the two geographical regions, with taller and slower growing species in Cedar Creek (North America) than in Jena (Central Europe). According to the community weighted means of single traits, communities in Jena were dominated by species that are shorter, have lower LDMC, but higher leaf N concentrations and SLA than those of the prairie grasslands at Cedar Creek. The dominant species of Central European grasslands are adapted to regular mowing or grazing disturbances, because the grasslands are anthropogenic and traditionally used for haymaking and fodder production (Ellenberg, 1988). Such sites are also typically fertilized (Blüthgen et al., 2012) and so short (compared to C4 dominated prairie grasslands) and fast growing plant strategies are well suited to these conditions (Grime, 2006).

The experimental plant communities of our study showed a stronger shift in their mean position in trait space over time, when they had initially higher functional or phylogenetic richness. This higher richness was generated in most cases by high sown plant diversity (in case of the Warming experiment, BioCON with elevated $CO_2$ and also in the Jena drought experiment) as we hypothesized. In case of the BioCON experiment with N enrichment, the functional trait space was even a better predictor of the temporal shift than plant diversity per se. It is important to note that, by chance, species-poor communities can also occupy a large trait space volume, if they contain functionally or phylogenetically very different species, and high-diversity communities can occupy a low trait space volume, if they contain many functionally similar species.

The direction of temporal changes implies a continuous shift towards dominance by tall and slow leaf economics trait species, independent of the global change drivers. The shift towards taller species suggests a response to increasing competition for light over successional time, that is consistent with earlier work at the site (Tilman, 1988), even though the plots were regularly burnt. In both Cedar Creek experiments, the shift towards 'tall and slow' communities is associated with the increasing dominance of *Andropogon gerardii*, a tall and slow growing C4 grass.

Global change drivers in general, had little effect on the functional composition of the communities in this study, and this lack of response is consistent with the analyses on functional and phylogenetic diversity (see above). The exceptions to this trend were warming and N addition. Warming favoured tall and slow growing species, but only diverse communities were able to make this shift. This shift was associated with an increase of legumes (e.g. *Dalea purpureum*) and the C4 grass *Andropogon gerardii*, while C3 grasses decreased relative to other functional groups over time (Cowles et al., 2016). N addition induced a stronger temporal shift towards 'tall and slow' species (which is again *Andropogon gerardii*), compared to control conditions in both diversity levels. This means that diverse communities selected for 'tall and slow' species without N addition, while at low diversity, the shift only occurred when N was added. This indicates that the general trend of the experimental communities at Cedar Creek to select for tall and slow-growing species was restricted in species poor communities due to N limitation (HilleRisLambers et al., 2004). In other grassland systems, nutrient addition typically benefits fast-growing species, also of tall stature (Allan et al., 2015; de Vries et al., 2012; Gazol et al., 2013; Nogueira et al., 2018; Price et al., 2014; Zhou et al., 2018). This suggests that light limitation is the primary selection force under N addition (Borer et al., 2014; Hautier et al., 2009), while 'fast-slow' traits are secondary. However, we cannot rule out the possibility to find different patterns, if other traits would have been measured at the site (Albert et al., 2011).

We found no significant effect of drought, elevated $CO_2$, or management intensification on the trait space occupied by the communities. However, intensive management weakly selected for small and fast species, independent of species richness. This is because nutrient addition is combined with increased mowing frequency in this experiment, potentially relaxing light limitation. The fact that we did not find stronger treatment effects here could be attributed to the relatively short duration of this experiment (4 years). Four years of experimental treatments might not sufficient to induce significant plant community shifts under the given environmental conditions and the mainly perennial species in the species pool.

One explanation for the lack of effects of drought or elevated $CO_2$ on functional composition could be that important traits were not quantified in our study. For drought, root traits might be more important than the growth strategy itself (Bardgett et al., 2014; Laliberté, 2017), because growth is reduced under drought (Wagg et al., 2017). Species capable of accessing soil nutrients and water under drought, of storing nutrients in belowground organs, or recovering rapidly after drought might be favoured in response to drought (Karlowsky et al., 2018; Muller et al., 2011; Tardieu et al., 2011). However, these traits may have little correlation with the fast-slow and size axes identified by our analysis. Furthermore, we cannot rule out the explanation that global change drivers in our study were not strong enough to cause significant changes in plant functional composition, or that they were experienced similarly across all diversity levels. Plant diversity can impact the physical microenvironment by increasing, for instance, soil water or nutrient content (Fischer et al., 2019; Fornara and Tilman, 2008; Lange et al., 2015). Global change drivers, such as drought or warming induced drought, are therefore less harsh at high diversity (Cowles et al., 2016; Vogel et al., 2013a). The impacts of global change on species interactions (Wright et al., 2014, 2015) and finally composition changes over time, consequently might therefore differ across diversity levels. Trait plasticity within species might also be important (Albert et al., 2011; Bjorkman et al., 2018;

Pontes et al., 2010; Siebenkäs et al., 2016), and if shifts towards favourable strategies were made this could limit species composition responses. For instance, within the BioCON experiment, it could be shown that elevated atmospheric $CO_2$ has strong effects on the photosynthetic capacity of species (Lee et al., 2011). This plasticity within species might have made compositional shifts unnecessary, and future experiments are needed that assess plant traits in the different experimental treatments, rather than referring to fixed literature traits and/or traits measured under ambient conditions in monocultures (Siebenkäs et al., 2016).

## 5. Conclusions

Our results show that plant communities shifted their functional composition over time, and in response to changes in environmental conditions. The direction and magnitude of the shift in functional composition depended on the inherent functional or phylogenetic richness of the communities, sown plant diversity, and partially on global change drivers. Our findings provide empirical support for the notion that species rich plant communities contain a wide range of ecological strategies. This large coverage of trait space may allow diverse communities to update community composition according to changes in environmental conditions. Species-poor communities, in contrast, are limited in their capacity to shift their functional composition because of the lower range of strategies that are present in these communities. Although responses to global change drivers were only seen in some cases, the results provide evidence for a mechanistic basis of the insurance effects of biodiversity, by showing that diverse communities have greater capacity to change more over time. Furthermore, it is notable that diverse communities shifted towards dominance of 'slow' traits, which are themselves related to more stable biomass production over time (Craven et al., 2018). Future studies should link such compositional shifts to temporal changes in ecosystem functioning (Guerrero-Ramírez et al., 2017; Meyer et al., 2016; Reich et al., 2012) to test the hypothesis that differences in the functioning between species-poor and species-rich communities can be explained by their distance from the ideal trait composition of a site, and thus the capacity of the community to perform well under the given environmental conditions. Based on the results of the present study, we conclude that fostering biodiversity should provide a critical insurance that will allow communities to respond to the changing environments they increasingly experience.

## Acknowledgements

This work was done on the basis of a workshop 'Community assembly', which was funded by the German research foundation (FOR 1451; Ei 862/13-1). The data from the Jena Experiment were obtained by funding from the Universities of Jena and Zurich (drought experiment) and German research foundation (FOR 456, FOR 1451, Management experiment). Funding for the BAC experiment was provided by the National Science Foundation DEB-062065, DEB-1234162 and the Minnesota Environment and Natural Resources Trust Fund as recommended by the Legislative– Citizen Commission on Minnesota Resources (LCCMR, project ID: 048-B1). BioCON was supported by the Department of Energy (DOE/DE-FG02-96ER62291) and the National Science Foundation (NSF Biocomplexity 0322057, NSF LTER DEB 9411972, DEB 0080382, DEB 0620652, NSF LTREB 0716587, DEB-1234162 and NSF LTREB 0716587), the DOE Office of Science, Biological and Environmental Research, through the Midwestern Regional Center of the National Institute for Climatic Change Research at Michigan Technological University (DE-FC02-06ER64158) and the Minnesota Environment and Natural Resources Trust Fund as recommended by the Legislative– Citizen Commission on Minnesota Resources. We gratefully acknowledge the help of the coordinators of the experimental sites as well as all gardeners, technicians and numerous student helpers, who supported the establishment of the experiments, maintained the long-term experimental plots and helped with data collections. Comments by two anonymous reviewers helped to improve our manuscript.

## References

Albert, C.H., Grassein, F., Schurr, F.M., Vieilledent, G., Violle, C., 2011. When and how should intraspecific variability be considered in trait-based plant ecology? Perspect. Plant Ecol. Evol. Syst. 13 (3), 217–225.

Allan, E., Weisser, W., Weigelt, A., Roscher, C., Fischer, M., Hillebrand, H., 2011. More diverse plant communities have higher functioning over time due to turnover in complementary dominant species. Proc. Natl. Acad. Sci. U. S. A. 108 (41), 17034–17039.

Allan, E., Jenkins, T., Fergus, A.J.F., Roscher, C., Fischer, M., Petermann, J., Weisser, W.W., Schmid, B., 2013. Experimental plant communities develop phylogenetically overdispersed abundance distributions during assembly. Ecology 94 (2), 465–477.

Allan, E., Manning, P., Alt, F., Binkenstein, J., Blaser, S., Blüthgen, N., Böhm, S., Grassein, F., Hölzel, N., Klaus, V.H., Kleinebecker, T., Morris, E.K., Oelmann, Y., Prati, D., Renner, S.C., Rillig, M.C., Schaefer, M., Schloter, M., Schmitt, B., Schöning, I., Schrumpf, M., Solly, E., Sorkau, E., Steckel, J., Steffen-Dewenter, I., Stempfhuber, B., Tschapka, M., Weiner, C.N., Weisser, W.W., Werner, M., Westphal, C., Wilcke, W., Fischer, M., 2015. Land use intensification alters ecosystem multifunctionality via loss of biodiversity and changes to functional composition. Ecol. Lett. 18 (8), 834–843.

Bardgett, R.D., Mommer, L., De Vries, F.T., 2014. Going underground: root traits as drivers of ecosystem processes. Trends Ecol. Evol. 29 (12), 692–699.

Bever, J.D., 1994. Feeback between plants and their soil communities in an old field community. Ecology 75 (7), 1965–1977.

Bjorkman, A.D., Myers-Smith, I.H., Elmendorf, S.C., Normand, S., Rüger, N., Beck, P.S.A., Blach-Overgaard, A., Blok, D., Cornelissen, J.H.C., Forbes, B.C., Georges, D., Goetz, S.J., Guay, K.C., Henry, G.H.R., HilleRisLambers, J., Hollister, R.D., Karger, D.N., Kattge, J., Manning, P., Prevéy, J.S., Rixen, C.,

Schaepman-Strub, G., Thomas, H.J.D., Vellend, M., Wilmking, M., Wipf, S., Carbognani, M., Hermanutz, L., Lévesque, E., Molau, U., Petraglia, A., Soudzilovskaia, N.A., Spasojevic, M.J., Tomaselli, M., Vowles, T., Alatalo, J.M., Alexander, H.D., Anadon-Rosell, A., Angers-Blondin, S., Beest, M.t., Berner, L., Björk, R.G., Buchwal, A., Buras, A., Christie, K., Cooper, E.J., Dullinger, S., Elberling, B., Eskelinen, A., Frei, E.R., Grau, O., Grogan, P., Hallinger, M., Harper, K.A., Heijmans, M.M.P.D., Hudson, J., Hülber, K., Iturrate-Garcia, M., Iversen, C.M., Jaroszynska, F., Johnstone, J.F., Jørgensen, R.H., Kaarlejärvi, E., Klady, R., Kuleza, S., Kulonen, A., Lamarque, L.J., Lantz, T., Little, C.J., Speed, J.D.M., Michelsen, A., Milbau, A., Nabe-Nielsen, J., Nielsen, S.S., Ninot, J.M., Oberbauer, S.F., Olofsson, J., Onipchenko, V.G., Rumpf, S.B., Semenchuk, P., Shetti, R., Collier, L.S., Street, L.E., Suding, K.N., Tape, K.D., Trant, A., Treier, U.A., Tremblay, J.-P., Tremblay, M., Venn, S., Weijers, S., Zamin, T., Boulanger-Lapointe, N., Gould, W.A., Hik, D.S., Hofgaard, A., Jónsdóttir, I.S., Jorgenson, J., Klein, J., Magnusson, B., Tweedie, C., Wookey, P.A., Bahn, M., Blonder, B., van Bodegom, P.M., Bond-Lamberty, B., Campetella, G., Cerabolini, B.E.L., Chapin, F.S., Cornwell, W.K., Craine, J., Dainese, M., de Vries, F.T., Díaz, S., Enquist, B.J., Green, W., Milla, R., Niinemets, Ü., Onoda, Y., Ordoñez, J.C., Ozinga, W.A., Penuelas, J., Poorter, H., Poschlod, P., Reich, P.B., Sandel, B., Schamp, B., Sheremetev, S., Weiher, E., 2018. Plant functional trait change across a warming tundra biome. Nature 562 (7725), 57–62.

Blüthgen, N., Dormann, C.F., Prati, D., 2012. A quantitative index of land-use intensity in grasslands: integrating mowing, grazing and fertilization. Basic Appl. Ecol. 13, 207–220.

Boeddinghaus, R., Marhan, S., Berner, D., Boch, S., Fischer, M., Hölzel, N., Kattge, J., Klaus, V., Kleinebecker, T., Oelmann, Y., Prati, D., Schäfer, D., Schöning, I., Schrumpf, M., Sorkau, E., Kandeler, E., Manning, P., 2019. Plant functional trait shifts explain concurrent changes in the structure and function of grassland soil microbial communities. J. Ecol. 1–14. https://doi.org/10.1111/1365-2745.13182.

Boersma, K.S., Dee, L.E., Miller, S.J., Bogan, M.T., Lytle, D.A., Gitelman, A.I., 2016. Linking multidimensional functional diversity to quantitative methods: a graphical hypothesis-evaluation framework. Ecology 97 (3), 583–593.

Borer, E.T., Seabloom, E.W., Gruner, D.S., Harpole, W.S., Hillebrand, H., Lind, E.M., Adler, P.B., Alberti, J., Anderson, T.M., Bakker, J.D., Biederman, L., Blumenthal, D., Brown, C.S., Brudvig, L.A., Buckley, Y.M., Cadotte, M., Chu, C., Cleland, E.E., Crawley, M.J., Daleo, P., Damschen, E.I., Davies, K.F., DeCrappeo, N.M., Du, G., Firn, J., Hautier, Y., Heckman, R.W., Hector, A., HilleRisLambers, J., Iribarne, O., Klein, J.A., Knops, J.M.H., La Pierre, K.J., Leakey, A.D.B., Li, W., MacDougall, A.S., McCulley, R.L., Melbourne, B.A., Mitchell, C.E., Moore, J.L., Mortensen, B., O'Halloran, L.R., Orrock, J.L., Pascual, J., Prober, S.M., Pyke, D.A., Risch, A.C., Schuetz, M., Smith, M.D., Stevens, C.J., Sullivan, L.L., Williams, R.J., Wragg, P.D., Wright, J.P., Yang, L.H., 2014. Herbivores and nutrients control grassland plant diversity via light limitation. Nature 508, 517.

Cadotte, M.W., Strauss, S.Y., 2011. Phylogenetic patterns of colonization and extinction in experimentally assembled plant communities. PLoS One 6 (5) e19363.

Cadotte, M.W., Tucker, C.M., 2017. Should environmental filtering be abandoned? Trends Ecol. Evol. 32 (6), 429–437.

Cadotte, M.W., Cardinale, B.J., Oakley, T.H., 2008. Evolutionary history and the effect of biodiversity on plant productivity. Proc. Natl. Acad. Sci. U. S. A. 105 (44), 17012–17017.

Cadotte, M.W., Cavender-Bares, J., Tilman, D., Oakley, T.H., 2009. Using phylogenetic, functional and trait diversity to understand patterns of plant community productivity. PLoS One 4 (5), 9.

Cadotte, M.W., Jonathan Davies, T., Regetz, J., Kembel, S.W., Cleland, E., Oakley, T.H., 2010. Phylogenetic diversity metrics for ecological communities: integrating species richness, abundance and evolutionary history. Ecol. Lett. 13 (1), 96–105.

Cadotte, M.W., Dinnage, R., Tilman, D., 2012. Phylogenetic diversity promotes ecosystem stability. Ecology 93 (sp8), S223–S233.

Cardinale, B.J., Duffy, J.E., Gonzalez, A., Hooper, D.U., Perrings, C., Venail, P., Narwani, A., Mace, G.M., Tilman, D., Wardle, D.A., Kinzig, A.P., Daily, G.C., Loreau, M., Grace, J.B., Larigauderie, A., Srivastava, D.S., Naeem, S., 2012. Biodiversity loss and its impact on humanity. Nature 486 (7401), 59–67.

Chao, A.N., Chiu, C.H., Jost, L., 2014. Unifying species diversity, phylogenetic diversity, functional diversity, and related similarity and differentiation measures through hill numbers. In: Futuyma, D.J. (Ed.), Annual Review of Ecology, Evolution, and Systematics. In: vol. 45. Annual Reviews, Palo Alto, pp. 297–324.

Chesson, P., 2000. Mechanisms of maintenance of species diversity. Annu. Rev. Ecol. Syst. 31, 343–366.

Chollet, S., Brabant, C., Tessier, S., Jung, V., 2018. From urban lawns to urban meadows: reduction of mowing frequency increases plant taxonomic, functional and phylogenetic diversity. Landsc. Urban Plan. 180, 121–124.

Cornwell, W.K., Ackerly, D.D., 2009. Community assembly and shifts in plant trait distributions across an environmental gradient in coastal California. Ecol. Monogr. 79 (1), 109–126.

Cowles, J.M., Wragg, P.D., Wright, A.J., Powers, J.S., Tilman, D., 2016. Shifting grassland plant community structure drives positive interactive effects of warming and diversity on aboveground net primary productivity. Glob. Chang. Biol. 22 (2), 741–749.

Craven, D., Eisenhauer, N., Pearse, W.D., Hautier, Y., Isbell, F., Roscher, C., Bahn, M., Beierkuhnlein, C., Bönisch, G., Buchmann, N., Byun, C., Catford, J.A., Cerabolini, B.E.L., Cornelissen, J.H.C., Craine, J.M., De Luca, E., Ebeling, A., Griffin, J.N., Hector, A., Hines, J., Jentsch, A., Kattge, J., Kreyling, J., Lanta, V., Lemoine, N., Meyer, S.T., Minden, V., Onipchenko, V., Polley, H.W., Reich, P.B., van Ruijven, J., Schamp, B., Smith, M.D., Soudzilovskaia, N.A., Tilman, D., Weigelt, A., Wilsey, B., Manning, P., 2018. Multiple facets of biodiversity drive the diversity–stability relationship. Nat. Ecol. Evol. 2, 1579–1587.

Crous, K.Y., Reich, P.B., Hunter, M.D., Ellsworth, D.S., 2010. Maintenance of leaf N controls the photosynthetic $CO_2$ response of grassland species exposed to 9 years of free-air $CO_2$ enrichment. Glob. Chang. Biol. 16 (7), 2076–2088.

De Deyn, G.B., Raaijmakers, C.E., Zoomer, H.R., Berg, M.P., de Ruiter, P.C., Verhoef, H.A., Bezemer, T.M., van der Putten, W.H., 2003. Soil invertebrate fauna enhances grassland succession and diversity. Nature 422, 711.

de Mazancourt, C., Isbell, F., Larocque, A., Berendse, F., De Luca, E., Grace, J.B., Haegeman, B., Wayne Polley, H., Roscher, C., Schmid, B., Tilman, D., van Ruijven, J., Weigelt, A., Wilsey, B.J., Loreau, M., 2013. Predicting ecosystem stability from community composition and biodiversity. Ecol. Lett. 16 (5), 617–625.

de Vries, F.T., Manning, P., Tallowin, J.R.B., Mortimer, S.R., Pilgrim, E.S., Harrison, K.A., Hobbs, P.J., Quirk, H., Shipley, B., Cornelissen, J.H.C., Kattge, J., Bardgett, R.D., 2012. Abiotic drivers and plant traits explain landscape-scale patterns in soil microbial communities. Ecol. Lett. 15 (11), 1230–1239.

Diaz, S., Cabido, M., 2001. Vive la différence: plant functional diversity matters to ecosystem processes. Trends Ecol. Evol. 16 (11), 646–655.

Díaz, S., Kattge, J., Cornelissen, J.H.C., Wright, I.J., Lavorel, S., Dray, S., Reu, B., Kleyer, M., Wirth, C., Colin Prentice, I., Garnier, E., Bönisch, G., Westoby, M., Poorter, H., Reich, P.B., Moles, A.T., Dickie, J., Gillison, A.N., Zanne, A.E., Chave, J., Joseph Wright, S., Sheremet'ev, S.N., Jactel, H., Baraloto, C.,

Cerabolini, B., Pierce, S., Shipley, B., Kirkup, D., Casanoves, F., Joswig, J.S., Günther, A., Falczuk, V., Rüger, N., Mahecha, M.D., Gorné, L.D., 2016. The global spectrum of plant form and function. Nature 529 (7585), 167–171.

Douma, J.C., de Haan, M.W.A., Aerts, R., Witte, J.-P.M., van Bodegom, P.M., 2012. Succession-induced trait shifts across a wide range of NW European ecosystems are driven by light and modulated by initial abiotic conditions. J. Ecol. 100 (2), 366–380.

Eisenhauer, N., Migunova, V.D., Ackermann, M., Ruess, L., Scheu, S., 2011a. Changes in plant species richness induce functional shifts in soil nematode communities in experimental grassland. PLoS One 6 (9), 9.

Eisenhauer, N., Milcu, A., Sabais, A.C.W., Bessler, H., Brenner, J., Engels, C., Klarner, B., Maraun, M., Partsch, S., Roscher, C., Schonert, F., Temperton, V.M., Thomisch, K., Weigelt, A., Weisser, W.W., Scheu, S., 2011b. Plant diversity surpasses plant functional groups and plant productivity as driver of soil biota in the long term. PLoS One 6 (1), 11.

Eisenhauer, N., Cesarz, S., Koller, R., Worm, K., Reich, P.B., 2012. Global change belowground: impacts of elevated CO2, nitrogen, and summer drought on soil food webs and biodiversity. Glob. Chang. Biol. 18 (2), 435–447.

Eisenhauer, N., Schielzeth, H., Barnes, A.D., Barry, K.E., Bonn, A., Brose, U., Bruelheide, H., Buchmann, N., Buscot, F., Ebeling, A., Ferlian, O., Freschet, G.T., Giling, D.P., Hättenschwiler, S., Hillebrand, H., Hines, J., Isbell, F., Koller-France, E., König-Ries, B., de Kroon, H., Meyer, S.T., Milcu, A., Müller, J., Nock, C.A., Petermann, J.S., Roscher, C., Scherber, C., Scherer-Lorenzen, M., Schmid, B., Schnitzer, S.A., Schuldt, A., Tscharntke, T., Türke, M., van Dam, N.M., van der Plas, F., Vogel, A., Wagg, C., Wardle, D.A., Weigelt, A., Weisser, W.W., Wirth, C., Jochum, M., 2019. A multitrophic perspective on biodiversity–ecosystem functioning research. Adv. Ecol. Res. 61, 1–54.

Ellenberg, H., 1988. Vegetation Ecology of Central Europe, fourth ed. Cambridge University Press, Cambridge.

Faith, D.P., 1992. Conservation evaluation and phylogenetic diversity. Biol. Conserv. 61 (1), 1–10.

Firn, J., McGree, J.M., Harvey, E., Flores-Moreno, H., Schütz, M., Buckley, Y.M., Borer, E.T., Seabloom, E.W., La Pierre, K.J., MacDougall, A.M., Prober, S.M., Stevens, C.J., Sullivan, L.L., Porter, E., Ladouceur, E., Allen, C., Moromizato, K.H., Morgan, J.W., Harpole, W.S., Hautier, Y., Eisenhauer, N., Wright, J.P., Adler, P.B., Arnillas, C.A., Bakker, J.D., Biederman, L., Broadbent, A.A.D., Brown, C.S., Bugalho, M.N., Caldeira, M.C., Cleland, E.E., Ebeling, A., Fay, P.A., Hagenah, N., Kleinhesselink, A.R., Mitchell, R., Moore, J.L., Nogueira, C., Peri, P.L., Roscher, C., Smith, M.D., Wragg, P.D., Risch, A.C., 2019. Leaf nutrients, not specific leaf area, are consistent indicators of elevated nutrient inputs. Nat. Ecol. Evol. 3, 400–406.

Fischer, C., Leimer, S., Roscher, C., Ravenek, J., de Kroon, H., Kreutziger, Y., Baade, J., Beßler, H., Eisenhauer, N., Weigelt, A., Mommer, L., Lange, M., Gleixner, G., Wilcke, W., Schröder, B., Hildebrandt, A., 2019. Plant species richness and functional groups have different effects on soil water content in a decade-long grassland experiment. J. Ecol. 107 (1), 127–141.

Flynn, D.F.B., Mirotchnick, N., Jain, M., Palmer, M.I., Naeem, S., 2011. Functional and phylogenetic diversity as predictors of biodiversity–ecosystem-function relationships. Ecology 92 (8), 1573–1581.

Folke, C., Holling, C.S., Perrings, C., 1996. Biological diversity, ecosystems, and the human scale. Ecol. Appl. 6 (4), 1018–1024.

Fornara, D.A., Tilman, D., 2008. Plant functional composition influences rates of soil carbon and nitrogen accumulation. J. Ecol. 96 (2), 314–322.

Fukami, T., Martijn Bezemer, T., Mortimer, S.R., van der Putten, W.H., 2005. Species divergence and trait convergence in experimental plant community assembly. Ecol. Lett. 8 (12), 1283–1290.

Gazol, A., Tamme, R., Price, J.N., 2013. A negative heterogeneity-diversity relationship found in experimental grassland communities. Oecologia 173, 545–555.

Gerhold, P., Price, J.N., Püssa, K., Kalamees, R., Aher, K., Kaasik, A., Pärtel, M., 2013. Functional and phylogenetic community assembly linked to changes in species diversity in a long-term resource manipulation experiment. J. Veg. Sci. 24 (5), 843–852.

Gerhold, P., Cahill, J.F., Winter, M., Bartish, I.V., Prinzing, A., 2015. Phylogenetic patterns are not proxies of community assembly mechanisms (they are far better). Funct. Ecol. 29 (5), 600–614.

Grime, J.P., 2006. Trait convergence and trait divergence in herbaceous plant communities: mechanisms and consequences. J. Veg. Sci. 17 (2), 255–260.

Guerrero-Ramírez, N.R., Craven, D., Reich, P.B., Ewel, J.J., Isbell, F., Koricheva, J., Parrotta, J.A., Auge, H., Erickson, H.E., Forrester, D.I., Hector, A., Joshi, J., Montagnini, F., Palmborg, C., Piotto, D., Potvin, C., Roscher, C., van Ruijven, J., Tilman, D., Wilsey, B., Eisenhauer, N., 2017. Diversity-dependent temporal divergence of ecosystem functioning in experimental ecosystems. Nat. Ecol. Evol. 1 (11), 1639–1642.

Hautier, Y., Niklaus, P.A., Hector, A., 2009. Competition for light causes plant biodiversity loss after eutrophication. Science 324 (5927), 636–638.

Hautier, Y., Tilman, D., Isbell, F., 2015. Anthropogenic environmental changes affect ecosystem stability via biodiversity. Science 348, 336–340.

HilleRisLambers, J., Harpole, W.S., Tilman, D., Knops, J., Reich, P.B., 2004. Mechanisms responsible for the positive diversity-productivity relationship in Minnesota grasslands. Ecol. Lett. 7 (8), 661–668.

Hoffmann, K., Bivour, W., Früh, B., Koßmann, M., Voß, P.-H., 2014. Klimauntersuchungen in Jena für die Anpassung an den Klimawandel und seine erwarteten Folgen. Ein Ergebnisbericht, Berichte des Deutschen Wetterdienstes. Selbstverlag des Deutschen Wetterdienstes, Offenbach am Main.

IPCC, 2014. Climate change 2014: synthesis report. In: Contribution of Working Groups I, II and III to the Fifth Assessment Report of the Intergovernmental Panel on Climate Change. IPCC, Geneva, Switzerland.

Isbell, F., Calcagno, V., Hector, A., Connolly, J., Harpole, W.S., Reich, P.B., Scherer-Lorenzen, M., Schmid, B., Tilman, D., van Ruijven, J., Weigelt, A., Wilsey, B.J., Zavaleta, E.S., Loreau, M., 2011. High plant diversity is needed to maintain ecosystem services. Nature 477 (7363), 199–U96.

Isbell, F., Reich, P.B., Tilman, D., Hobbie, S.E., Polasky, S., Binder, S., 2013. Nutrient enrichment, biodiversity loss, and consequent declines in ecosystem productivity. Proc. Natl. Acad. Sci. U. S. A. 110 (29), 11911–11916.

Isbell, F., Craven, D., Connolly, J., Loreau, M., Schmid, B., Beierkuhnlein, C., Bezemer, T.M., Bonin, C., Bruelheide, H., de Luca, E., Ebeling, A., Griffin, J.N., Guo, Q.F., Hautier, Y., Hector, A., Jentsch, A., Kreyling, J., Lanta, V., Manning, P., Meyer, S.T., Mori, A.S., Naeem, S., Niklaus, P.A., Polley, H.W., Reich, P.B., Roscher, C., Seabloom, E.W., Smith, M.D., Thakur, M.P., Tilman, D., Tracy, B.F., van der Putten, W.H., van Ruijven, J., Weigelt, A., Weisser, W.W., Wilsey, B., Eisenhauer, N., 2015. Biodiversity increases the resistance of ecosystem productivity to climate extremes. Nature 526 (7574), 574–U263.

Isbell, F., Gonzalez, A., Loreau, M., Cowles, J., Diaz, S., Hector, A., Mace, G.M., Wardle, D., O'Connor, M.I., Duffy, J., Turnbull, L.A., Thompson, P.L., Larigauderie, A., 2017. Linking the influence and dependence of people on biodiversity across scales. Nature 546, 65–72.

Janeček, Š., de Bello, F., Horník, J., Bartoš, M., Černý, T., Doležal, J., Dvorský, M., Fajmon, K., Janečková, P., Jiráská, Š., Mudrák, O., Klimešová, J., 2013. Effects of land-use changes on plant functional and taxonomic diversity along a productivity gradient in wet meadows. J. Veg. Sci. 24 (5), 898–909.

Johnson, N.C., Zak, D.R., Tilman, D., Pfleger, F.L., 1991. Dynamics of vesicular-arbuscular mycorrhizae during old field succession. Oecologia 86 (3), 349–358.

Kahmen, S., Poschlod, P., 2004. Plant functional trait responses to grassland succession over 25 years. J. Veg. Sci. 15 (1), 21–32.

Karlowsky, S., Augusti, A., Ingrisch, J., Hasibeder, R., Lange, M., Lavorel, S., Bahn, M., Gleixner, G., 2018. Land use in mountain grasslands alters drought response and recovery of carbon allocation and plant-microbial interactions. J. Ecol. 106 (3), 1230–1243.

Kattge, J., Diaz, S., Lavorel, S., Prentice, C., Leadley, P., Bonisch, G., Garnier, E., Westoby, M., Reich, P.B., Wright, I.J., Cornelissen, J.H.C., Violle, C., Harrison, S.P., van Bodegom, P.M., Reichstein, M., Enquist, B.J., Soudzilovskaia, N.A., Ackerly, D.D., Anand, M., Atkin, O., Bahn, M., Baker, T.R., Baldocchi, D., Bekker, R., Blanco, C.C., Blonder, B., Bond, W.J., Bradstock, R., Bunker, D.E., Casanoves, F., Cavender-Bares, J., Chambers, J.Q., Chapin, F.S., Chave, J., Coomes, D., Cornwell, W.K., Craine, J.M., Dobrin, B.H., Duarte, L., Durka, W., Elser, J., Esser, G., Estiarte, M., Fagan, W.F., Fang, J., Fernandez-Mendez, F., Fidelis, A., Finegan, B., Flores, O., Ford, H., Frank, D., Freschet, G.T., Fyllas, N.M., Gallagher, R.V., Green, W.A., Gutierrez, A.G., Hickler, T., Higgins, S.I., Hodgson, J.G., Jalili, A., Jansen, S., Joly, C.A., Kerkhoff, A.J., Kirkup, D., Kitajima, K., Kleyer, M., Klotz, S., Knops, J.M.H., Kramer, K., Kuhn, I., Kurokawa, H., Laughlin, D., Lee, T.D., Leishman, M., Lens, F., Lenz, T., Lewis, S.L., Lloyd, J., Llusia, J., Louault, F., Ma, S., Mahecha, M.D., Manning, P., Massad, T., Medlyn, B.E., Messier, J., Moles, A.T., Muller, S.C., Nadrowski, K., Naeem, S., Niinemets, U., Nollert, S., Nuske, A., Ogaya, R., Oleksyn, J., Onipchenko, V.G., Onoda, Y., Ordonez, J., Overbeck, G., Ozinga, W.A., Patino, S., Paula, S., Pausas, J.G., Penuelas, J., Phillips, O.L., Pillar, V., Poorter, H., Poorter, L., Poschlod, P., Prinzing, A., Proulx, R., Rammig, A., Reinsch, S., Reu, B., Sack, L., Salgado-Negre, B., Sardans, J., Shiodera, S., Shipley, B., Siefert, A., Sosinski, E., Soussana, J.F., Swaine, E., Swenson, N., Thompson, K., Thornton, P., Waldram, M., Weiher, E., White, M., White, S., Wright, S.J., Yguel, B., Zaehle, S., Zanne, A.E., Wirth, C., 2011. TRY—a global database of plant traits. Glob. Chang. Biol. 17 (9), 2905–2935.

Kimball, B.A., Conley, M.M., Wang, S., Lin, X., Luo, C., Morgan, J., Smith, D., 2008. Infrared heater arrays for warming ecosystem field plots. Glob. Chang. Biol. 14 (2), 309–320.

Kraft, N.J.B., Ackerly, D.D., 2010. Functional trait and phylogenetic tests of community assembly across spatial scales in an Amazonian forest. Ecol. Monogr. 80 (3), 401–422.

Kraft, N.J.B., Adler, P.B., Godoy, O., James, E.C., Fuller, S., Levine, J.M., 2015. Community assembly, coexistence and the environmental filtering metaphor. Funct. Ecol. 29 (5), 592–599.

Laliberté, E., 2017. Below-ground frontiers in trait-based plant ecology. New Phytol. 213 (4), 1597–1603.

Laliberte, E., Legendre, P., 2010. A distance-based framework for measuring functional diversity from multiple traits. Ecology 91 (1), 299–305.

Lange, M., Eisenhauer, N., Sierra, C.A., Bessler, H., Engels, C., Griffiths, R.I., Mellado-Vázquez, P.G., Malik, A.A., Roy, J., Scheu, S., Steinbeiss, S., Thomson, B.C., Trumbore, S.E., Gleixner, G., 2015. Plant diversity increases soil microbial activity and soil carbon storage. Nat. Commun. 6, 6707.

Lange, M., Koller-France, E., Hildebrandt, A., Oelmann, Y., Wilcke, W., Gleixner, G., 2019. How plant diversity impacts the coupled water, nutrient and carbon cycles. Adv. Ecol. Res. 61, 185–219.

Lee, T.D., Barrot, S.H., Reich, P.B., 2011. Photosynthetic responses of 13 grassland species across 11 years of free-air $CO_2$ enrichment is modest, consistent and independent of N supply. Glob. Chang. Biol. 17 (9), 2893–2904.

Lenth, R.V., 2016. Least-squares means: the R package lsmeans. J. Stat. Softw. 69 (1), 1–33.

Lepš, J., de Bello, F., Smilauer, P., Dolezal, J., 2011. Community trait response to environment: disentangling species turnover vs intraspecific trait variability effects. Ecography 34 (5), 856–863.

Lepš, J., Májeková, M., Vítová, A., Doležal, J., de Bello, F., 2018. Stabilizing effects in temporal fluctuations: management, traits, and species richness in high-diversity communities. Ecology 99 (2), 360–371.

Li, S.-P., Cadotte, M.W., Meiners, S.J., Hua, Z.-S., Jiang, L., Shu, W.-S., 2015. Species colonisation, not competitive exclusion, drives community overdispersion over long-term succession. Ecol. Lett. 18 (9), 964–973.

Lin, B.B., Flynn, D.F.B., Bunker, D.E., Uriarte, M., Naeem, S., 2011. The effect of agricultural diversity and crop choice on functional capacity change in grassland conversions. J. Appl. Ecol. 48 (3), 609–618.

Liu, D.J., Penuelas, J., Ogaya, R., Estiarte, M., Tielborger, K., Slowik, F., Yang, X.H., Bilton, M.C., 2018. Species selection under long-term experimental warming and drought explained by climatic distributions. New Phytol. 217 (4), 1494–1506.

Manning, P., Newington, J.E., Robson, H.R., Saunders, M., Eggers, T., Bradford, M.A., Bardgett, R.D., Bonkowski, M., Ellis, R.J., Gange, A.C., Grayston, S.J., Kandeler, E., Marhan, S., Reid, E., Tscherko, D., Godfray, H.C.J., Rees, M., 2006. Decoupling the direct and indirect effects of nitrogen deposition on ecosystem function. Ecol. Lett. 9 (9), 1015–1024.

Manning, P., Morrison, S.A., Bonkowski, M., Bardgett, R.D., 2008. Nitrogen enrichment modifies plant community structure via changes to plant–soil feedback. Oecologia 157 (4), 661–673.

Meyer, S.T., Ebeling, A., Eisenhauer, N., Hertzog, L., Hillebrand, H., Milcu, A., Pompe, S., Abbas, M., Bessler, H., Buchmann, N., De Luca, E., Engels, C., Fischer, M., Gleixner, G., Hudewenz, A., Klein, A.-M., de Kroon, H., Leimer, S., Loranger, H., Mommer, L., Oelmann, Y., Ravenek, J.M., Roscher, C., Rottstock, T., Scherber, C., Scherer-Lorenzen, M., Scheu, S., Schmid, B., Schulze, E.-D., Staudler, A., Strecker, T., Temperton, V., Tscharntke, T., Vogel, A., Voigt, W., Weigelt, A., Wilcke, W., Weisser, W.W., 2016. Effects of biodiversity strengthen over time as ecosystem functioning declines at low and increases at high biodiversity. Ecosphere 7 (12)e01619-n/a.

Mouchet, M.A., Villeger, S., Mason, N.W.H., Mouillot, D., 2010. Functional diversity measures: an overview of their redundancy and their ability to discriminate community assembly rules. Funct. Ecol. 24 (4), 867–876.

Muller, B., Pantin, F., Génard, M., Turc, O., Freixes, S., Piques, M., Gibon, Y., 2011. Water deficits uncouple growth from photosynthesis, increase C content, and modify the relationships between C and growth in sink organs. J. Exp. Bot. 62 (6), 1715–1729.

Naeem, S., Li, S.B., 1997. Biodiversity enhances ecosystem reliability. Nature 390 (6659), 507–509.

Nogueira, C., Nunes, A., Bugalho, M.N., Branquinho, C., McCulley, R.L., Caldeira, M.C., 2018. Nutrient addition and drought interact to change the structure and decrease the functional diversity of a Mediterranean grassland. Front. Ecol. Evol. 6 (155).

Pakeman, R.J., Lennon, J.J., Brooker, R.W., 2011. Trait assembly in plant assemblages and its modulation by productivity and disturbance. Oecologia 167 (1), 209–218.

Pearse, W.D., Purvis, A., 2013. phyloGenerator: an automated phylogeny generation tool for ecologists. Methods Ecol. Evol. 4 (7), 692–698.
Pinheiro, J., Bates, D., DebRoy, S., Sarkar, D., R Core Team, 2018. Nlme: Linear and Nonlinear Mixed Effects Models. Springer Verlag, New York.
Pontes, L.D., Louault, F., Carrere, P., Maire, V., Andueza, D., Soussana, J.F., 2010. The role of plant traits and their plasticity in the response of pasture grasses to nutrients and cutting frequency. Ann. Bot. 105 (6), 957–965.
Price, J.N., Gazol, A., Tamme, R., Hiiesalu, I., Pärtel, M., 2014. The functional assembly of experimental grasslands in relation to fertility and resource heterogeneity. Funct. Ecol. 28 (2), 509–519.
Purschke, O., Schmid, B.C., Sykes, M.T., Poschlod, P., Michalski, S.G., Durka, W., Kühn, I., Winter, M., Prentice, H.C., 2013. Contrasting changes in taxonomic, phylogenetic and functional diversity during a long-term succession: insights into assembly processes. J. Ecol. 101 (4), 857–866.
R Core Team, 2018. R: A Language and Environment for Statistical Computing. R Foundation for Statistical Computing, Vienna, Austria.
R Development Core Team, 2011. R: A Language and Environment for Statistical Computing. R Foundation for Statistical Computing, Vienna, Austria.
Reich, P.B., 2014. The world-wide 'fast–slow' plant economics spectrum: a traits manifesto. J. Ecol. 102 (2), 275–301.
Reich, P.B., Knops, J., Tilman, D., Craine, J., Ellsworth, D., Tjoelker, M., Lee, T., Wedin, D., Naeem, S., Bahauddin, D., Hendrey, G., Jose, S., Wrage, K., Goth, J., Bengston, W., 2001. Plant diversity enhances ecosystem responses to elevated $CO_2$ and nitrogen deposition. Nature 410 (6839), 809–812.
Reich, P.B., Tilman, D., Isbell, F., Mueller, K., Hobbie, S.E., Flynn, D.F.B., Eisenhauer, N., 2012. Impacts of biodiversity loss escalate through time as redundancy fades. Science 336 (6081), 589–592.
Revelle, W., 2018. Psych: Procedures for Personality and Psychological Research. Northwestern University, Evanston, Illinois, USA.
Roscher, C., Schumacher, J., Baade, J., Wilcke, W., Gleixner, G., Weisser, W.W., Schmid, B., Schulze, E.D., 2004. The role of biodiversity for element cycling and trophic interactions: an experimental approach in a grassland community. Basic Appl. Ecol. 5 (2), 107–121.
Roscher, C., Schumacher, J., Gubsch, M., Lipowsky, A., Weigelt, A., Buchmann, N., Schmid, B., Schulze, E.-D., 2012. Using plant functional traits to explain diversity-productivity relationships. PLoS One 7 (5) e36760.
Roscher, C., Schumacher, J., Lipowsky, A., Gubsch, M., Weigelt, A., Pompe, S., Kolle, O., Buchmann, N., Schmid, B., Schulze, E.D., 2013. A functional trait-based approach to understand community assembly and diversity-productivity relationships over 7 years in experimental grasslands. Perspect. Plant Ecol. Evol. Syst. 15 (3), 139–149.
Roscher, C., Schumacher, J., Gerighausen, U., Schmid, B., 2014. Different assembly processes drive shifts in species and functional composition in experimental grasslands varying in sown diversity and community history. PLoS One 9 (7) e101928.
Roscher, C., Gerighausen, U., Schmid, B., Schulze, E.-D., 2015. Plant diversity and community history shift colonization success from early- to mid-successional species. J. Plant Ecol. 8 (3), 231–241.
Roscher, C., Schumacher, J., Petermann, J.S., Fergus, A.J.F., Gerighausen, U., Michalski, S.G., Schmid, B., Schulze, E.D., 2016. Convergent high diversity in naturally colonized experimental grasslands is not related to increased productivity. Perspect. Plant Ecol. Evol. Syst. 20, 32–45.
Siebenkäs, A., Schumacher, J., Roscher, C., 2016. Trait variation in response to resource availability and plant diversity modulates functional dissimilarity among species in experimental grasslands. J. Plant Ecol. 10 (6), 981–993.

Srivastava, D.S., Cadotte, M.W., MacDonald, A.A.M., Marushia, R.G., Mirotchnick, N., 2012. Phylogenetic diversity and the functioning of ecosystems. Ecol. Lett. 15 (7), 637–648.
Steinbeiss, S., Bessler, H., Engels, C., Temperton, V.M., Buchmann, N., Roscher, C., Kreutziger, Y., Baade, J., Habekost, M., Gleixner, G., 2008. Plant diversity positively affects short-term soil carbon storage in experimental grasslands. Glob. Chang. Biol. 14 (12), 2937–2949.
Tardieu, F., Granier, C., Muller, B., 2011. Water deficit and growth. Co-ordinating processes without an orchestrator? Curr. Opin. Plant Biol. 14 (3), 283–289.
Tilman, D., 1988. Plant Strategies and the Dynamics and Structure of Plant Communities, Monographs in Population Biology. Princeton University Press, Princeton, NJ..
Tilman, D., Reich, P.B., Knops, J., Wedin, D., Mielke, T., Lehman, C., 2001. Diversity and productivity in a long-term grassland experiment. Science 294 (5543), 843–845.
Tilman, D., Reich, P.B., Knops, J.M.H., 2006. Biodiversity and ecosystem stability in a decade-long grassland experiment. Nature 441 (7093), 629–632.
Török, P., Matus, G., Tóth, E., Papp, M., Kelemen, A., Sonkoly, J., Tóthmérész, B., 2018. Both trait-neutrality and filtering effects are validated by the vegetation patterns detected in the functional recovery of sand grasslands. Sci. Rep. 8 (1), 13703.
Tucker, C.M., Cadotte, M.W., Carvalho, S.B., Davies, T.J., Ferrier, S., Fritz, S.A., Grenyer, R., Helmus, M.R., Jin, L.S., Mooers, A.O., Pavoine, S., Purschke, O., Redding, D.W., Rosauer, D.F., Winter, M., Mazel, F., 2016. A guide to phylogenetic metrics for conservation, community ecology and macroecology. Biol. Rev. 92, 698–715.
Valladares, F., Bastias, C.C., Godoy, O., Granda, E., Escudero, A., 2015. Species coexistence in a changing world. Front. Plant Sci. 6, 16.
Van der Putten, W.H., Van Dijk, C., Peters, B.A.M., 1993. Plant-specific soil-borne diseases contribute to succession in foredune vegetation. Nature 362 (6415), 53–56.
Vile, D., Shipley, B., Garnier, E., 2006. A structural equation model to integrate changes in functional strategies during old-field succession. Ecology 87 (2), 504–517.
Villeger, S., Mason, N.W.H., Mouillot, D., 2008. New multidimensional functional diversity indices for a multifaceted framework in functional ecology. Ecology 89 (8), 2290–2301.
Vogel, A., Scherer-Lorenzen, M., Weigelt, A., 2012. Grassland resistance and resilience after drought depends on management intensity and species richness. PLoS One 7 (5), e36992. https://doi.org/10.1371/journal.pone.0036992.
Vogel, A., Eisenhauer, N., Weigelt, A., Scherer-Lorenzen, M., 2013a. Plant diversity does not buffer drought effects on early-stage litter mass loss rates and microbial properties. Glob. Chang. Biol. 19 (9), 2795–2803.
Vogel, A., Fester, T., Eisenhauer, N., Scherer-Lorenzen, M., Schmid, B., Weisser, W.W., Weigelt, A., 2013b. Separating drought effects from roof artifacts on ecosystem processes in a grassland drought experiment. PLoS One 8 (8) e70997.
Vogel, A., Ebeling, A., Gleixner, G., Roscher, C., Scheu, S., Ciobanu, M., Koller-France, E., Lange, M., Lochner, A., Meyer, S.T., Oelmann, Y., Wilcke, W., Schmid, B., Eisenhauer, N., 2019. A new experimental approach to test why biodiversity effects strengthen as ecosystems age. Adv. Ecol. Res. 61, 221–264.
Wagg, C., O'Brien, M.J., Vogel, A., Scherer-Lorenzen, M., Eisenhauer, N., Schmid, B., Weigelt, A., 2017. Plant diversity maintains long-term ecosystem productivity under frequent drought by increasing short-term variation. Ecology 98 (11), 2952–2961.
Webb, C.O., 2000. Exploring the phylogenetic structure of ecological communities: an example for rain forest trees. Am. Nat. 156 (2), 145–155.
Webb, C.O., Ackerly, D.D., McPeek, M.A., Donoghue, M.J., 2002. Phylogenies and community ecology. Annu. Rev. Ecol. Syst. 33, 475–505.

Weigelt, A., Weisser, W.W., Buchmann, N., Scherer-Lorenzen, M., 2009. Biodiversity for multifunctional grasslands: equal productivity in high-diversity low-input and low-diversity high-input systems. Biogeosciences 6 (8), 1695–1706.
Weisser, W.W., Roscher, C., Meyer, S.T., Ebeling, A., Luo, G., Allan, E., Beßler, H., Barnard, R.L., Buchmann, N., Buscot, F., Engels, C., Fischer, C., Fischer, M., Gessler, A., Gleixner, G., Halle, S., Hildebrandt, A., Hillebrand, H., de Kroon, H., Lange, M., Leimer, S., Le Roux, X., Milcu, A., Mommer, L., Niklaus, P.A., Oelmann, Y., Proulx, R., Roy, J., Scherber, C., Scherer-Lorenzen, M., Scheu, S., Tscharntke, T., Wachendorf, M., Wagg, C., Weigelt, A., Wilcke, W., Wirth, C., Schulze, E.-D., Schmid, B., Eisenhauer, N., 2017. Biodiversity effects on ecosystem functioning in a 15-year grassland experiment: patterns, mechanisms, and open questions. Basic Appl. Ecol. 23 (Suppl. C), 1–73.
Wright, A., Schnitzer, S.A., Reich, P.B., 2014. Living close to your neighbors: the importance of both competition and facilitation in plant communities. Ecology 95 (8), 2213–2223.
Wright, A., Schnitzer, S.A., Reich, P.B., 2015. Daily environmental conditions determine the competition–facilitation balance for plant water status. J. Ecol. 103 (3), 648–656.
Yachi, S., Loreau, M., 1999. Biodiversity and ecosystem productivity in a fluctuating environment: the insurance hypothesis. Proc. Natl. Acad. Sci. U. S. A. 96, 1463–1468.
Zanne, A.E., Tank, D.C., Cornwell, W.K., Eastman, J.M., Smith, S.A., FitzJohn, R.G., McGlinn, D.J., O'Meara, B.C., Moles, A.T., Reich, P.B., Royer, D.L., Soltis, D.E., Stevens, P.F., Westoby, M., Wright, I.J., Aarssen, L., Bertin, R.I., Calaminus, A., Govaerts, R., Hemmings, F., Leishman, M.R., Oleksyn, J., Soltis, P.S., Swenson, N.G., Warman, L., Beaulieu, J.M., Ordonez, A., 2013. Data From: Three Keys to the Radiation of Angiosperms Into Freezing Environments. Dryad Digital Repository. https://doi.org/10.5061/dryad.63q27.2, Dryad.
Zanne, A.E., Tank, D.C., Cornwell, W.K., Eastman, J.M., Smith, S.A., FitzJohn, R.G., McGlinn, D.J., O'Meara, B.C., Moles, A.T., Reich, P.B., Royer, D.L., Soltis, D.E., Stevens, P.F., Westoby, M., Wright, I.J., Aarssen, L., Bertin, R.I., Calaminus, A., Govaerts, R., Hemmings, F., Leishman, M.R., Oleksyn, J., Soltis, P.S., Swenson, N.G., Warman, L., Beaulieu, J.M., 2014. Three keys to the radiation of angiosperms into freezing environments. Nature 506 (7486), 89–92.
Zhou, X., Guo, Z., Zhang, P., Du, G., 2018. Shift in community functional composition following nitrogen fertilization in an alpine meadow through intraspecific trait variation and community composition change. Plant Soil 431 (1), 289–302.

CHAPTER FOUR

# Terrestrial laser scanning reveals temporal changes in biodiversity mechanisms driving grassland productivity

Claudia Guimarães-Steinicke[a,*], Alexandra Weigelt[a,b], Anne Ebeling[c], Nico Eisenhauer[b,d], Joaquín Duque-Lazo[e], Björn Reu[f], Christiane Roscher[b,g], Jens Schumacher[h], Cameron Wagg[i,j], Christian Wirth[a,b,k]

[a]Systematic Botany and Functional Biodiversity, Institute of Biology, Leipzig University, Leipzig, Germany
[b]German Centre for Integrative Biodiversity Research (iDiv) Halle-Jena-Leipzig, Leipzig, Germany
[c]Institute of Ecology and Evolution, Friedrich Schiller University Jena, Jena, Germany
[d]Institute of Biology, Leipzig University, Leipzig, Germany
[e]Department of Forestry, School of Agriculture and Forestry, University of Córdoba, Córdoba, Spain
[f]School of Biology, Industrial University of Santander, Bucaramanga, Colombia
[g]UFZ, Helmholtz Centre for Environmental Research, Physiological Diversity, Leipzig, Germany
[h]Institute of Mathematics, Friedrich Schiller University Jena, Jena, Germany
[i]Department of Evolutionary Biology and Environmental Studies, University of Zürich, Zürich, Switzerland
[j]Fredericton Research and Development Centre, Agriculture and Agri-Food Canada, Fredericton, NB, Canada
[k]Max Planck Institute for Biogeochemistry, Jena, Germany
*Corresponding author: e-mail address: claudia.steinicke@uni-leipzig.de

## Contents

| | | |
|---|---|---|
| 1. | Introduction | 134 |
| 2. | Material and methods | 138 |
| | 2.1 Study site and trait based-experiment | 138 |
| | 2.2 Terrestrial laser scanning: Data acquisition and processing | 139 |
| | 2.3 Diversity drivers: Functional diversity, functional identity, and species richness | 141 |
| | 2.4 Data analyses | 142 |
| 3. | Results | 144 |
| | 3.1 Intra- and inter-annual variation in mean height as a proxy for aboveground biomass | 144 |
| | 3.2 Intra-annual diversity and identity effects on plant development | 147 |
| 4. | Discussion | 149 |
| | 4.1 Intra-annual changes in functional diversity effects on plant development | 150 |
| | 4.2 Intra-annual changes in identity effects on plant development | 151 |
| | 4.3 Species richness effects and inter-annual differences | 152 |
| | 4.4 A new method for biodiversity-ecosystem functioning research in grasslands | 154 |

5. Conclusions 155
Acknowledgements 155
References 155

## Abstract

Biodiversity often enhances ecosystem functioning likely due to multiple, often temporarily separated drivers. Yet, most studies are based on one or two snapshot measurements per year. We estimated productivity using bi-weekly estimates of high-resolution canopy height in 2 years with terrestrial laser scanning (TLS) in a grassland diversity experiment. We measured how different facets of plant diversity (functional dispersion [FDis], functional identity [PCA species scores], and species richness [SR]) predict aboveground biomass over time. We found strong intra- and inter-annual variability in the relative importance of different mechanisms underlying the diversity effects on mean canopy height, i.e., resource partitioning (via FDis) and identity effects (via species scores), respectively. TLS is a promising tool to quantify community development non-destructively and to unravel the temporal dynamics of biodiversity-ecosystem functioning mechanisms. Our results show that harvesting at estimated peak biomass—as done in most grassland experiments—may miss important variation in underlying mechanisms driving cumulative biomass production.

## 1. Introduction

The deterioration of global ecosystems due to intensifying human activities reduces species diversity and thereby impairs ecosystem functioning (Eisenhauer et al., 2019, in this issue; Loreau et al., 2002; Millennium Ecosystem Assessments, 2005; Naeem et al., 2012). Biodiversity experiments have repeatedly revealed a positive relationship between biodiversity and ecosystem functioning (BEF) (Bunker, 2005; Cardinale et al., 2007a, 2012; Reich et al., 2012). Yet, the underlying mechanisms are still subject to debate (Hooper et al., 2005; Tilman et al., 2014; Vogel et al., 2019, in this issue). Ecologists have tried to separate BEF mechanisms such as above- and below-ground resource partitioning and sampling effects experimentally (Bachmann et al., 2015; Jesch et al., 2018; Kahmen et al., 2006; Tilman et al., 2001; von Felten et al., 2009) and statistically (Fox, 2005; Loreau and Hector, 2001; Tilman et al., 1997). However, merging the potential joint effects of multiple, often temporally separated drivers of ecosystem functions (Everwand et al., 2014), into a single annual measurement is unlikely to reveal the full spectrum of mechanisms underlying the effect of BEF relationships.

Different facets of plant diversity, e.g., functional diversity based on plant traits, or plant species richness, point to different underlying mechanisms (Cadotte, 2017), and their relative importance as predictors of aboveground productivity may change during the growing season. If functional identity, quantified as community mean trait expression (Cingolani et al., 2007; Fonseca et al., 2000), is a strong predictor of functioning, we may conclude that an *identity effect* is operating. In this case, increasing diversity increases the chance of including species with specific trait values that have strong effects on functioning (Aarssen, 1997; Grime, 1998; Huston, 1997). If in turn, functional diversity of grassland communities, a facet of biodiversity determined by the dispersion of species traits in multidimensional trait space (Laliberté and Legendre, 2010), emerges as a strong predictor of plant development, we might infer that *resource partitioning* is an important driver of BEF relationships. Resource partitioning is a mechanism that requires being functionally dissimilar as a prerequisite for more efficient resource acquisition in mixtures. Both facets of diversity require that the selected traits comprehensively represent the functional dimensions relevant to the target ecosystem function, e.g., resource partitioning was shown to depend on the presence of species varying in rooting depth (Oram et al., 2018). In addition, if mere species richness is a significant predictor—instead of or in addition to functional diversity or identity—this may point to alternative mechanisms, e.g., *pathogen dilution* or *facilitation* (Eisenhauer et al., 2012; Guerrero-Ramírez and Eisenhauer, 2017; Wright et al., 2017a).

Despite considerable efforts to unravel the importance of resource partitioning or identity effects for BEF relationships, we still know remarkably little about their relative importance. Indeed, these mechanisms may vary during the growing season (Bachmann et al., 2018; Cardinale et al., 2007a). Temperate grassland communities, the target communities of our study undergo substantial microclimatic, biogeochemical, and structural changes throughout the growing season (Gibson, 2009). Here, we develop three possible scenarios for extensively managed European grasslands for how the relative importance of BEF mechanisms may change during the growing season:

**(1)** *Functional diversity as an indicator of resource partitioning:* After early spring, increasing temperature and radiation enhance plant growth, generally leading to denser canopies and light limitation, especially for small-statured species (Daßler et al., 2008; Roscher et al., 2011), and nutrients are progressively depleted by plant uptake from the soil (Regan et al., 2014). Increased competition for light and nutrients as plant communities approach peak biomass should enhance the relative importance of functional diversity of resource uptake strategies, allowing for resource partitioning and, as a consequence, higher productivity. Although to a lesser extent, resource partitioning is expected to increase biomass after recovery following mowing, as biomass packing is less intense in the late summer sward in most years. This pattern is reflected by the fact that the second mowing often only yields half of the biomass compared to the first mowing in spring (Jongen et al., 2011; Regan et al., 2014; Weigelt et al., 2010; Zeeman et al., 2019), canopy height is lower and light attenuation to the ground reaches higher values (Bachmann et al., 2018; Lorentzen et al., 2008). The effects of functional trait diversity on ecosystem functioning should manifest where functional diversity increases along a gradient optimizing spatial resource partitioning (light, water, nutrient) above- and belowground.

(2) *Functional identity as an indicator for identity effects:* Additive partitioning of net biodiversity effects (Loreau and Hector (2001) revealed that in young, freshly-seeded grassland experiments the sampling effect for fast-growing species is the dominant mechanism underlying the positive BEF relationship, particularly in terms of biomass accumulation (Marquard et al., 2009; Reich et al., 2012; Wagg et al., 2017a,b). Over time, the relative importance of the complementarity effect increases (Cardinale et al., 2007b; Marquard et al., 2009; van Ruijven and Berendse, 2005; Vogel et al., 2019, in this issue). A similar situation may arise during the recovery after winter. In early spring, energy is limiting, but water and nutrients are available in excess (Eviner et al., 2006). Community productivity does not depend on the efficient sharing of these resources via resource complementarity (Fargione and Tilman, 2006), but rather on the occurrence of early-growing species (Mamolos et al., 2011). Diverse mixtures have a higher likelihood of containing species with early phenology and may thus perform better at the beginning of the growing season (identity effect). A similar situation arises after mowing when competition for light, water, and nutrients declines markedly (Humbert et al., 2012; Schippers and Joenje, 2002). As before, community re-growth may not depend on efficient resource sharing under these conditions, but rather on the presence of species with ample storage compounds in the roots and the capability of rapid resprouting (Eisenhauer et al., 2009) to allow a fast sward recovery (Roscher et al., 2012). The functional identity effect should be maximized, where diversity increases along a phenological trait gradient of early and late growing species.

**(3)** *Species richness as an indicator of alternative mechanisms:* In addition to functional identity and diversity, species richness per se could determine an ecosystem functioning. This may be the case where biotic interactions that are not directly linked to productivity-related plant traits are involved, e.g., pathogen or herbivory dilution (Civitello et al., 2015; Hendriks et al., 2013) or mutual *facilitation* via mycorrhizal networks (Eisenhauer et al., 2012; Wagg et al., 2015; Wright et al., 2017a). Biotic interactions might be particularly important during times of increased disturbance (Callaway et al., 2002; Maalouf et al., 2012; Michalet et al., 2006) or abiotic stress such as drought (Craven et al., 2016; De Boeck et al., 2011; Isbell et al., 2015; Wagg et al., 2017a,b).

We examine the effects of these three different facets of grassland diversity, i.e., functional diversity, functional identity, and species richness, on biomass production throughout the growing season of two subsequent years in a long-term biodiversity experiment (Trait-Based Biodiversity Experiment; Ebeling et al., 2014). The experiment is managed as a mown grassland with two cuts per year. We approximate biomass production as height growth measured at high spatial and temporal resolutions using biweekly and non-destructive high-resolution scans of vegetation canopy height with a terrestrial laser scanner throughout two growing seasons. Thereby, we cover different environmental conditions related to seasonal changes in water availability, temperature, and competition for light. According to Ebeling et al. (2014), the Trait-Based Biodiversity Experiment manipulates plant trait composition using two plant species pools. The first pool reflects plant size spanning the range from small to large species and, thus, expressing different strategies of spatial resource use (Campbell et al., 1991). The second pool represents a phenological trait gradient that ranges from early-growing species to late-growing species and, therefore, captures a spectrum of temporal resource-use strategies. These two pools are ideally suited to address differences between functional identity and functional diversity effects throughout the growing season. We calculate community means of traits (CM) and functional dispersion (FDis) based on the presence-absence of species to reflect the potential importance of the different facets of diversity to control biomass production in communities along 12 dates throughout the growing season. Using this data, we test the following hypotheses:

**Hypothesis 1.** We expect that there are intra-annual differences in the relative importance of the three plant diversity facets, namely, functional identity, diversity, and species richness.

**Hypothesis 2.** Resource partitioning (as indicated by functional trait dispersion) is a key driver during the peak of biomass. This effect is strongest in the first pool (plant size), where we expect positive resource partitioning effects at times of high biomass.

**Hypothesis 3.** Functional identity (as indicated by community means of traits) is a key driver during initial stages of the growing season and directly after mowing. This effect is strongest in the second pool (phenology), where we expect a negative identity effect at the start of the growing season (selection for early species) and a positive identity effect after the mowing (selection for late species).

**Hypothesis 4.** Species richness emerges as an additional predictor of alternative mechanisms, such as pathogen dilution or facilitation, which may be particularly stronger during times of disturbances.

## 2. Material and methods
### 2.1 Study site and trait based-experiment

This study was conducted in 2014 and 2015 within the Trait-Based Biodiversity Experiment (TBE; Ebeling et al., 2014) at the field site of the Jena Experiment (Thuringia, Germany; 50°55′ N, 11°35′E, 130m above sea level) (Roscher et al., 2004; Weisser et al., 2017). The TBE was established in 2010 and manipulates trait diversity of plant communities, i.e., plant height, leaf size, rooting depth, root length density, onset of flowering, and start of the growing period (see Ebeling et al., 2014) as well as plant species richness using a subset of the 48 non-legume species of the Jena Experiment species pool (Roscher et al., 2004). A principal component analysis assessed the species trait variation of the 48 non-legume species relevant for resource acquisition in space (PCA 1) and time (PCA2). The first ordination axis was positively correlated with leaf size, plant height, and rooting depth and negatively to root length density. Thus, species along this axis vary from short plants with small leaves and shallow, dense roots (i.e. many grasses) to taller plants with large leaves and deeper roots (mostly tall herbs). The second axis spans a phenological gradient, which was negatively correlated with species with earlier flowering and start of growth and positively correlated with species with late flowering and onset of growth (Ebeling et al., 2014).

PCA axes were separated into four sectors, and two species from each sector were selected to build two experimental species pools of eight species each. Therefore, it was established: pool 1 by selecting species from sectors along axis 1 and hereafter referred to as 'spatial' pool for simplicity—containing species varying in size (e.g., plant height, rooting depth, root length density, and leaf size); pool 2 by selecting species from sectors along axis 2—and hereafter as 'temporal' pool—spanning a phenological gradient. Overall, on 46 plots per species pool (3.5 m × 3.5 m), the TBE design covers a gradient of plant species richness ranging from 1 to 8 species. In central Europe, grassland management covers a gradient of intensities, depending on site conditions and economy. In Thuringia, where the Jena Experiment is located, the most common practice of extensively managed grasslands is no fertilization with 2–3 cuts per year (Roscher et al., 2004). The Jena Experiment mimics this practice and maintains experimental plots with biannual mowing (June, September) and no fertilization. Plots are weeded three times per year (April, July, and October) to maintain the sown species richness gradient. We conducted our communities scanning measurements in 16 monocultures, 32 two-species mixtures, 24 three-species mixtures, 18 four-species mixtures, and 2 eight-species mixtures.

## 2.2 Terrestrial laser scanning: Data acquisition and processing

To obtain a non-destructive proxy of plant height at high temporal resolution during the growing season, we used a terrestrial laser scanner (TLS) Faro Focus 3D X330 (FARO Technologies Inc, 2011). We scanned 92 plots biweekly from April to September in 2014 and 2015, resulting in 12 time-steps each year. The TLS was mounted upside-down on a tripod that was elevated 3.35 m above soil level. The legs of the tripod were centred on permanent survey markers to guarantee identical scanning areas over time. For each plot, we extracted an area of $3.75\,m^2$ (1.5 m × 2.5 m) below the scanner to reduce the effect of shadows within scans. The laser scanner measures the distance between the surface of an object and the scanner. The discrete returns of laser beams registered by the laser scanning produce a point cloud image of the surface of the grassland vegetation. The laser device emits an infrared pulse of 1550 nm with a beam divergence of $0.011°$ (0.19 mrad) at a range of 0.6–130 m. The scanning parameters used for resolution was 44.4 million points (full scan) and the first level of quality (1×) to maximize the efficiency of the scanning processes (FARO Technologies Inc, 2011). This yielded 3D point clouds with a scan size of 10,154 × 1138 points. This configuration provided a three-dimensional representation of the community structure and canopy cover (Fig. 1). All scan data were transformed from a point cloud into XYZ coordinates by using the proprietary software 'Scene' (version 5.2.0, Faro Technologies, Inc., Lake Mary, Florida, USA).

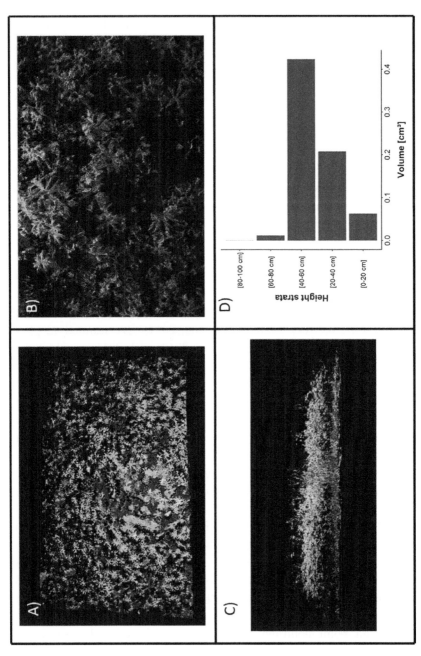

**Fig. 1** An example of a point cloud plot from the terrestrial laser scanner. (A) Top view from the monoculture plot of *Geranium pratense*; (B) zoomed view; (C) sided view and, (D) volume estimated for each height profile (0–10, 10–20, 20–30, 30–40, 40–50, 50–60, and 60–80 cm).

To reduce errors and increase accuracy, we applied two filtering processes to the point clouds using 'CloudCompare 2.8.1' (http://www.cloudcompare.org). First, we used statistical outlier removal, which aims to decrease source error by computing the average distance of every sixth point to its neighbours and rejects points that are farther than the average distance ($N=6$, Sigma $=1.5$). Second, we applied a noise filter that computes a plane with an approximately 2 mm circumference around each point and removes the points far away from the fitted plane. By applying both filters, we decreased occlusion problems that may have occurred due to the single perspective scanner position of each plot (scanner was positioned in upside down view).

Further, data processing was conducted with the LAStools suite of processing scripts (http://www.lastools.org, Isenburg, 2014). The XYZ coordinates of point clouds were transformed into standard LAS format (LASer—the interchange of three-dimensional point cloud data). We classified the points from the soil surface to 5 cm height as 'ground points' and all remaining points up to the upper canopy layer as 'non-ground points'. The ground points were interpolated to produce a digital terrain model by applying a triangulated network (TIN) approach. To obtain the height above ground level, we used the digital terrain model to normalize the Z coordinates of vegetation returns. The vegetation points were exported to FUSION/LDV (McGaughey, 2011) to produce structural metrics (using function cloud metrics), describing components such as canopy height, canopy relief, mean return height and variation in return height (Table S2 in Supplementary Material in the online version at https://doi.org/10.1016/bs.aecr.2019.06.003). We obtained 25 response variables: mean, maximum, standard deviation, the coefficient of variation, variance, skewness, and kurtosis of canopy height, percentile values of 15 layers of height from minimum to maximum height values, canopy relief ratio, height quadratic mean and height cubic mean. We then used these variables in the full model to estimate aboveground biomass with the goal to test the suitability of mean height as the key variable for the approximation of biomass (see Section 2.4.1).

### 2.3 Diversity drivers: Functional diversity, functional identity, and species richness

We tested the effects of different aspects of diversity, e.g., functional identity, functional diversity, and species richness, on our proxy of plant community biomass estimated by a terrestrial laser scanner (TLS) during the two growing seasons. We used two different indices to capture differences in the distribution of trait values within a community. First, we calculated functional dispersion (FDis), which describes the mean of the distance of each species to the centre of all species in multidimensional trait space (Laliberté and Legendre, 2010). The calculation of FDis permits a small number of species with continuous as well as qualitative trait measures (Laliberté and Legendre, 2010; Tobner et al., 2016). The second index was the community mean of traits (CM), which is the mean trait value of all species in a community (Cingolani et al., 2007; Fonseca et al., 2000).

To calculate FDis and CM, we used species scores from axis 1 and 2 of the experimental design principal component analysis (PCA) to describe the functional species position in the trait space (Ebeling et al., 2014). As we only had a measurement of abundance data of plant communities at two times of scanning campaigns (two peaks of biomass during the season), we used target species presence based on the experimental design of sown species to calculate FDis and CM, which results in unweighted means (Anderson, 2006). As a consequence, our results are independent of seasonal shifts in species-specific abundance distributions. Lastly, sown species richness was included as the third facet of biodiversity.

## 2.4 Data analyses
### 2.4.1 Estimation of plant biomass from terrestrial laser scanning metrics

To obtain a common proxy of community biomass for both years and further assess the effects of diversity drivers during the growing season, we regressed the aboveground biomass of the mowing in May and August of both years on the selected 25 TLS-derived metrics. Biomass was mowed in two randomly selected $20 \times 50$ cm areas per plot, dried at $70\,°C$ for $72\,h$ and weighed. The two samples per plot and sampling campaign were averaged and scaled to $1\,m^2$. The model selection aimed to identify the most parsimonious model (lowest AIC) using stepwise forward selection. To avoid highly correlated explanatory variables, we controlled for multicollinearity using the variation inflation factor with a cut-off value of vif $< 4$ (O'Brien, 2007). The step-wise multiple linear models revealed that TLS-derived metrics accounted for 54.2% of the variation of the dry aboveground biomass of $1\,m^2$ for both years (Adj. $R^2 = 0.53$, MAE $= 2.47\,g$; Fig. S1 in Supplementary Material in the online version at https://doi.org/10.1016/bs.aecr.2019.06.003). Even though all plots were mowed at the peak biomass and therefore variation in biomass is rather low (in contrast to comparing very low biomass right after mowing or after winter with very high biomass at peak biomass). The best model included mean canopy height, maximum height, and canopy relief ratio and height variance as explanatory variables. The mean height explained 45% of the variance, and the maximum height explained 4.75%, canopy relief, and height variance explained 3.71% and 0.74%, respectively. The residuals retained 45.8% of the explained variance. Further, we selected the best predictor of the multiple linear regression models (estimation of biomass) as the response variable for both years to investigate how diversity influenced the growth of plant communities intra-annually (see below) Therefore, we used mean height as a proxy for community aboveground biomass in all subsequent analyses.

### 2.4.2 Temporal changes in diversity effects on plant productivity

To investigate temporal changes in diversity effects on plant productivity, we fit linear mixed-effects models using the *lme* function in the *nlme* package of the statistical software R (Pinheiro and Bates, 2000; R Core Team, 2018; Zuur et al., 2009). We examined the relationship of diversity and mean height (as a proxy for community aboveground biomass) using TLS measurements obtained on 12 dates throughout the growing season. We fit mixed-effects models for each species pool separately. Treating the pools separately takes advantage of the design of the Trait-Based Biodiversity Experiment, in which we explicitly maximized diversity strategies throughout the growing season (Ebeling et al., 2014). For the spatial pool, we had 576 and for the temporal pool 528 individual measurements of mean height. To check for the effects of the year of measurement, we first ran an analysis of the effects of diversity (FDis, CM of species scores in the trait space, and species richness) on canopy height pooling the 2 years with the variable 'year' as an interaction term. In this analysis, the mixed-effect models resulted in highly significant differences between years (see Table S1 in Supplementary Material in the online version at https://doi.org/10.1016/bs.aecr.2019.06.003). As our research questions focused on potential factors that vary intra-annually and may influence productivity, such as environmental conditions and nutrient availability, we used independent models for each year. This decision was supported by the evidence of differences in climate during the growing season. For instance, seasonal precipitation better correlates with community productivity than total precipitation (Robinson et al., 2013); the same was observed with air temperature (Craine et al., 2012). Thus, we separated the dataset by year to obtain independent models for each year and investigate their separate relationships with diversity based on the 12 TLS measurements of community growth per year.

Analysis of the residuals using the auto-correlation function (ACF) indicated the presence of temporal autocorrelation within the 12 TLS measurements per year (Pinheiro and Bates, 2000). Therefore, the mixed-effects models were fitted with a correlation structure function. First, we assessed the standardized residuals using an empirical autocorrelation plot and compared the performance of models for different auto-correlation structures functions such as AR1 and ARMA(1,0) and ARMA(2,0) with distinct error structure based on AIC values (Pinheiro and Bates, 2000). We thus applied in the final mixed effect model the autoregressive correlation structure, which displayed the lowest AIC value (R function correlation = corARMA(2,0); package nlme, version 3.1–128) for the times of measurement (Chi and Reinsel, 1989; Pinheiro and Bates, 2000).

In the mixed-effects models, we treated block and plot as nested random factors to allow responses to varying randomly between blocks and plots. The variables mean plant height and species richness were log-transformed, and FDis and CM of species scores were z-transformed to stabilize the variance. We treated the sampling times as factors and the three drivers related to diversity (FDis, CM of species scores in the trait space, species richness) interacting with time as fixed effects. We chose a parameterization without intercept so that the individual coefficients at each sampling time vary around zero and indicate the occurrence of significant positive or negative effects (instead of expressing the difference to some arbitrary reference sampling time). We fitted the final models with REML estimation. We assessed the homogeneity of residuals with residuals vs. fitted values plots and Q-Q plots for data normality using '*Pearson*' correlation (Zuur et al., 2009). We used the *anova* function for mixed-effects models using the *F-statistic* (likelihood ratio test). To evaluate the mixed-effect model predictions for each time slice throughout the growing season, we regressed the predicted values of each species pool and year models versus the corresponding observed mean height obtained by the TLS measurements. Statistical analyses were performed with R v. 3.5.3 (R Core Team, 2018).

## 3. Results

### 3.1 Intra- and inter-annual variation in mean height as a proxy for aboveground biomass

The mean canopy height obtained from terrestrial laser scanning varied strongly within each growing season (Fig. 2 for 2014 and 3 for 2015). This is also reflected by the fact that time was a significant predictor in both years

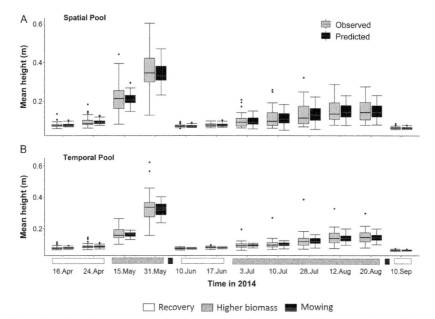

**Fig. 2** Boxplot of mean canopy height obtained from terrestrial laser scanning of plant communities in 2014 and the canopy height predicted by the best mixed-effects model accounting for diversity effects on plant development over the growing season. Note that the functional identity used in this model was calculated based on the community means of target species. Boxplots show 95% prediction interval for the mean of predicted values based on best model predictors and a 95% confidence interval of observed values at each time slot. (A) Data from Spatial and (B) Temporal pools in 2014.

and species pools (Tables 1 and 2). In 2014, plant canopy mean height peaked on May 31st and after the first mowing on August 12th, in which the vegetation reached approximately half of the mean height of May (Fig. 2). In 2015, the plant canopy mean height peaked on May 19th and reached a plateau of higher canopy height in the later season starting on July 7th (Fig. 3). In 2015, the mean height was overall lower than in 2014, and this effect was most pronounced after the first mowing (Figs 2A and 3A). Model predictions per time slice of effects of diversity variables on observed mean canopy height presented best fit for the year 2014 in both pools, compared to 2015 (Fig. S2 in Supplementary Material in the online version at https://doi.org/10.1016/bs.aecr.2019.06.003). For dates during the summer, model predictions could explain up to 79% on July 28th in 2014 (Fig. S2A in Supplementary Material in the online version at https://doi.org/10.1016/bs.aecr.2019.06.003) and 70% on July 7th in 2015 (Fig. S2B in Supplementary Material in the online version at

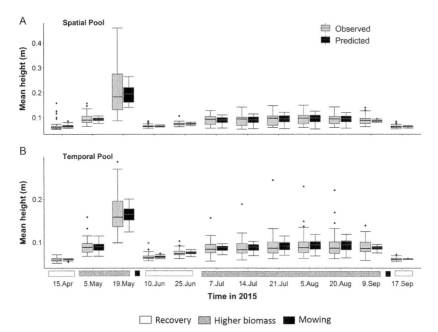

**Fig. 3** Boxplot of mean canopy height obtained from terrestrial laser scanning of plant communities in 2015 and height predicted by the best mixed-effects model accounting for diversity effects on plant development over the growing season. Note that the functional identity used in this model was calculated based on the community means of target species. Boxplots show 95% prediction interval for the mean of predicted values based on best model predictors and 95% confidence interval of observed values at each time slot. Data from (A) Spatial and (B) Temporal pools in 2015.

**Table 1** Analysis of variance for 2014 with the fixed-effects part of the mixed-effects models considering the effects of drivers of diversity on mean canopy height over the growing season.

| | Spatial pool | | | | Temporal pool | | | |
|---|---|---|---|---|---|---|---|---|
| | numDF | denDF | F-test | P value | | numDF | denDF | F-test | P value |
| Time | 12 | 481 | 890.11 | **<0.0001** | Time | 12 | 437 | 1790.9 | **<0.0001** |
| FDis x time | 12 | 481 | 3.10 | **0.0003** | FDis x time | 12 | 437 | 2.94 | **0.0006** |
| PCA1x time | 12 | 481 | 18.93 | **<0.0001** | PCA2x time | 12 | 437 | 12.25 | **<0.0001** |
| SR x time | 12 | 481 | 1.44 | 0.141 | SR x time | 12 | 437 | 1.71 | 0.060 |

Results are separated for spatial pool (different spatial resource use traits) and temporal pool (different phenological traits). Abbreviations: FDis, functional dispersion; Functional identity is represented by PCA axis 1 and 2 which represent the position of species in a trait space to each corresponding axes of the PCA design of Trait-Based Experiment; SR, species richness. Predictors with 'x' describe the interaction with time (as a factor). Restricted maximum likelihood estimated the variance components of the mixed effect models. Values in the columns are numerator degrees of freedom (numDF), denominator degrees of freedom (denDF), variance ratio (F-test), and P-values for Wald tests. Significant relationships at level 0.05 are in bold.

**Table 2** Analysis of variance for 2015 with the fixed-effects part of the mixed-effects models considering the effects of drivers of diversity on mean canopy height over the growing season.

|  | Spatial pool | | | | Temporal pool | | | |
| --- | --- | --- | --- | --- | --- | --- | --- | --- |
|  | numDF | denDF | F-test | P value |  | numDF | denDF | F-test | P value |
| time | 12 | 481 | 1490.1 | **<0.0001** | time | 12 | 440 | 3068.94 | **<0.0001** |
| FDis x time | 12 | 481 | 4.57 | **<0.0001** | FDis x time | 12 | 440 | 1.09 | 0.360 |
| PCA1 x time | 12 | 481 | 8.27 | **<0.0001** | PCA2 x time | 12 | 440 | 9.70 | **<0.0001** |
| SR x time | 12 | 481 | 3.93 | **<0.0001** | SR x time | 12 | 440 | 0.68 | 0.768 |

Results are separated for spatial pool (different spatial resource use traits) and temporal pool (different phenological traits). Abbreviations and model parameters as described in Table 1.

https://doi.org/10.1016/bs.aecr.2019.06.003). Although the temporal pool presented a lower fit between predicted and observed data compared to the spatial pool in both years, it also showed a good fit in dates close to the second peak of biomass in 2014 (Fig. S2C in Supplementary Material in the online version at https://doi.org/10.1016/bs.aecr.2019.06.003) and reasonable fit during some dates in the recovery after first mowing period in 2015 (Fig. S2D in Supplementary Material in the online version at https://doi.org/10.1016/bs.aecr.2019.06.003).

### 3.2 Intra-annual diversity and identity effects on plant development

Our results show differences between the two species pools in the effects of functional dispersion, identity (PCA1 and 2) and species richness on mean plant height over the growing season (Tables 1 and 2). These differences between species pools were most pronounced in 2015. As a general pattern for the spatial pool, in both years FDis and, in particular, identity effects varied in strength and similarly in direction between individual dates as indicated by significant interactions with time (Figs 4 and 5). Effects of species richness (hereafter SR) varied less (significant interaction with time only in 2015—Tables 1 and 2) but were strongly positive shortly before the first mowing (Figs 4 and 5 see SR). In the temporal pool, FDis, although significant over time only in 2014, was not important at any specific date in both years. As in the spatial pool, the effects of identity—here concerning PCA2—varied strongly in strength and direction in both years, as also shown by a highly significant interaction with time. SR, in the temporal pool, was

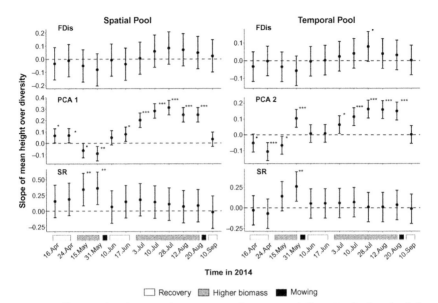

**Fig. 4** Coefficient plot showing the variation of estimates (effect sizes) of each of the diversity drivers and their contrast to zero with 95% of confidence interval during the growing season based on the best mixed-effects in 2014. The plot shows Spatial and Temporal pools data separately. Abbreviations: FDis = Functional Dispersion, PCA1 and 2 = principal component axes 1 and 2, SR = species richness).

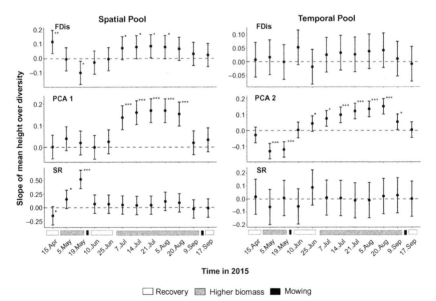

**Fig. 5** Coefficient plot showing the variation of estimates (effect sizes) of each diversity drivers and their contrast to zero with 95% of confidence interval during the growing season based on the best mixed-effects in 2015. The plot shows Spatial and Temporal pools data separately. For abbreviations see Fig. 4.

neither significant in 2014 nor 2015 but interacted significantly with time before first mowing in 2014 (Table 1 and Fig. 4).

To test our main hypothesis of changing mechanisms over time, as indicated by varying strength and direction of the effects of different facets of diversity over time, we extracted the slopes for individual dates from the mixed effects models (Figs 4 and 5 for 2014 and 2015, respectively). In the temporal pool, we found distinct intra-annual patterns of the identity (PCA2) effects in both years. Height was negatively and significantly related to PCA2 at the beginning of the growing season (24th and 15th in 2014 and May 5th and 19th in 2015), showing that mixtures with earlier species performed better in spring. For the rest of the growing season, especially during peak biomass before the second mowing (most of July and August in both years), the effect reversed, i.e., mixtures with late species performed better towards mid and late summer. Even though there was a significant interaction of FDis and time in 2014, FDis did not emerge as a significant predictor for single dates. SR, in contrast, increased mean height on one date during peak biomass before the first mowing in 2014 despite the marginal interaction of SR and time.

For the spatial pool, we found differences in diversity effects between the 2 years for individual dates. While in 2014 only identity (PCA1) and SR effects could be detected, in the drier year of 2015 all diversity drivers, FDis, PCA1, and SR emerged as significant predictors for individual dates (Figs 4 and 5). The outcome in 2014 was not directly compatible with the results of the anova analysis, where SR effects (interaction with time) were not significant (Table 1). In 2014, there was a reversal of the identity effect. Mixtures with taller species with large leaves (high mean PCA1 score) reached highest mean heights early in the growing season (April 16th and 24th) and during summer from July onwards (Figs 4 and 5), while the opposite was true for the time of peak biomass before the first mowing (May 15th and 31st, Fig. 4). In 2015, the signal of PCA1 was only positive in midsummer (starting on July 7th, Fig. 5). Moreover, still in 2015, species richness and FDis effects emerged as significant before and shortly after the first mowing (only for FDis), the signs of their effects were opposed, positive for SR and negative for FDis (Figs 4 and 5).

## 4. Discussion

We studied the effects of three facets of plant diversity on a proxy of plant aboveground biomass, namely, mean canopy height. For this, we employed the non-destructive method of laser scanning at a high spatial and temporal resolution over two growing seasons in a temperate grassland. We took advantage of a biodiversity experiment with varying plant functional trait composition to investigate how the different facets of diversity influence productivity in mixtures, which represent functional trait gradients related to spatial (spatial pool) and temporal resource use (temporal pool). We expected strong intra-annual variability in the relative importance of biodiversity-ecosystem functioning (BEF) mechanisms reflected by temporal changes in the importance of different diversity facets for predicting productivity (Hypothesis 1). We hypothesized the primary importance of functional dispersion (FDis) during the peak of biomass (Hypothesis 2) and of functional identity during initial growth periods (Hypothesis 3).

The results support our first hypothesis (Hypothesis 1). The mechanisms underlying the diversity effects on productivity, i.e., resource partitioning (via FDis) and identity effect (via community species scores means), respectively, varied intra- and inter-annually. The two biomass mowings carried out in the Jena Experiment each integrate over growth dynamics of 3–9 months if one considered the slow growth in the winter time. Even within these comparatively short periods, the type and strength of dominating mechanisms changed markedly. Most prominently, before the first mowing, April (early growth) and May (towards peak biomass) differed in the underlying drivers of productivity. Species richness effects could only be detected in May in both years, and in 2014 the direction of the identity effects reversed between April and May in both pools. After the first mowing, identity effects were generally weak in June and became only strong towards mid-July (exception for the temporal pool in 2015). It seems that mowing the biomass at peak biomass, as done in most grassland experiments, may miss important variation in resource partitioning and identity effects as mechanisms driving biomass production.

## 4.1 Intra-annual changes in functional diversity effects on plant development

Our Hypothesis 2 stated that, during periods of peak biomass, resource scarcity strengthens the importance of functional diversity (FDis) in controlling productivity, because plants will be driven towards resource partitioning (Allan et al., 2011; Roscher et al., 2011, 2016; Tilman, 1997). In contrast, we did not expect resource partitioning to play a significant role during the recovery phases after winter or mowing, when above- and belowground resources are abundant. Despite the significant interaction between functional dispersion and time in both plant species pools and years for the spatial pool, we only found one date where functional diversity affected mean height in the expected positive direction (April 15th, 2015 in the spatial pool).

Yet, FDis exhibited a negative effect for one date in parallel with a positive effect of species richness. In 2014, we did not find a single date with a significant effect of FDis, and we can only speculate about the cause of these counterintuitive results. It is conceivable that the significant interaction with time in both pools in 2014 and the spatial pool in 2015 has a sustained influence over the entire growing season, but is never strong enough to emerge as significant when focusing on single dates. This is true at least for the spatial pool, where species with deep roots and tall stature suited to capture space (tall herbs) are mixed with species that can densely fill space above- and belowground, i.e., mostly grasses with high specific root length and thin leaves (Hooper, 1998).

The negative effect of FDis on biomass productivity in 2015 may arise as an artefact created by collinearity with species richness (the correlation between FDis and SR was 0.64 for spatial pool and 0.51 for the temporal pool in 2015) (Venail et al., 2015) or because species richness takes over the representation of the complementarity signal. This may happen when the traits underlying FDis do not sufficiently represent the true mechanism underlying the resource partitioning. For instance, we may have missed essential root traits for nutrient uptake from the soil, as these may be more important for plant performance than leaf traits (Schroeder-Georgi et al., 2016). FDis in the temporal pool did not emerge as a significant predictor at single dates; it was only marginally significant on July 28th (again, despite a significant interaction of FDis with time in 2014) (Table 1). This is less surprising as one would expect that temporal resource partitioning arising from mixing species with different phenologies would not be detectable at individual dates but rather requires performance data integrated over longer time periods (Cardinale et al., 2007b; Kahmen et al., 2006; Ravenek et al., 2014; Tilman et al., 2001).

## 4.2 Intra-annual changes in identity effects on plant development

We found that identity effects dominated at the recovery after winter in the year 2014 for the spatial pool and recovery after mowing, as well as in both years and species pools when recovery after winter or mowing may require particular characteristics of plant species. These findings support our Hypothesis 3, which stated that identity effects would emerge during the recovery after winter and mowing. It was not or only weakly supported for the time after the first mowing. As predicted in Hypothesis 3, functional identity effects were strongest in the temporal pool (PCA2), representing the phenology gradient from early (negative values) to late species (positive values). Our results revealed that in both years mixtures dominated by early species had higher productivity in early spring (negative PCA2 effect on biomass), whereas mixtures with late species had an advantage in mid to late summer (strong positive PCA2 effect later in the year). Examples for plant species driving these dynamics are the early grass *Anthoxanthum odoratum*, and in the late summer, the grass *Holcus lanatus* (Dolezal et al., 2019). It is important to note here that the identity effect of PCA2 involved a change in the sign of the coefficient within the growing season. Thus, when analysed and modelled with a lower temporal resolution—as usually done in these experiments—the opposing effects of PCA2 could cancel out each other, potentially leading to the false conclusion that identity effects are not relevant.

In addition to the expected identity effect in the temporal pool, we found that identity in the spatial pool (PCA1) did not support our expectations. We would have expected that mixtures dominated by taller species grow taller irrespective of the season (i.e. the significant interaction between time and PCA1—see Tables 1 and 2). However, spatial pool (PCA1) was only a significant and positive predictor of mean height mainly towards the end of the growing season in 2015. We explain this by the fact that grasses that tend to have negative values on PCA1 (lower maximum height and dense, thin roots) dominate the early growing seasons, while taller herbs with positive values on PCA1 become more dominant and may leave their imprint on productivity only during the summer period (Martínková et al., 2002).

## 4.3 Species richness effects and inter-annual differences

We found that (SR) was generally less important as a predictor of mean height compared to the other two diversity facets (Figs 4 and 5). It was a significant predictor only in the spatial pool in 2015 (Table 2). Yet, we found three periods during which SR was positively influencing mean height, in these cases during peak biomass before the first mowing (spatial and temporal pools, 2014: Fig. 4; and spatial pool, 2015: Fig. 5). This points to a potential role for SR during crowded conditions. Given that the effect of FDis was not clearly pronounced during the highest biomass periods (or exhibits an opposite sign), we conclude that diversity mechanisms other than spatial or temporal resource partitioning—such as facilitation or pathogen dilution—might play a role for plant growth (de Kroon et al., 2012; Eisenhauer, 2012; Wright et al., 2017b). Alternatively, functional traits other than the ones considered in this study may drive resource partitioning, e.g., traits reflecting different forms of nitrogen preferences, which might change with diversity (Dolezal et al., 2019; Sauheitl et al., 2010). Moreover, we consider it likely that species richness may capture diversity mechanisms related to plant-soil feedback effects in addition to, or rather than, resource partitioning. This is corroborated by an increasing number of studies relating the shape and strength of the biodiversity-ecosystem functioning relationship to soil biotic feedback effects (Eisenhauer et al., 2012; de Kroon et al., 2012; Guerrero-Ramírez and Eisenhauer, 2017; Vogel et al., 2019, in this issue). There is growing evidence that the deterioration of plant communities at low diversity is related to the accumulation of specific plant antagonists generating negative feedback effects (de Kroon et al., 2012; Hendriks et al., 2013; Kulmatiski et al., 2012; Maron et al., 2011; Schnitzer et al., 2011). On the other hand, positive feedback effects, (e.g. by mycorrhizal fungi, biocontrol bacteria) might increase plant performance at high diversity (Eisenhauer, 2012; Latz et al., 2012; Wagg et al., 2011).

Our results indicate that the importance of SR effects during the highest biomass period might have been more pronounced during 2014 as compared to 2015 (three significant individual dates in 2014, whereas in 2015 only one). This might be related to inter-annual climatic differences between these 2 years. The year 2015 was characterized by particularly low productivity compared to 2014 (Figs 2 and 3). This was partly due to lower than average precipitation and high daily mean air temperatures ($>20\,°C$) during July and August, which caused a sustained period of summer drought (Fig. S3 in Supplementary Material in the online version at https://doi.org/10.1016/bs.aecr.2019.06.003). It is possible that the slight influence of SR in 2015 is a consequence of the more stressful conditions, favouring the emergence of facilitation according to the stress gradient hypothesis (Greenlee and Callaway, 1996; Pugnaire et al., 1996; Wagg et al., 2017a,b; Wright et al., 2017b).

## 4.4 A new method for biodiversity-ecosystem functioning research in grasslands

Terrestrial laser scanning (TLS) allowed us to follow the development of swards throughout the growing season at a high temporal resolution. The 3D point cloud produced by the TLS represented the canopy surface with a spatial resolution below 2 mm and was well suited to describe subtle changes in the canopy during the growing season. Much of the theory we are invoking (e.g. resource partitioning) refers to biomass production. The mean canopy height derived from TLS measurements by pooling years together explained 54% of dry mowed community biomass (Fig. S1 in Supplementary Material in the online version at https://doi.org/10.1016/bs.aecr.2019.06.003). It is important to note that we could only calibrate against high values of biomass corresponding to peak biomass at the time of the mowing (two times in the year) and that the biomass samples cover a considerable lower area than those studied with the TLS. This narrowed the range of values towards high values only and thus restricted our calibration precision. Far higher values of $R^2$ may have been obtained if we also had included biomass measurements during low biomass periods.

A major shortcoming of the method is that the TLS is not able to differentiate between species. It is therefore not possible to quantify relative abundances based on TLS data (at least for grassland communities), which would be important for differentiating between the contributions of different species to overyielding in mixtures (Loreau and Hector, 2001). Also, a visual inspection of the RGB pictures taken alongside the scanning does not yield reliable species abundance data. We, therefore, base our analysis on presence-absence data, although the trait-based indices could, in principle, be abundance-weighted. Thus, we can only assess the functional *potential* of our community to express diversity or identity control. It should be noted that basing the calculation of FDis and functional identity on presence-absence is equivalent to assuming perfect evenness and will yield results that maximize the range of FDis and minimize it for functional identity. This implies that identity effects may occur despite an intrinsic bias against identity effects. For future applications, we thus recommend recording species cover values based on visual inspection alongside the TLS measurements. Temporal constraints during fieldwork and image processing prevented the scanning of individual plots from different perspectives. The one perspective scan of 92 plots every second week provided a good overview of sward development and a feasible data storage potential. However, we understand that other scan perspectives could have considerably increased point information for the same position and decreased occlusion.

## 5. Conclusions

Our approach of estimating ecosystem function using terrestrial laser scanning at a high temporal resolution is a promising method for repeated, non-destructive biomass estimation. The present results place a caveat on the interpretation of previous studies measuring biomass production once or twice a year at estimated peak biomass (Cadotte, 2017; Cardinale et al., 2012; Eisenhauer et al., 2011). We believe that an assessment of changes in community composition alongside the TLS measurements may reveal even stronger seasonal changes in diversity control as detected in this study, further emphasizing the temporal dynamics of resource partitioning, identity effects, and species richness. Our work may stimulate the evaluation of inferences drawn from measurements conducted at a coarser temporal resolution and to reveal how the interplay between different biodiversity-ecosystem functioning mechanisms unfolds over time and responds to climate variation and disturbances. Methods of automated identification of species or functional groups applying, e.g., multi- or hyperspectral images along with laser scanning may greatly enhance the potential of the approach presented here.

## Acknowledgements

Special thanks to Dr Shaun Levick for introducing LAStools and Dr David Eichenberg for inputs regarding statistics. All the coordinators, the gardening team from the Jena Experiment and many field assistants for maintaining the field, and student helpers Nataliia Kodash and Arthur Ferrari for helping on scanning days. The Jena Experiment is funded by the German Research Foundation (DFG, FOR 1451). Further support came from the German Centre for Integrative Biodiversity Research (iDiv) Halle-Jena-Leipzig, funded by the German Research Foundation (FZT 118).

## References

Aarssen, L.W., 1997. High productivity in grassland ecosystems: effected by species diversity or productive species? Oikos 80, 183–184.

Allan, E., Weisser, W., Weigelt, A., Roscher, C., Fischer, M., Hillebrand, H., 2011. More diverse plant communities have higher functioning over time due to turnover in complementary dominant species. Proc. Natl. Acad. Sci. U.S.A. 108, 17034–17039.

Anderson, M.J., 2006. Distance-based tests for homogeneity of multivariate dispersions. Biometrics 62, 245–253.

Bachmann, D., Gockele, A., Ravenek, J.M., Roscher, C., Strecker, T., Weigelt, A., Buchmann, N., 2015. No evidence of complementary water use along a plant species richness gradient in temperate experimental grasslands. PLoS One 10, e0116367.

Bachmann, D., Roscher, C., Buchmann, N., 2018. How do leaf trait values change spatially and temporally with light availability in a grassland diversity experiment? Oikos 127 (7), 935–948.

Bunker, D.E., 2005. Species loss and aboveground carbon storage in a tropical Forest. Science 310, 1029–1031.

Cadotte, M.W., 2017. Functional traits explain ecosystem function through opposing mechanisms. Ecol. Lett. 20, 989–996.
Callaway, R.M., Brooker, R.W., Choler, P., Kikvidze, Z., Lortie, C.J., Michalet, R., Paolini, L., Pugnaire, F.I., Newingham, B., Aschehoug, E.T., Armas, C., Kikodze, D., Cook, B.J., 2002. Positive interactions among alpine plants increase with stress. Nature 417, 844.
Campbell, B.D., Grime, J.P., Mackey, J.M.L., 1991. A trade-off between scale and precision in resource foraging. Oecologia 87, 532–538.
Cardinale, B.J., Wright, J.P., Cadotte, M.W., Carroll, I.T., Hector, A., Srivastava, D.S., Loreau, M., Weis, J.J., 2007a. Impacts of plant diversity on biomass production increase through time because of species complementarity. Proc. Natl. Acad. Sci. U. S. A. 104, 18123–18128.
Cardinale, B.J., Wright, J.P., Cadotte, M.W., Carroll, I.T., Hector, A., Srivastava, D.S., Loreau, M., Weis, J.J., 2007b. Impacts of plant diversity on biomass production increase through time because of species complementarity. Proc. Natl. Acad. Sci. U. S. A. 104, 18123–18128.
Cardinale, B.J., Duffy, J.E., Gonzalez, A., Hooper, D.U., Perrings, C., Venail, P., Narwani, A., Mace, G.M., Tilman, D., Wardle, D.A., 2012. Biodiversity loss and its impact on humanity. Nature 486, 59–67.
Chi, E.M., Reinsel, G.C., 1989. Models for longitudinal data with random effects and AR(1) errors. J. Am. Stat. Assoc. 84, 452–459.
Cingolani, A.M., Cabido, M., Gurvich, D.E., Renison, D., Díaz, S., 2007. Filtering processes in the assembly of plant communities: are species presence and abundance driven by the same traits? J. Veg. Sci. 18, 911–920.
Civitello, D.J., Cohen, J., Fatima, H., Halstead, N.T., Liriano, J., McMahon, T.A., Ortega, C.N., Sauer, E.L., Sehgal, T., Young, S., Rohr, J.R., 2015. Biodiversity inhibits parasites: broad evidence for the dilution effect. Proc. Natl. Acad. Sci. U. S. A. 112, 8667–8671.
Craine, J.M., Nippert, J.B., Elmore, A.J., Skibbe, A.M., Hutchinson, S.L., Brunsell, N.A., 2012. Timing of climate variability and grassland productivity. Proc. Natl. Acad. Sci. U. S. A. 109, 3401–3405.
Craven, D., Isbell, F., Manning, P., Connolly, J., Bruelheide, H., Ebeling, A., Roscher, C., van Ruijven, J., Weigelt, A., Wilsey, B., Beierkuhnlein, C., de Luca, E., Griffin, J.N., Hautier, Y., Hector, A., Jentsch, A., Kreyling, J., Lanta, V., Loreau, M., Meyer, S.T., Mori, A.S., Naeem, S., Palmborg, C., Polley, H.W., Reich, P.B., Schmid, B., Siebenkäs, A., Seabloom, E., Thakur, M.P., Tilman, D., Vogel, A., Eisenhauer, N., 2016. Plant diversity effects on grassland productivity are robust to both nutrient enrichment and drought. Philos. Trans. R. Soc. B 371, 20150277.
Daßler, A., Roscher, C., Temperton, V.M., Schumacher, J., Schulze, E.-D., 2008. Adaptive survival mechanisms and growth limitations of small-stature herb species across a plant diversity gradient. Plant Biol. 10, 573–587.
De Boeck, H.J., Dreesen, F.E., Janssens, I.A., Nijs, I., 2011. Whole-system responses of experimental plant communities to climate extremes imposed in different seasons. New Phytol. 189, 806–817.
de Kroon, H., Hendriks, M., van Ruijven, J., Ravenek, J., Padilla, F.M., Jongejans, E., Visser, E.J.W., Mommer, L., 2012. Root responses to nutrients and soil biota: drivers of species coexistence and ecosystem productivity: root responses and ecosystem productivity. J. Ecol. 100, 6–15.
Dolezal, J., Lanta, V., Mudrák, O., Lepš, J., 2019. Seasonality promotes grassland diversity: interactions with mowing, fertilization and removal of dominant species. J. Ecol. 107, 203–215.

Ebeling, A., Pompe, S., Baade, J., Eisenhauer, N., Hillebrand, H., Proulx, R., Roscher, C., Schmid, B., Wirth, C., Weisser, W.W., 2014. A trait-based experimental approach to understand the mechanisms underlying biodiversity–ecosystem functioning relationships. Basic Appl. Ecol. 15, 229–240.

Eisenhauer, N., 2012. Aboveground–belowground interactions as a source of complementarity effects in biodiversity experiments. Plant and Soil 351, 1–22.

Eisenhauer, N., Milcu, A., Sabais, A.C.W., Scheu, S., 2009. Earthworms enhance plant regrowth in a grassland plant diversity gradient. Eur. J. Soil Biol. 45, 455–458.

Eisenhauer, N., Bohan, D.A., Dumbrell, A.J., 2019. Mechanistic links between biodiversity and ecosystem functioning. Adv. Ecol. Res. 61, xix–xxviii.

Eisenhauer, N., Milcu, A., Sabais, A.C., Bessler, H., Brenner, J., Engels, C., Klarner, B., Maraun, M., Partsch, S., Roscher, C., 2011. Plant diversity surpasses plant functional groups and plant productivity as driver of soil biota in the long term. PLoS One 6, e16055.

Eisenhauer, N., Reich, P.B., Scheu, S., 2012. Increasing plant diversity effects on productivity with time due to delayed soil biota effects on plants. Basic Appl. Ecol. 13, 571–578.

Everwand, G., Fry, E.L., Eggers, T., Manning, P., 2014. Seasonal variation in the capacity for plant trait measures to predict grassland carbon and water fluxes. Ecosystems 17, 1095–1108.

Eviner, V.T., Chapin III, F.S., Vaughn, C.E., 2006. Seasonal variation in plant species effects on soil N and P dynamics. Ecology 87, 974–986.

Fargione, J., Tilman, D., 2006. Plant species traits and capacity for resource reduction predict yield and abundance under competition in nitrogen-limited grassland. Funct. Ecol. 20, 533–540.

FARO Technologies Inc. 2011. 250 Technology Park Lake Mary, FL 32746.

Fonseca, C.R., Overton, J.M.C., Collins, B., Westoby, M., 2000. Shifts in trait-combinations along rainfall and phosphorus gradients. J. Ecol. 88, 964–977.

Fox, J.W., 2005. Interpreting the 'selection effect' of biodiversity on ecosystem function. Ecol. Lett. 8, 846–856.

Gibson, D.J., 2009. Grasses and Grassland Ecology. University Press, Oxford.

Greenlee, J.T., Callaway, R.M., 1996. Abiotic stress and the relative importance of interference and facilitation in montane bunchgrass communities in Western Montana. Am. Nat. 148, 386–396.

Grime, J.P., 1998. Benefits of plant diversity to ecosystems: immediate, filter and founder effects. J. Ecol. 86, 902–910.

Guerrero-Ramírez, N.R., Eisenhauer, N., 2017. Trophic and non-trophic interactions influence the mechanisms underlying biodiversity-ecosystem functioning relationships under different abiotic conditions. Oikos 126 (12), 1748–1759.

Hendriks, M., Mommer, L., de Caluwe, H., Smit-Tiekstra, A.E., van der Putten, W.H., de Kroon, H., 2013. Independent variations of plant and soil mixtures reveal soil feedback effects on plant community overyielding. J. Ecol. 101, 287–297.

Hooper, D.U., 1998. The role of complementarity and competition in ecosystem responses to variation in plant diversity. Ecology 79, 704–719.

Hooper, D.U., Chapin, F.S., Ewel, J.J., Hector, A., Inchausti, P., Lavorel, S., Lawton, J.H., Lodge, D.M., Loreau, M., Naeem, S., et al., 2005. Effects of biodiversity on ecosystem functioning: a consensus of current knowledge. Ecol. Monogr. 75, 3–35.

Humbert, J.-Y., Pellet, J., Buri, P., Arlettaz, R., 2012. Does delaying the first mowing date benefit biodiversity in meadowland? Environ. Evid. 1, 1.

Huston, M.A., 1997. Hidden treatments in ecological experiments: re-evaluating the ecosystem function of biodiversity. Oecologia 110, 449–460.

Isbell, F., Craven, D., Connolly, J., Loreau, M., Schmid, B., Beierkuhnlein, C., Bezemer, T.M., Bonin, C., Bruelheide, H., de Luca, E., Ebeling, A., Griffin, J.N., Guo, Q., Hautier, Y., Hector, A., Jentsch, A., Kreyling, J., Lanta, V., Manning, P.,

Meyer, S.T., Mori, A.S., Naeem, S., Niklaus, P.A., Polley, H.W., Reich, P.B., Roscher, C., Seabloom, E.W., Smith, M.D., Thakur, M.P., Tilman, D., Tracy, B.F., van der Putten, W.H., van Ruijven, J., Weigelt, A., Weisser, W.W., Wilsey, B., Eisenhauer, N., 2015. Biodiversity increases the resistance of ecosystem productivity to climate extremes. Nature 526, 574–577.

Isenburg, M., 2014. LAStools. Retrieved from http://lastools.org. Version 190604, Academic.

Jesch, A., Barry, K.E., Ravenek, J.M., Bachmann, D., Strecker, T., Weigelt, A., Buchmann, N., de Kroon, H., Gessler, A., Mommer, L., Roscher, C., Scherer-Lorenzen, M., 2018. Below-ground resource partitioning alone cannot explain the biodiversity-ecosystem function relationship: a field test using multiple tracers. J. Ecol. 106, 2002–2018.

Jongen, M., Pereira, J.S., Aires, L.M.I., Pio, C.A., 2011. The effects of drought and timing of precipitation on the inter-annual variation in ecosystem-atmosphere exchange in a Mediterranean grassland. Agric. For. Meteorol. 151, 595–606.

Kahmen, A., Renker, C., Unsicker, S.B., Buchmann, N., 2006. Niche complementarity for nitrogen: an explanation for the biodiversity and ecosystem functioning relationship?? Ecology 87, 1244–1255.

Kulmatiski, A., Beard, K.H., Heavilin, J., 2012. Plant-soil feedbacks provide an additional explanation for diversity-productivity relationships. Proc. R. Soc. B Biol. Sci. 279, 3020–3026.

Laliberté, E., Legendre, P., 2010. A distance-based framework for measuring functional diversity from multiple traits. Ecology 91, 299–305.

Latz, E., Eisenhauer, N., Rall, B.C., Allan, E., Roscher, C., Scheu, S., Jousset, A., 2012. Plant diversity improves protection against soil-borne pathogens by fostering antagonistic bacterial communities: Plant diversity improves protection against soil-borne pathogens. J. Ecol. 100, 597–604.

Loreau, M., Hector, A., 2001. Partitioning selection and complementarity in biodiversity experiments. Nature 412, 72–76.

Loreau, M., Naeem, S., Inchausti, P. (Eds.), 2002. Biodiversity and Ecosystem Functioning: Synthesis and Perspectives. University Press, Oxford.

Lorentzen, S., Roscher, C., Schumacher, J., Schulze, E.-D., Schmid, B., 2008. Species richness and identity affect the use of aboveground space in experimental grasslands. Perspect. Plant Ecol. Evol. Syst. 10, 73–87.

Maalouf, J.-P., Le Bagousse-Pinguet, Y., Marchand, L., Touzard, B., Michalet, R., 2012. The interplay of stress and mowing disturbance for the intensity and importance of plant interactions in dry calcareous grasslands. Ann. Bot. 110, 821–828.

Mamolos, A.P., Vasilikos, C.V., Veresoglou, D.S., 2011. Temporal patterns of growth and nutrient accumulation of plant species in a Mediterranean mountainous grassland. Ecol. Res. 26, 583–593.

Maron, J.L., Marler, M., Klironomos, J.N., Cleveland, C.C., 2011. Soil fungal pathogens and the relationship between plant diversity and productivity: soil pathogens, productivity and invasibility. Ecol. Lett. 14, 36–41.

Marquard, E., Weigelt, A., Temperton, V.M., Roscher, C., Schumacher, J., Buchmann, N., Fischer, M., Weisser, W.W., Schmid, B., 2009. Plant species richness and functional composition drive overyielding in a six-year grassland experiment. Ecology 90, 3290–3302.

Martínková, J., Smilauer, P., Mihulka, S., 2002. Phenological pattern of grassland species: relation to the ecological and morphological traits. Flora Morphol. Distrib. Funct. Ecol. Plants 197, 290–302.

McGaughey, R., 2011. FUSION/LDV: Software for LiDAR Data Analysis and Visualization. U.S. Department of Agriculture, Forest Service, Pacific Northwest Research Station, University of Washington, Seattle, WA.

Michalet, R., Brooker, R.W., Cavieres, L.A., Kikvidze, Z., Lortie, C.J., Pugnaire, F.I., Valiente-Banuet, A., Callaway, R.M., 2006. Do biotic interactions shape both sides of the humped-back model of species richness in plant communities? Ecol. Lett. 9, 767–773.

Millennium Ecosystem Assessments, 2005, *Ecosystems and Human Well-being: Biodiversity Synthesis*, World Resource Institute.

Naeem, S., Duffy, J.E., Zavaleta, E.S., 2012. The functions of biological diversity in an age of extinction. Science 336, 1401–1406.

O'Brien, R.M., 2007. A caution regarding rules of thumb for variance inflation factors. Qual. Quant. 41, 673–690.

Oram, N.J., Ravenek, J.M., Barry, K.E., Weigelt, A., Chen, H., Gessler, A., Gockele, A., de Kroon, H., van der Paauw, J.W., Scherer-Lorenzen, M., Smit-Tiekstra, A., van Ruijven, J., Mommer, L., 2018. Below-ground complementarity effects in a grassland biodiversity experiment are related to deep-rooting species. J. Ecol. 106, 265–277.

Pinheiro, J.C., Bates, D.M., 2000. Mixed-Effects Models in Sand S-PLUS. Springer New York, New York, NY.

Pugnaire, F.I., Haase, P., Puigdefabregas, J., 1996. Facilitation between higher plant species in a semiarid environment. Ecology 77, 1420–1426.

R Core Team, 2018. R: A Language and Environment for Statistical Computing. R Foundation for Statistical Computing, Vienna, Austria.

Ravenek, J.M., Bessler, H., Engels, C., Scherer-Lorenzen, M., Gessler, A., Gockele, A., De Luca, E., Temperton, V.M., Ebeling, A., Roscher, C., 2014. Long-term study of root biomass in a biodiversity experiment reveals shifts in diversity effects over time. Oikos 123, 1528–1536.

Regan, K.M., Nunan, N., Boeddinghaus, R.S., Baumgartner, V., Berner, D., Boch, S., Oelmann, Y., Overmann, J., Prati, D., Schloter, M., Schmitt, B., Sorkau, E., Steffens, M., Kandeler, E., Marhan, S., 2014. Seasonal controls on grassland microbial biogeography: are they governed by plants, abiotic properties or both? Soil Biol. Biochem. 71, 21–30.

Reich, P.B., Tilman, D., Isbell, F., Mueller, K., Hobbie, S.E., Flynn, D.F., Eisenhauer, N., 2012. Impacts of biodiversity loss escalate through time as redundancy fades. Science 336, 589–592.

Robinson, T.M.P., La Pierre, K.J., Vadeboncoeur, M.A., Byrne, K.M., Thomey, M.L., Colby, S.E., 2013. Seasonal, not annual precipitation drives community productivity across ecosystems. Oikos 122, 727–738.

Roscher, C., Schumacher, J., Baade, J., Wilcke, W., Gleixner, G., Weisser, W.W., Schmid, B., Schulze, E.-D., 2004. The role of biodiversity for element cycling and trophic interactions: an experimental approach in a grassland community. Basic Appl. Ecol. 5, 107–121.

Roscher, C., Kutsch, W.L., Kolle, O., Ziegler, W., Schulze, E.-D., 2011. Adjustment to the light environment in small-statured forbs as a strategy for complementary resource use in mixtures of grassland species. Ann. Bot. 107, 965–979.

Roscher, C., Schumacher, J., Gubsch, M., Lipowsky, A., Weigelt, A., Buchmann, N., Schmid, B., Schulze, E.-D., 2012. Using plant functional traits to explain diversity–productivity relationships. PLoS One 7, e36760.

Roscher, C., Schmid, B., Kolle, O., Schulze, E.-D., 2016. Complementarity among four highly productive grassland species depends on resource availability. Oecologia 181, 571–582.

Sauheitl, L., Glaser, B., Dippold, M., Leiber, K., Weigelt, A., 2010. Amino acid fingerprint of a grassland soil reflects changes in plant species richness. Plant and Soil 334, 353–363.

Schippers, P., Joenje, W., 2002. Modelling the effect of fertiliser, mowing, disturbance and width on the biodiversity of plant communities of field boundaries. Agr Ecosyst Environ 93, 351–365.

Schnitzer, S.A., Klironomos, J.N., HilleRisLambers, J., Kinkel, L.L., Reich, P.B., Xiao, K., Rillig, M.C., Sikes, B.A., Callaway, R.M., Mangan, S.A., van Nes, E.H., Scheffer, M., 2011. Soil microbes drive the classic plant diversity–productivity pattern. Ecology 92, 296–303.

Schroeder-Georgi, T., Wirth, C., Nadrowski, K., Meyer, S.T., Mommer, L., Weigelt, A., 2016. From pots to plots: hierarchical trait-based prediction of plant performance in a Mesic grassland. J. Ecol. 104, 206–218.

Tilman, D., 1997. The influence of functional diversity and composition on ecosystem processes. Science 277, 1300–1302.

Tilman, D., Lehman, C., Thompson, K., 1997. Plant diversity and ecosystem productivity: theoretical considerations. Proc. Natl. Acad. Sci. U. S. A. 94, 1857–1861.

Tilman, D., Reich, P.B., Knops, J., Wedin, D., Mielke, T., Lehman, C., 2001. Diversity and productivity in a long-term grassland experiment. Science 294, 843–845.

Tilman, D., Isbell, F., Cowles, J.M., 2014. Biodiversity and ecosystem functioning. Annu. Rev. Ecol. Evol. Syst. 45, 471–493.

Tobner, C.M., Paquette, A., Gravel, D., Reich, P.B., Williams, L.J., Messier, C., 2016. Functional identity is the main driver of diversity effects in young tree communities. Ecol. Lett. 19, 638–647.

van Ruijven, J., Berendse, F., 2005. Diversity–productivity relationships: initial effects, long-term patterns, and underlying mechanisms. Proc. Natl. Acad. Sci. U. S. A. 102, 695–700.

Venail, P., Gross, K., Oakley, T.H., Narwani, A., Allan, E., Flombaum, P., Isbell, F., Joshi, J., Reich, P.B., Tilman, D., van Ruijven, J., Cardinale, B.J., 2015. Species richness, but not phylogenetic diversity, influences community biomass production and temporal stability in a re-examination of 16 grassland biodiversity studies. Funct. Ecol. 29, 615–626.

Vogel, A., Ebeling, A., Gleixner, G., Roscher, C., Scheu, S., Ciobanu, M., Koller-France, E., Lange, M., Lochner, A., Meyer, S.T., Oelmann, Y., Wilcke, W., Schmid, B., Eisenhauer, N., 2019. A new experimental approach to test why biodiversity effects strengthen as ecosystems age. Adv. Ecol. Res. 61, 221–264.

von Felten, S., Hector, A., Buchmann, N., Niklaus, P.A., Schmid, B., Scherer-Lorenzen, M., 2009. Belowground nitrogen partitioning in experimental grassland plant communities of varying species richness. Ecology 90, 1389–1399.

Wagg, C., Jansa, J., Schmid, B., van der Heijden, M.G.A., 2011. Belowground biodiversity effects of plant symbionts support aboveground productivity: biodiversity effects of soil symbionts. Ecol. Lett. 14, 1001–1009.

Wagg, C., Veiga, R., van der Heijden, M.G.A., 2015. Facilitation and antagonism in mycorrhizal networks. In: Horton, T.R. (Ed.), Mycorrhizal Networks. Springer Netherlands, Dordrecht, pp. 203–226.

Wagg, C., Ebeling, A., Roscher, C., Ravenek, J., Bachmann, D., Eisenhauer, N., Mommer, L., Buchmann, N., Hillebrand, H., Schmid, B., Weisser, W.W., 2017a. Functional trait dissimilarity drives both species complementarity and competitive disparity. Funct. Ecol. 31, 2320–2329.

Wagg, C., O'Brien, M.J., Vogel, A., Scherer-Lorenzen, M., Eisenhauer, N., Schmid, B., Weigelt, A., 2017b. Plant diversity maintains long-term ecosystem productivity under frequent drought by increasing short-term variation. Ecology 98, 2952–2961.

Weigelt, A., Marquard, E., Temperton, V.M., Roscher, C., Scherber, C., Mwangi, P.N., Felten, S., Buchmann, N., Schmid, B., Schulze, E.-D., Weisser, W.W., 2010. The Jena experiment: six years of data from a grassland biodiversity experiment: ecological archives E091-066. Ecology 91, 930–931.

Weisser, W.W., Roscher, C., Meyer, S.T., Ebeling, A., Luo, G., Allan, E., Beßler, H., Barnard, R.L., Buchmann, N., Buscot, F., Engels, C., Fischer, C., Fischer, M., Gessler, A., Gleixner, G., Halle, S., Hildebrandt, A., Hillebrand, H., de Kroon, H., Lange, M., Leimer, S., Le Roux, X., Milcu, A., Mommer, L., Niklaus, P.A.,

Oelmann, Y., Proulx, R., Roy, J., Scherber, C., Scherer-Lorenzen, M., Scheu, S., Tscharntke, T., Wachendorf, M., Wagg, C., Weigelt, A., Wilcke, W., Wirth, C., Schulze, E.-D., Schmid, B., Eisenhauer, N., 2017. Biodiversity effects on ecosystem functioning in a 15-year grassland experiment: patterns, mechanisms, and open questions. Basic Appl. Ecol. 23, 1–73.

Wright, A.J., Wardle, D.A., Callaway, R., Gaxiola, A., 2017a. The overlooked role of facilitation in biodiversity experiments. Trends Ecol. Evol. 32, 383–390.

Wright, A.J., Wardle, D.A., Callaway, R., Gaxiola, A., 2017b. The overlooked role of facilitation in biodiversity experiments. Trends Ecol. Evol. 32, 383–390.

Zeeman, M.J., Shupe, H., Baessler, C., Ruehr, N.K., 2019. Productivity and vegetation structure of three differently managed temperate grasslands. Agr Ecosyst Environ 270–271, 129–148.

Zuur, A.F., Ieno, E.N., Walker, N., Saveliev, A.A., Smith, G.M., 2009. Mixed Effects Models and Extensions in Ecology with R. Springer New York, New York, NY.

CHAPTER FIVE

# Plant functional trait identity and diversity effects on soil meso- and macrofauna in an experimental grassland

Rémy Beugnon[a,b,*], Katja Steinauer[a,b,c], Andrew D. Barnes[a,b,d], Anne Ebeling[e], Christiane Roscher[a,f], Nico Eisenhauer[a,b]

[a]German Centre for Integrative Biodiversity Research (iDiv) Halle-Jena-Leipzig, Leipzig, Germany
[b]Institute of Biology, Leipzig University, Leipzig, Germany
[c]Department of Terrestrial Ecology, Netherlands Institute of Ecology, Wageningen, The Netherlands
[d]School of Science, University of Waikato, Hamilton, New Zealand
[e]Institute of Ecology and Evolution, Friedrich Schiller University Jena, Jena, Germany
[f]UFZ, Helmholtz Centre for Environmental Research, Physiological Diversity, Leipzig, Germany
*Corresponding author: e-mail address: remy.beugnon@idiv.de

## Contents

| | |
|---|---|
| 1. Introduction | 164 |
| 2. Material and methods | 167 |
|    2.1 Experimental design | 167 |
|    2.2 Plant community indices | 169 |
|    2.3 Statistical analyses | 169 |
| 3. Results | 171 |
|    3.1 Plant species richness effects | 171 |
|    3.2 Trait-based models | 172 |
|    3.3 Comparison of plant species richness-based and trait-based models | 174 |
| 4. Discussion | 174 |
|    4.1 Plant species richness has a weak effect on soil communities | 176 |
|    4.2 Plant traits as more powerful predictors of soil fauna communities | 176 |
|    4.3 The importance of plant trait identity effects across soil fauna groups | 177 |
|    4.4 Soil fauna responses to spatial resource acquisition traits | 178 |
|    4.5 Soil fauna responses to temporal resource acquisition traits | 179 |
| 5. Conclusions | 179 |
| Acknowledgements | 180 |
| References | 180 |
| Further reading | 184 |

## Abstract

Understanding aboveground-belowground linkages and their consequences for ecosystem functioning is a major challenge in soil ecology. It is already well established that soil communities drive essential ecosystem processes, such as nutrient cycling, decomposition, or carbon storage. However, knowledge of how plant diversity affects belowground community structure is limited. Such knowledge can be gained from studying the main plant functional traits that modulate plant community effects on soil fauna. Here, we used a grassland experiment manipulating plant species richness and plant functional diversity to explore the effects of community-level plant traits on soil meso- and macrofauna and the trophic structure of soil fauna by differentiating predators and prey. The functional composition of plant communities was described by six plant traits related to spatial and temporal resource use: plant height, leaf area, rooting depth, root length density, growth start, and flowering start. Community-Weighted Means (CWMs), Functional Dissimilarity (FDis), and Functional Richness (FRic) were calculated for each trait. Community-level plant traits better explained variability in soil fauna than did plant species richness. Notably, each soil fauna group was affected by a unique set of plant traits. Moreover, the identity of plant traits (CWM) explained more variance of soil fauna groups than trait diversity. The abundances of soil fauna at the lower trophic levels were better explained by community-level plant traits than higher trophic levels soil fauna groups. Taken together, our results highlight the importance of the identity of different plant functional traits in driving the diversity and trophic structure of soil food communities.

## 1. Introduction

Over the past decades, ecologists have extensively studied aboveground-belowground linkages (Bardgett and Wardle, 2010) and their effects on ecosystem properties (Wardle et al., 2004). Plants have been identified as a major ecological link between these compartments (Grime, 2001; Wardle et al., 2004), by providing carbon resources and nutrients to both aboveground and belowground consumer communities (Bardgett and Wardle, 2010) that are connected through feeding relationships with plants (e.g. Johnson et al., 2012; McKenzie et al., 2013). Given these aboveground-belowground interactions, it is not surprising that aboveground and belowground diversity were reported to be positively linked (De Deyn and Van Der Putten, 2005; Scherber et al., 2010).

During the last years, it has been highlighted that changes in plant diversity can have significant consequences for the structure and functioning of above- and belowground consumer communities (Ebeling et al., 2018b; Eisenhauer et al., 2013; Giling et al., 2019; Haddad et al., 2009; Hertzog et al., 2017; Hines et al., 2019; Hooper et al., 2000; Meyer et al., 2017; Schuldt et al., 2019). For aboveground consumer communities, recent studies in experimental grasslands have shown that a loss of plant species causes a reduction in the functional richness and composition of herbivores and omnivores (Ebeling et al., 2018a), as well as a shift in food web structure (Giling et al., 2019). Although plant diversity effects on belowground consumer communities have received less attention, the few existing papers reported a positive relationship between plant diversity and soil microbial biomass (Eisenhauer et al., 2010a,b; Lange et al., 2019; Strecker et al., 2016) as well as between abundance and diversity of soil meso- and macrofauna (Eisenhauer et al., 2011a, 2013; Milcu et al., 2013; Scherber et al., 2010). Identifying the underlying mechanisms of these relationships are subject to current research (e.g. Eisenhauer et al., 2019; Mellado-Vázquez et al., 2016).

In the past decade, considerable progress has been made to identify the mechanisms behind plant diversity effects on ecosystem properties using plant functional traits (Diaz and Cabido, 2001; Flynn et al., 2011; Reich et al., 2012). The significance of plant traits for selected ecosystem functions is now well described (Lavorel and Garnier, 2002; Roscher et al., 2012). For instance, early studies found strong effects of leaf traits on net primary productivity (Violle et al., 2007), litter decomposability (Kazakou et al., 2006), and the species richness of aboveground arthropods (Symstad et al., 2000). More recently, research on soil community responses to plant traits has accelerated (e.g. Eisenhauer and Powell, 2017; Laliberté, 2016; Milcu et al., 2013; Steinauer et al., 2017). These studies have provided evidence that functionally and phylogenetically diverse plant communities enhance the density and diversity of soil fauna (Milcu et al., 2013), and that soil microbial communities and associated functions are mainly driven by plant traits related to spatial resource acquisition (Steinauer et al., 2017). These previous findings were related to effects of dominant plant traits on ecosystem properties. Such dominant plant effects were expressed in the 'biomass ratio hypothesis' by Grime (1998), which predicts that effects of specific plant functional traits on ecosystem properties (e.g. soil biota) should be largely determined by the species dominating the biomass of the plant community (Steinauer et al., 2017). Moreover, those first examples highlight the potential that plant traits have for improving our understanding of plant community effects on soil biota (Eisenhauer and Powell, 2017; Laliberté, 2016).

Despite these findings, understanding the effects of specific above- and belowground plant traits on soil communities remains limited. To address this gap in knowledge, we studied the importance of single plant functional traits related to spatial and temporal resource acquisition on soil meso- and macrofauna in the framework of the so-called Trait-Based Biodiversity Experiment (TBE; Ebeling et al., 2014). This experiment manipulates species richness and functional diversity of plant communities based on spatial- and temporal resource acquisition traits in a crossed factorial design (Ebeling et al., 2014). It is therefore suited to calculate trait diversity (here: Functional Richness and Functional Dispersion; Laliberté et al., 2014; Villéger et al., 2008), and the expression of single plant traits at the community level (here: Community Weighted Mean: CWM, Garnier et al., 2004; Lavorel et al., 2008; Fig. S1 in Supplementary Material in the online version at https://doi.org/10.1016/bs.aecr.2019.06.004). We sampled soil fauna communities (7796 individuals from 68 morphospecies) on 138 experimental plots, and we grouped individuals either based on their general size class (soil macrofauna and mesofauna), or feeding strategy (predator and prey). For each of these fauna groups, we quantified the abundance and species richness.

We tested the following hypotheses:

**Hypothesis 1.** Plant species richness and diversity of plant functional traits will positively affect the abundance and species richness of soil fauna communities. Our hypothesis is based on earlier findings of a positive relationship between plant species richness and soil fauna communities (Eisenhauer et al., 2011a; Milcu et al., 2013). Similarly, the diversity of plant traits might increase the diversity of microenvironments and/or resources (e.g. root traits; Hooper et al., 2000; Kuzyakov and Blagodatskaya, 2015; Postma and Lynch, 2012).

**Hypothesis 2.** A high community-level expression (i.e. high CWM) and diversity of plant traits related to high root productivity are particularly important for the abundance and diversity of soil fauna communities. This hypothesis is based on previous studies showing significant effects of root inputs (Eisenhauer et al., 2017; Kuzyakov and Blagodatskaya, 2015) and root length density on soil microorganisms (Steinauer et al., 2017), and the CWM root length density and rooting depth, that have been shown to be related to the quantity of carbon inputs to decomposers (Bardgett et al., 2014). Also, this hypothesis is in line with the concept of niche complementarity through different resource foraging strategies (Tilman, 1982, 1988), suggesting that the diversity of habitats and resources in soil (i.e. diversity of plant traits) will facilitate the coexistence of a high number of consumer species (Hooper et al., 2000)

**Hypothesis 3.** The community-level expression (CWM) and diversity of plant traits related to temporal resource acquisition have a strong effect on shaping soil fauna communities. For example, a plant community containing species differing in their peak growth (i.e. high phenological diversity) might provide resources to the belowground compartment evenly throughout the growing season, thereby supporting diverse soil fauna communities (Kuzyakov and Blagodatskaya, 2015), as well as, increasing community level expression (CWM) of temporal traits may delay species population dynamics in the season. By testing our Hypotheses 2 and 3, we thus sought to identify which community-level plant traits drive soil fauna community composition.

**Hypothesis 4.** The diversity and expression of certain traits within a plant community have strong effects on consumers at lower trophic levels (herbivores and decomposers), with attenuating bottom-up effects on higher trophic levels (Kaunzinger and Morin, 1998; Scherber et al., 2010). This is in line with a previously shown bottom-up effect of plant diversity on consumers (Eisenhauer et al., 2013; Haddad et al., 2001; Scherber et al., 2010).

## 2. Material and methods
### 2.1 Experimental design

This study was conducted in the Trait-Based Biodiversity Experiment (TBE; Ebeling et al., 2014) established in 2010 within the framework of a long-term grassland biodiversity experiment (Jena Experiment, Roscher et al., 2014). The experimental site is located in the floodplain of the Saale river close to the city of Jena (Germany; 50°55′ N, 11°35′ E, 130 m a.s.l.). Mean annual air temperature is 9.9 °C, and mean annual precipitation is 610 mm (1981–2010; Hoffmann et al., 2014) in the region. Before the establishment of the experiment, the area had been an unfertilized mown grassland for 8 years. In 2010, the previous grassland community was removed and new plant communities were sown on 138 plots (3.5 × 3.5 m) to cover a gradient of plant species richness (1, 2, 3, 4, and 8) and plant functional diversity (1, 2, 3, and 4) (see Ebeling et al., 2014).

The functional diversity gradient was formed by the selection of six resource acquisition traits: two aboveground spatial traits [maximum plant height (MH) and leaf area (LA)], two belowground spatial traits [rooting depth (RD) and root length density (RLD)], and two temporal traits [growth start (GS) and flowering start (FS)]. Those traits were analysed by a standardized Principal Component Analysis (PCA) (Ebeling et al., 2014) including all 48 non-legume species of the species pool of the Jena Experiment. The PCA axis 1 spans a gradient of spatial resource acquisition traits, and the PCA axis 2 displays a gradient of temporal resource acquisition traits. The PCA axes were divided into four sectors, and two species from each sector were selected to create three plant species pools each comprising eight species: species pool 1 covers species along the entire axis 1 with an intermediate position on axis 2; species pool 2 covers species along the entire axis 2 with an intermediate position on axis 1; and species pool 3 is the combination of the extremes of both axes (Ebeling et al., 2014). Plant communities were assembled in order to show a gradient in trait dissimilarity between species according to their assignment to different sectors along the two leading axes of the PCA. The experimental plots were arranged in three blocks accounting for variation in soil properties (see Ebeling et al., 2014).

### 2.1.1 Soil fauna sampling

In September 2014, i.e., 4 years after establishment of the plant communities, soil cores for soil mesofauna (5 cm deep, 5 cm diameter, Macfadyen, 1961) and soil macrofauna (10 cm deep, 25 cm diameter, Kempson et al., 1963) were sampled, taking one sample per plot for each method. Soil arthropods were extracted by a gradual heating, collected in glycol, and then stored in 70% ethanol until identification. For both mesofauna and macrofauna, we only recorded taxonomic groups that were adequately assessed by these extraction methods. For mesofauna samples, we identified mites (Krantz and Walter, 2009) and collembolans (Hopkin, 2007). For macrofauna samples, we separated chilopods, symphylans, diplopods, hemipterans (Aphidoidea), and beetles (Staphylinidae) (Coleman et al., 2004; Table S1 in Supplementary Material in the online version at https://doi.org/10.1016/bs.aecr.2019.06.004). Other taxonomic groups were excluded from analyses, because the method of extraction was considered inappropriate (e.g. for Diptera larvae). All extracted fauna from the target taxonomic groups were assigned to morphospecies based on consistent morphological characteristics. They were further assigned to trophic groups by using information from the literature on their respective taxonomic groups (Coleman et al., 2004; Table S1 in Supplementary Material in the online version at https://doi.org/10.1016/bs.aecr.2019.06.004). We defined all lower trophic level consumers (i.e. herbivores and decomposers) as 'prey', and all higher trophic levels as 'predators'.

The dataset contained 121 samples for each meso- and macrofauna as some samples were lost during the extraction procedure.

### 2.1.2 Plant cover measurement

Plant-specific cover (%) of sown plant species in each plot was estimated by using a decimal scale (modified after Londo, 1976) on the entire plot area (3.5 × 3.5 m) in mid-August 2014. The realized plant community composition and species relative abundances were used to calculate abundance-weighted plant community indices (see below).

## 2.2 Plant community indices

Two indices of plant trait diversity were calculated: Functional Richness (FRic) and Functional Dispersion (FDis) (Fig. S1 in Supplementary Material in the online version at https://doi.org/10.1016/bs.aecr.2019.06.004). All calculations were performed for each trait separately, which means that the functional diversity indices only refer to one dimension. In one-dimensional space (i.e. one trait), Functional Richness (FRic) is defined as the range between the maximum and minimum of the trait values within the plant community (Villéger et al., 2008), and Functional Dispersion (FDis) is defined as the weighted variance of the trait values within the plant community (Laliberté and Legendre, 2010). These indices have been identified to be more sensitive to processes of community assembly than species richness (Laliberté and Legendre, 2010; Roscher et al., 2014), and they describe trait distributions in two complementary ways: Functional Richness describes the range of trait values within a community, while Functional Dispersion informs on how evenly species trait values are distributed along this range. Further, we calculated CWM of each trait, based on their species-specific cover for each plant community in 2014 (Garnier et al., 2004; Lavorel et al., 2008; Roscher et al., 2012). To avoid differences of too many orders of magnitude between the explanatory variables, they were rescaled based on range to fit between $-1$ and 1. The calculations were performed with the R package FD (Laliberté et al., 2014).

## 2.3 Statistical analyses

We used linear mixed-effects models to test the effects of plant species richness on the abundance and species richness of all soil fauna, mesofauna, macrofauna, corresponding subgroups of predators and prey. Species abundance and richness were modelled using a Poisson distribution with observation-level random effects to take in account over-dispersion (Blolker, 2019; Elston et al., 2001). For each model, 'block' was specified as a random effect to account for the spatial arrangement of the plots. Linear mixed-effects models were performed using the 'lme4' package (Bates et al., 2012) within the R statistical environment (R Development Core Team, 2010).

To identify the effect of plant trait-based indices on each soil fauna group, we proceeded in two steps: (1) for each index (CWM, FDis, FRic), we selected a subset of important traits, and (2) we built a full trait-based model composed of the traits selected into the subsets of each index (CWM, FDis, and FRic). (1) *Trait selection for each of the three indices describing community-level plant traits (FRic, FDis, and CWM):* we employed a model selection approach following Burnham and Anderson (2002) and Grueber et al. (2011). For a given index, we used linear mixed-effect models to test the effect of the six plant traits (growth start, flowering start, leaf area, maximum height, root length density, and rooting depth) on soil fauna groups. All possible combinations of the six plant traits were modelled for each response variable (abundance and species richness) of each fauna group (total fauna, macrofauna, mesofauna, mesofauna predators, mesofauna prey, macrofauna predators, and macrofauna prey). A set of best candidate models was defined by all models with a maximum $\Delta$AICc of 2 compared to the model with the lowest AICc (Bolker et al., 2009; Burnham and Anderson, 2002). For each response variable, we selected all explanatory variables included in the set of best candidate models (Grueber et al., 2011; Nakagawa and Freckleton, 2011). Trait selection was performed using the R package 'MuMIn' (Barton, 2015). (2) *Trait-based model:* we used linear mixed-effect models to test the effect of all traits previously selected for all indices on each response variable. For both steps and as done in the plant species richness models, species abundances and richness were modelled with a Poisson distribution with observation-level random effects, and 'block' was specified as a random effect. Finally, for each group of soil fauna, we calculated the proportion of total variance explained by the trait-based models using marginal $R^2$. $R^2$ were calculated following Nakagawa and Schielzeth (2013).

To compare plant species richness and trait-based models, we compared the AICc of plants species richness-based model predictions (Eq. 1) or trait-based model predictions (Eq. 2) and the AICc of the model with both plants species richness and trait-based model predictions as explanatory variables:

$$\Delta AICc_1 = AICc(y \sim trait.prediction) - AICc(y \sim trait.prediction + PD) \quad (1)$$
$$\Delta AICc_2 = AICc(y \sim PD) - AICc(y \sim PD + trait.prediction) \quad (2)$$

where, $y$ is our response variable, $PD$ is the plant species richness, and *trait. prediction* is the prediction of the corresponding trait-based model. We considered the models distinct when $|\Delta AICc| > 2$. If $AICc_1 > 2$, adding plant species richness improved our model predictions, while if $AICc_2 > 2$, adding traits indices improved our model predictions.

## 3. Results
### 3.1 Plant species richness effects

Plant species richness did not have any significant effect on neither abundance nor species richness of our studied fauna groups (Fig. 1; Supplementary

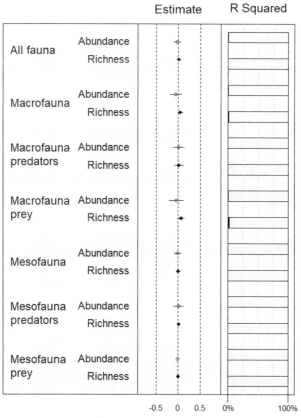

**Fig. 1** Plant species richness effects on soil fauna abundance and species richness. Circles indicate the estimates of plant species richness effects on the different indices of soil fauna groups with a confidence interval of 95%.

Material in the online version at https://doi.org/10.1016/bs.aecr.2019.06.004 S1). The variance explained by plant species richness in our model was extremely low (from $R^2 < 0.01$ to $R^2 = 0.016$; Fig. 1; Supplementary Material in the online version at https://doi.org/10.1016/bs.aecr.2019.06.004 S1).

## 3.2 Trait-based models

For each trait index, we selected a subset of the most relevant traits. However, FDis and FRic indices were highly correlated (Pearson's correlation: $0.97 +/- 0.01$, Fig. S2 in Supplementary Material in the online version at https://doi.org/10.1016/bs.aecr.2019.06.004), and the subsets of traits selected were similar (see Supplementary Material in the online version at https://doi.org/10.1016/bs.aecr.2019.06.004 S2 and S3A). Given these similarities, we only used the traits selected for FDis to build out trait-based models.

For soil fauna abundance, we observed that the amount of variance explained by our trait-based models differed between groups of soil fauna. More specifically, our models explained only a small fraction of all fauna variability ($R^2 = 0.04$; Fig. 2; Supplementary Material in the online version at https://doi.org/10.1016/bs.aecr.2019.06.004 S3B), while size based groups were better explained (macrofauna $R^2 = 0.16$ and mesofauna $R^2 = 0.05$; Fig. 2; Supplementary Material in the online version at https://doi.org/10.1016/bs.aecr.2019.06.004 S3B). Moreover, the abundance of prey was better explained than that of predators for a given size-based group (macrofauna or mesofauna, Fig. 2). A unique set of explanatory variables was selected for each fauna group (Supplementary Material in the online version at https://doi.org/10.1016/bs.aecr.2019.06.004 S3A). For example, both macrofauna and mesofauna abundance models included CWM root length density, maximum height, and growth start, while the model for mesofauna abundance also included CWM flowering start and FDis of all six traits (see Supplementary Material in the online version at https://doi.org/10.1016/bs.aecr.2019.06.004 S3A and Fig. 2).

Overall, our models revealed that CWM indices had a higher explanatory power than FDis indices. None of the FDis indices had a significant effect on any soil fauna group (Fig. 2; Supplementary Material in the online version at https://doi.org/10.1016/bs.aecr.2019.06.004 S3B). Of the CWM indices, two traits had strong effects on soil fauna abundances. First, CWM root length density had a positive effect on macrofauna (estimate $= 0.44$, P-value $= 0.04$),

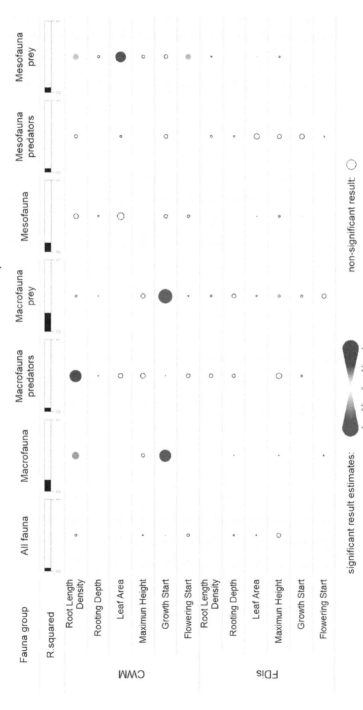

**Fig. 2** Plant functional trait effects on soil fauna abundance. Circle size and colour indicate the estimates of a given trait effect on the abundance of soil fauna groups. Blue circles denote positive estimates, red circles denote negative estimates and empty circles denote a non-significant relationship (i.e. P-value <0.05).

macrofauna predator species (estimate = 0.74, *P*-value = 0.002), and mesofauna prey (estimate = 0.35, *P*-value = 0.05; Fig. 2; Supplementary Material in the online version at https://doi.org/10.1016/bs.aecr.2019.06.004 S3B). Second, CWM growth start had a strong negative effect on macrofauna (estimate = − 0.74, *P*-value < 0.001) and macrofauna prey (estimate = − 0.86, *P*-value = 0.03; Fig. 2; Supplementary Material in the online version at https://doi.org/10.1016/bs.aecr.2019.06.004 S3B). The other traits had more inconsistent and weaker effects across fauna groups (e.g. a positive effect of CWM leaf area on mesofauna prey and negative effect of CWM flowering start on mesofauna prey; Fig. 2). Our models did not show any significant trend for species richness of soil fauna groups, although the explained variance was within the same order of magnitude (Fig. S3 in Supplementary Material in the online version at https://doi.org/10.1016/bs.aecr.2019.06.004).

## 3.3 Comparison of plant species richness-based and trait-based models

For abundance of all fauna groups, adding plant species richness to the plant trait-based models did not improve the model ($\Delta AICc_1 \leq 2$; Fig. 3; Supplementary Material in the online version at https://doi.org/10.1016/bs.aecr.2019.06.004 S4), while adding plant traits to a plant species richness-based model significantly improved the model predictions ($\Delta AICc_2$ from 213 to 45′714; Fig. 3; Supplementary Material in the online version at https://doi.org/10.1016/bs.aecr.2019.06.004 S4). These results indicate that variance explained by the plant trait-based model already accounts for the variance explained by plant species richness. By contrast, the plant trait-based model explained a higher proportion of variance than the plant species richness model and was not improved by including plant species richness in the model (Fig. 3). We observed the same for the species richness of soil fauna groups; however, the overall explanatory power of the models was low (Fig. S4 in Supplementary Material in the online version at https://doi.org/10.1016/bs.aecr.2019.06.004).

## 4. Discussion

Our results revealed that plant traits play a significant role in structuring soil communities (Hypothesis 1), whereas plant species richness appeared to be of relatively minor importance in the present study. Importantly, we found that different feeding groups (predator and prey) and size classes

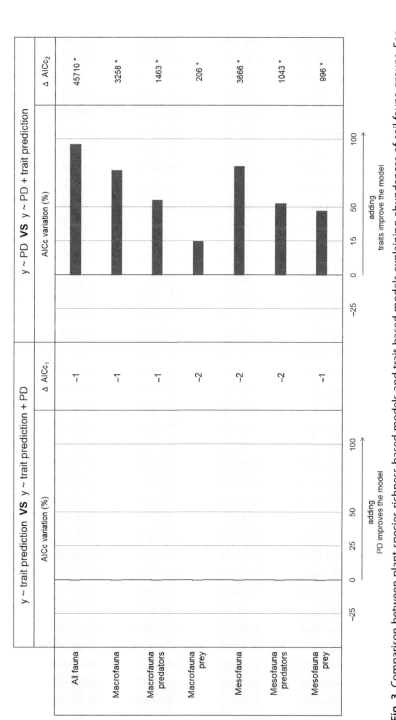

**Fig. 3** Comparison between plant species richness-based models and trait-based models explaining abundances of soil fauna groups. For each fauna group, a positive variation of AICc (blue line) and a ΔAICc >2 indicate that the information added to the model (i.e. plant species richness—PD—or trait prediction) increased the model prediction.

(meso- and macrofauna) were affected by different sets of community-level plant traits. Taken together, our results indicate that the expression of certain plant traits at the community level governs the structure of soil fauna communities.

## 4.1 Plant species richness has a weak effect on soil communities

Surprisingly, we found that plant species richness had no significant effect on the abundance and species richness of any soil fauna group (Fig. 1). These results contradict our Hypothesis 1 and findings from previous studies on both aboveground and belowground communities (e.g. Ebeling et al., 2018a; Eisenhauer et al., 2011; Haddad et al., 2009; Milcu et al., 2013; Scherber et al., 2010; Schuldt et al., 2019) that showed positive plant diversity effects on the abundance and diversity of consumers. Moreover, our results are in contrast to the 'More Individuals Hypothesis', which predicts an increase of plant productivity with increasing plant species richness, and consequently an increase of herbivorous and detritivorous species (Srivastava and Lawton, 1998). Non-significant plant diversity effects on soil fauna have been observed before (e.g. Eisenhauer et al., 2009; Milcu et al., 2008), which is possibly explained by the short-term duration of most previous experiments as plant diversity effects need several years to manifest (Eisenhauer et al., 2010a,b, 2012). This is because plant community-specific organic matter gradually accumulates following the establishment of the experimental grassland plots (Habekost et al., 2008), which in turn drives the assembly of specific soil communities (Eisenhauer et al., 2011b). As a consequence, these results should be treated with care as longer-term studies are required to test plant diversity-ecosystem functioning relationships (Cardinale et al., 2007; Reich et al., 2012; Thakur et al., 2015).

## 4.2 Plant traits as more powerful predictors of soil fauna communities

Although only a small proportion of the variance of soil fauna groups was explained by plant functional traits (max. 16%), they generally explained more variance than plant species richness. In fact, functional traits have often been argued to underlie significant plant diversity effects (e.g. Milcu et al., 2013; Mouillot et al., 2013). Previous research from the Jena Experiment, however, has shown that long-term effects of plant species richness on soil organisms are more important than plant functional group effects (e.g. Eisenhauer et al., 2010a,b, 2011a). Notably, these previous studies used broad categories of plant functional groups, but did not consider finer gradients in spatial and temporal resource use traits (Ebeling et al., 2014). Moreover, the present study did not include any legumes, which have repeatedly been shown to play a major role for soil communities and processes by fixing nitrogen through their mutualistic relationship with rhizobia (Eisenhauer et al., 2009; Lange et al., 2015; Milcu et al., 2008; Spehn et al., 2000). In addition, it should be noted that species richness of soil fauna was poorly explained by our models, which may be the result of a low variability of species richness within each given fauna group (Table S1 in Supplementary Material in the online version at https://doi.org/10.1016/bs.aecr.2019.06.004).

## 4.3 The importance of plant trait identity effects across soil fauna groups

We observed that our plant trait-based model varied in its explanatory power for soil fauna abundance depending on the fauna group. While only a small fraction of the variance of all soil fauna was explained by the model ($R^2 = 0.04$), 16% of the variance in soil macrofauna abundance was explained. Moreover, for both macrofauna and mesofauna, explanatory power for prey abundance was always higher than for predators (providing support for our Hypothesis 4). These findings are in line with previously found bottom-up effects of plant community properties on consumer species (see Eisenhauer et al., 2010a,b, 2013; Scherber et al., 2010), whereby the strength of plant community effects decreased with increasing trophic level (Kaunzinger and Morin, 1998; Scherber et al., 2010).

While these results broadly confirm previous findings, our study provides particularly novel insights into how different traits exert significant effects on meso- and macrofauna predators and prey. Thus, our study emphasizes that different facets of plant community traits likely affect soil food web structure (see Schuldt et al., 2019 for aboveground invertebrate food webs) and that some plant community effects on predators may be mediated by variations in their prey populations, while there may be other simultaneous direct effects of vegetation structure. For instance, macrofauna predators were mainly affected by CWM root length density, while macrofauna prey were mainly affected by CWM growth start. In our final plant trait-based model, CWM traits showed stronger effects on soil fauna than plant trait diversity (Fig. 2). These observations suggest that dominant plant trait values are more important than the diversity of plant traits for specific groups of soil fauna. This is in line with the 'biomass ratio hypothesis' (Grime, 1998) that predicts a stronger effect of dominant species within a community. Nevertheless, the finding of dissimilar traits influencing different groups of soil fauna might provide a mechanism underlying the often-observed positive plant diversity effects on soil communities (e.g. Eisenhauer et al., 2013; Scherber et al., 2010).

## 4.4 Soil fauna responses to spatial resource acquisition traits

Our models highlighted the importance of two spatial resource acquisition traits: root length density and leaf area. These results are in the line with Hypothesis 2—that both CWM and diversity of plant traits related to high root productivity should influence soil community structure—and confirm previous findings for soil microbial communities (Steinauer et al., 2017). Root length density (RLD) affected several fauna groups (Hypothesis 2). In particular, macrofauna predator abundance and mesofauna prey abundance increased with CWM root length density. CWM root length density may be related to an increase of plant-derived carbon inputs to the soil that are available for primary consumers (i.e. herbivores or decomposers; Bardgett et al., 2014). Therefore, we suspect that the effect of CWM root length density on prey abundance and macrofauna predator abundance was due to enhanced belowground plant biomass (Barry et al., 2019; Milcu et al., 2008; Eisenhauer et al., 2010a,b; Scherber et al., 2010). Moreover, macrofauna predators seem to be more related to the response of mesofauna prey to changes in root length density than to that of mesofauna predators. This could be due to the fact that mesofauna predators in the sampled communities of this study were mainly comprised of Gamasina mites, which may preferably feed on nematodes, insect larvae, and Collembola with various degrees of specialization (Koehler, 1999).

The other resource acquisition trait, leaf area (LA), also had a positive effect on mesofauna prey abundance. It has been shown that traits associated with the leaf economic spectrum (e.g. specific leaf area or leaf dry matter content) can be related to soil functioning (e.g. decomposition; Garnier et al., 2004; Lavorel and Garnier, 2002; or, nitrification; Laughlin, 2011). In the species pool of the TBE, grass species tended to have a smaller leaf area than forb species (Ebeling et al., 2014); grass species in the Jena Experiment have a higher leaf dry matter content than forbs (Bachmann et al., 2018), which could explain the relationship of leaf size with the traits of the leaf economics spectrum. Based on these previous findings, it stands to reason that leaf traits should have indirect effects on soil communities through changes in soil properties and processes, such that increases in soil carbon or nitrification positively influence resources of detritivorous mesofauna.

## 4.5 Soil fauna responses to temporal resource acquisition traits

An important plant trait for soil fauna identified by our analysis was growth start (GS, Hypothesis 3). We found that a later start of plant growth (i.e. an increase in CWM growth start) had a negative effect on macrofauna abundance and species richness. While it is important to note that our assessment of soil fauna was limited to a single sampling event in early fall (September), variation in temporal plant traits such as growth start may still reliably indicate differences in the productivity of the plant community across the growing season. There are multiple examples from temperate grasslands to tropical forests showing that soil community composition, population dynamics, and plant community effects on soil biota vary over time (e.g. Eisenhauer et al., 2009; Moche et al., 2015). Bearing that in mind, a single snapshot measurement may not be able to clearly determine if temporal plant traits modify soil communities or just shift community dynamics in time. However, the significant effect of temporal plant traits found in this study suggests that multiple assessments of soil communities within and across seasons and years are required to more comprehensively study plant community effects on soil fauna (Berg and Bengtsson, 2007; Eisenhauer et al., 2018; Moche et al., 2015).

## 5. Conclusions

This study suggests that soil fauna abundance and diversity are better explained by plant trait identity and, to a lesser extent, trait diversity than by plant species richness. However, the effects of plant traits were not always consistent and depended on the soil fauna group in question. Our results further suggest that future studies should take into account multiple root traits (Laliberté, 2016) as well as their plasticity in responses to abiotic and biotic drivers (Eisenhauer and Powell, 2017) to better predict plant community effects on soil biota and functions. This study reveals, for the first time, the importance of temporal plant traits for soil fauna, highlighting the need for repeated assessments that cover the temporal dynamics of communities across different seasons (Dombos et al., 2017; Eisenhauer et al., 2018).

## Acknowledgements

We thank Ulrich Pruschitzki, Alfred Lochner, Julia Friese and Josephine Grenzer for their help during sampling and soil fauna extraction processes. We would also like to thank Prof Dr Stefan Scheu for his assistance during soil fauna extraction. We thank the gardeners, technicians, and managers for their work in maintaining the field site and also many student helpers for weeding the experimental plots. Comments by two anonymous reviewers helped us to improve this paper. The Jena Experiment was funded by the Deutsche Forschungsgemeinschaft (German Research Foundation; FOR 1451). NE acknowledges funding by the German Research Foundation (Ei 862/3-2). Additional support came from the German Centre for Integrative Biodiversity Research (iDiv) Halle-Jena-Leipzig funded by the German Research Foundation (FZT 118).

## References

Bachmann, D., Roscher, C., Buchmann, N., 2018. How do leaf trait values change spatially and temporally with light availability in a grassland diversity experiment? Oikos 127 (7), 935–948. https://doi.org/10.1111/oik.04533.

Bardgett, R.D., Wardle, D.A., 2010. Aboveground-Belowground Linkages, first ed. Oxford University Press.

Bardgett, R.D., Mommer, L., De Vries, F.T., 2014. Going underground: root traits as drivers of ecosystem processes. Trends Ecol. Evol. 29 (12), 692–699. Elsevier Ltd. https://doi.org/10.1016/j.tree.2014.10.006.

Barry, K.E., et al., 2019. Above- and belowground overyielding are related at the community and species level in a grassland biodiversity experiment. Adv. Ecol. Res. 61, 55–89.

Barton, K., 2015. MuMIn: Multi—Model Inference (R package version 1.13. 4). http://CRAN.R-project.org/package=MuMIn.

Bates, D., et al., 2012. Package 'lme4'. In: CRAN. R Foundation for Statistical Computing, Vienna, Austria.

Berg, M.P., Bengtsson, J., 2007. Temporal and spatial variability in soil food web structure. Oikos 116, 1789–1804. https://doi.org/10.1111/j.2007.0030-1299.15748.x.

Blolker, B., 2019. GLMM FAQ. Available at: https://bbolker.github.io/mixedmodels-misc/glmmFAQ.html#overdispersion.

Bolker, B.M., et al., 2009. Generalized linear mixed models: a practical guide for ecology and evolution. Trends Ecol. Evol. 24 (3), 127–135. https://doi.org/10.1016/j.tree.2008.10.008.

Burnham, K.P., Anderson, D.R., 2002. Model Selection and Multimodel Inference—A Practical Information—Theoretic Approach, second ed. Springer-Verlag New York.

Cardinale, B.J., et al., 2007. Impacts of plant diversity on biomass production increase through time because of species complementarity. Proc. Natl. Acad. Sci. U.S.A. 104 (46), 18123–18128. https://doi.org/10.1073/pnas.0709069104.

Coleman, D.C., Crossley, D.A., Hendrix, P.F., 2004. Fundamentals of Soil Ecology, second ed. Academic Press.

De Deyn, G.B., Van Der Putten, W.H., 2005. Linking aboveground and belowground diversity. Trends Ecol. Evol. 20 (11), 625–633. https://doi.org/10.1016/j.tree.2005.08.009.

Diaz, S., Cabido, M., 2001. Vive la difference: plant functional diversity matters to ecosystem processes: plant functional diversity matters to ecosystem processes. Trends Ecol. Evol. 16 (11), 646–655.

Dombos, M., et al., 2017. EDAPHOLOG monitoring system: automatic, real-time detection of soil microarthropods. Methods Ecol. Evol. 8, 313–321. https://doi.org/10.1111/2041-210X.12662.

Ebeling, A., et al., 2014. A trait-based experimental approach to understand the mechanisms underlying biodiversity–ecosystem functioning relationships. Basic Appl. Ecol. 15 (3), 229–240. Elsevier GmbH. https://doi.org/10.1016/j.baae.2014.02.003.

Ebeling, A., Rzanny, M., et al., 2018a. Plant diversity induces shifts in the functional structure and diversity across trophic levels. Oikos 127 (2), 208–219. https://doi.org/10.1111/oik.04210.

Ebeling, A., Hines, J., et al., 2018b. Plant diversity effects on arthropods and arthropod-dependent ecosystem functions in a biodiversity experiment. Basic Appl. Ecol. 26, 50–63. Elsevier GmbH. https://doi.org/10.1016/j.baae.2017.09.014.

Eisenhauer, N., Powell, J.R., 2017. Plant trait effects on soil organisms and functions. Pedobiologia 65, 1–4. https://doi.org/10.1016/j.pedobi.2017.11.001.

Eisenhauer, N., et al., 2009. Impacts of earthworms and arbuscular mycorrhizal fungi (Glomus intraradices) on plant performance are not interrelated. Soil Biol. Biochem. 41 (3), 561–567. https://doi.org/10.1016/j.soilbio.2008.12.017.

Eisenhauer, A.N., et al., 2010. Plant diversity effects on soil microorganisms support the singular hypothesis. Ecology 91 (2), 485–496. Available at http://www.jstor.org/stable/25661074.

Eisenhauer, N., et al., 2011. Plant diversity surpasses plant functional groups and plant productivity as driver of soil biota in the long term. PLoS ONE 6 (1), e16055. Edited by A. Hector. https://doi.org/10.1371/journal.pone.0016055.

Eisenhauer, N., Reich, P.B., Scheu, S., 2012. Increasing plant diversity effects on productivity with time due to delayed soil biota effects on plants. Basic Appl. Ecol. 13 (7), 571–578. https://doi.org/10.1016/j.baae.2012.09.002.

Eisenhauer, N., et al., 2013. Plant diversity effects on soil food webs are stronger than those of elevated CO2 and N deposition in a long-term grassland experiment. Proc. Natl. Acad. Sci. U.S.A. 110, 6889–6894. https://doi.org/10.1073/pnas.1217382110.

Eisenhauer, N., et al., 2017. Root biomass and exudates link plant diversity with soil bacterial and fungal biomass. Sci. Rep. 7, 1–8. https://doi.org/10.1038/srep44641Nature Publishing Group.

Eisenhauer, N., et al., 2018. The dark side of animal phenology. Trends Ecol. Evol. 33 (12), 898–901. Elsevier Ltd. https://doi.org/10.1016/j.tree.2018.09.010.

Eisenhauer, N., et al., 2019. A multitrophic perspective on biodiversity–ecosystem functioning research. Adv. Ecol. Res. 61, 1–54.

Elston, D.A., et al., 2001. Analysis of aggregation, a worked example: numbers of ticks on red grouse chicks. Parasitology 122 (Pt. 5), 563–569. Available at http://www.ncbi.nlm.nih.gov/pubmed/11393830.

Flynn, D.F.B., et al., 2011. Functional and phylogenetic diversity as predictors of biodiversity—ecosystem-function relationships. Ecology 92 (8), 1573–1581.

Garnier, E., et al., 2004. Plant functional markers capture ecosystem properties during secondary succession. Ecology 85 (9), 2630–2637.

Giling, D.P., et al., 2019. Plant diversity alters the representation of motifs in food webs. Nat. Commun. 10 (1), 1226. 2019 10:1. Springer US. https://doi.org/10.1038/s41467-019-08856-0.

Grime, J.P., 1998. Benefits of plant diversity to ecosystems: immediate, filter and founder effects. J. Ecol. 86, 902–910.

Grime, J.P., 2001. Plant Strategies Vegetation Processes and Ecosystem Properties, second ed. Wiley.

Grueber, C.E., et al., 2011. Multimodel inference in ecology and evolution: challenges and solutions. J. Evol. Biol. 24 (4), 699–711. https://doi.org/10.1111/j.1420-9101.2010.02210.x.

Habekost, M., et al., 2008. Seasonal changes in the soil microbial community in a grassland plant diversity gradient four years after establishment. Soil Biol. Biochem. 40 (10), 2588–2595. https://doi.org/10.1016/j.soilbio.2008.06.019.

Haddad, N.M., et al., 2001. Contrasting effects of plant richness and composition on insect communities: a field experiment contrasting effects of plant richness and composition on insect communities: a field experiment. Am. Nat. 158, 17–35.

Haddad, N.M., et al., 2009. Plant species loss decreases arthropod diversity and shifts trophic structure. Ecol. Lett. 12, 1029–1039. https://doi.org/10.1111/j.1461-0248.2009.01356.x.

Hertzog, L.R., et al., 2017. Plant diversity increases predation by ground-dwelling invertebrate predators. Ecosphere 8 (11), 1–14. https://doi.org/10.1002/ecs2.1990/full.

Hines, J., et al., 2019. Mapping change in biodiversity and ecosystem function research: food webs foster integration of experiments and science policy. Adv. Ecol. Res. 61, 297–322.

Hoffmann, K., et al., 2014. Klimauntersuchungen in Jena für die Anpassung an den Klimawandel und seine erwarteten Folgen. Selbstverlag des Deutschen Wetterdienstes, Offenbach am Main.

Hooper, D.U., et al., 2000. Interactions between aboveground and belowground biodiversity in terrestrial ecosystems: patterns, mechanisms, and feedbacks. Bioscience 50 (12), 1049–1061.

Hopkin, S.P., 2007. A Key to the Collembola (Springtails) of Britain and Ireland. Field Studies Council.

Johnson, S.N., et al., 2012. Aboveground—belowground herbivore interactions: a meta-analysis. Ecology 93 (10), 2208–2215.

Kaunzinger, C.M.K., Morin, P.J., 1998. Productivity controls food-chain properties in microbial communities. Nature 395, 495–497.

Kazakou, E., et al., 2006. Co-Variations in Litter Decomposition, Leaf Traits and Plant Growth in Species from a Mediterranean Old-Field Succession. Published by: British Ecological Society Linked references are available on JSTOR for this article: Co-variations in litter decompoFunct. Ecol. 20 (1), 21–30.

Kempson, D., Lloyd, M., Ghelardi, R., 1963. A new extractor for woodland litter. Pedobiologia 3 (1), 21.

Koehler, H.H., 1999. Predatory mites (Gamasina, Mesostigmata). Agric. Ecosyst. Environ. 74 (1–3), 395–410. https://doi.org/10.1016/S0167-8809(99)00045-6.

Krantz, G.W., Walter, D.E., 2009. Manual of Acarology, third ed. Texas Tech University Press.

Kuzyakov, Y., Blagodatskaya, E., 2015. Microbial hotspots and hot moments in soil: concept & review. Soil Biol. Biochem. 83, 184–199. Elsevier Ltd. https://doi.org/10.1016/j.soilbio.2015.01.025.

Laliberté, E., 2016. Below-ground frontiers in trait-based plant ecology. New Phytol. 213 (4), 1597–1603. https://doi.org/10.1111/nph.14247.

Laliberté, E., Legendre, P., 2010. A distance-based framework for measuring functional diversity from multiple traits. Ecology 91 (1), 299–305. https://doi.org/10.1890/08-2244.1.

Laliberté, E., Legendre, P., Shipley, B., Laliberté, M.E., 2014. Package 'FD': Measuring functional diversity from multiple traits, and other tools for functional ecology.

Lange, M., et al., 2015. Plant diversity increases soil microbial activity and soil carbon storage. Nat. Commun. 6, 1–8. https://doi.org/10.1038/ncomms7707.

Lange, M., et al., 2019. How plant diversity impacts the coupled water, nutrient and carbon cycles. Adv. Ecol. Res. 61, 185–219.

Laughlin, D.C., 2011. Nitrification is linked to dominant leaf traits rather than functional diversity. J. Ecol. 99 (5), 1091–1099. https://doi.org/10.1111/j.1365-2745.2011.01856.x.

Lavorel, S., Garnier, E., 2002. Predicting changes in community composition and ecosystem functioning from plant traits: revisiting the holy grail. Funct. Ecol. 16, 545–556.

Lavorel, S., et al., 2008. Assessing functional diversity in the field—methodology matters!. Funct. Ecol. 22 (1), 134–147. https://doi.org/10.1111/j.1365-2435.2007.01339.x071124124908001–???.

Londo, 1976. The decimal scale for releves of permanent quadrats. Vegetatio 33 (1954), 61–64.

Macfadyen, A., 1961. Improved funnel-type extractors for soil arthropods. J. Anim. Ecol. 30 (1), 171–184. Available at: http://www.jstor.org/stable/2120 REF.

McKenzie, S.W., et al., 2013. Reciprocal feeding facilitation between above- and below-ground herbivores. Biol. Lett. 9 (5), 1–5. https://doi.org/10.1098/rsbl.2013.0341.

Mellado-Vázquez, P.G., et al., 2016. Plant diversity generates enhanced soil microbial access to recently photosynthesized carbon in the rhizosphere. Soil Biol. Biochem. 94, 122–132. https://doi.org/10.1016/j.soilbio.2015.11.012.

Meyer, S.T., et al., 2017. Consistent increase in herbivory along two experimental plant diversity gradients over multiple years. Ecosphere 8 (7), 1–19. https://doi.org/10.1002/ecs2.1876.

Milcu, A., et al., 2008. Earthworms and legumes control litter decomposition in a plant diversity gradient. published by: Wiley on behalf of the ecological Society of Amer, Ecology 89 (7), 1872–1882.

Milcu, A., et al., 2013. Functionally and phylogenetically diverse plant communities key to soil biota. Ecology 94 (8), 1878–1885.

Moche, M., et al., 2015. Soil biology & biochemistry monthly dynamics of microbial community structure and their controlling factors in three floodplain soils. Soil Biol. Biochem. 90, 169–178. Elsevier Ltd. https://doi.org/10.1016/j.soilbio.2015.07.006.

Mouillot, D., et al., 2013. A functional approach reveals community responses to disturbances. Trends Ecol. Evol. 28 (3), 167–177. https://doi.org/10.1016/j.tree.2012.10.004.

Nakagawa, S., Freckleton, R.P., 2011. Model averaging, missing data and multiple imputation: a case study for behavioural ecology. Behav. Ecol. Sociobiol. 65, 103–116.

Nakagawa, S., Schielzeth, H., 2013. A general and simple method for obtaining R2 from generalized linear mixed-effects models. Methods Ecol. Evol. 4 (2), 133–142. https://doi.org/10.1111/j.2041-210x.2012.00261.x.

Postma, J.A., Lynch, J.P., 2012. Complementarity in root architecture for nutrient uptake in ancient maize/bean and maize/bean/squash polycultures. Ann. Bot. 110, 521–534. https://doi.org/10.1093/aob/mcs082.

R Development Core Team, R, 2010. A Language and Environment for Statistical Computing. R foundation for Statistical Computing, Vienna.

Reich, P.B., et al., 2012. Impacts of biodiversity loss escalate through time as redundancy fades. Science 336 (6081), 589–592.

Roscher, C., et al., 2012. Using plant functional traits to explain diversity–productivity relationships. PLoS ONE 7 (5), e36760. Edited by H. Y. H. Chen. Public Library of Science. https://doi.org/10.1371/journal.pone.0036760.

Roscher, C., et al., 2014. Different assembly processes drive shifts in species and functional composition in experimental grasslands varying in sown diversity and community history. PLoS One 9 (7), 1–12. https://doi.org/10.1371/journal.pone.0101928.

Scherber, C., et al., 2010. Bottom-up effects of plant diversity on multitrophic interactions in a biodiversity experiment. Nature 469, 553–556. https://doi.org/10.1038/nature09492.

Schuldt, A., et al., 2019. Multiple plant diversity components drive consumer communities across ecosystems. Nat. Commun. 10, 1–11.

Spehn, E.M., et al., 2000. Plant diversity and soil heterotrophic activity in experimental grassland systems. Plant and Soil 224, 217–230. https://doi.org/10.1023/A.

Srivastava, D., Lawton, J., 1998. Why more productive sites have more species: an experimental test of theory using tree-hole communities. Am. Nat. 152, 510–529.

Steinauer, K., et al., 2017. Spatial plant resource acquisition traits explain plant community effects on soil microbial properties. Pedobiologia 65 (February), 50–57. Elsevier. https://doi.org/10.1016/j.pedobi.2017.07.005.

Strecker, T., et al., 2016. Functional composition of plant communities determines the spatial and temporal stability of soil microbial properties in a long-term plant diversity experiment. Oikos 125 (12), 1743–1754. https://doi.org/10.1111/oik.03181.

Symstad, A.J., Siemann, E., Haarstad, J., 2000. An experimental test of the effect of plant functional group diversity on arthropod diversity. Oikos 89 (2), 243–253. https://doi.org/10.1034/j.1600-0706.2000.890204.x.

Thakur, M.P., et al., 2015. Plant diversity drives soil microbial biomass carbon in grasslands irrespective of global environmental change factors. Glob. Chang. Biol. 21 (11), 4076–4085. https://doi.org/10.1111/gcb.13011.

Tilman, D., 1982. Resource Competition and Community Structure. In: Monographs in Population Biology. Princeton University Press, Princeton.

Tilman, D., 1988. Dynamics and Structure of Plant Communities. In: Monographs in Population Biology. Princeton University Press, Princeton.

Villéger, S., Mason, N.W.H., Mouillot, D., 2008. New multidimentional functional diversity indices for a multifaceted framework in functional ecology. Ecology 89, 2290–2301.

Violle, C., et al., 2007. Let the concept of trait be functional!. Oikos 116 (5), 882–892. https://doi.org/10.1111/j.0030-1299.2007.15559.x.

Wardle, D.A., et al., 2004. Ecological linkages between aboveground and belowground biota. Science 304, 1629–1633. https://doi.org/10.1126/science.1094875.

## Further reading

Allan, E., et al., 2015. Land use intensification alters ecosystem multifunctionality via loss of biodiversity and changes to functional composition. Ecol. Lett. 18, 834–843. https://doi.org/10.1111/ele.12469.

Loreau, M., Hector, A., 2001. Partitioning selection and complementarity in biodiversity experiments. Nature 412, 72–76.

Steinbeiss, S., et al., 2008. Plant diversity positively affects short-term soil carbon storage in experimental grasslands. Glob. Chang. Biol. 14, 2937–2949. 2008. https://doi.org/10.1111/j.1365-2486.2008.01697.x.

CHAPTER SIX

# How plant diversity impacts the coupled water, nutrient and carbon cycles

**Markus Lange[a],\*, Eva Koller-France[b,c], Anke Hildebrandt[a,d,e,f], Yvonne Oelmann[b], Wolfgang Wilcke[c], Gerd Gleixner[a]**

[a]Department of Biogeochemical Processes, Max Planck Institute for Biogeochemistry, Jena, Germany
[b]Geoecology, University of Tübingen, Tübingen, Germany
[c]Institute of Geography and Geoecology, Karlsruhe Institute of Technology (KIT), Karlsruhe, Germany
[d]UFZ-Helmholtz Centre for Environmental Research, Leipzig, Germany
[e]Institute of Geosciences, Friedrich-Schiller-University Jena, Jena, Germany
[f]German Centre for Integrative Biodiversity Research (iDiv) Halle-Jena-Leipzig, Leipzig, Germany
\*Corresponding author: e-mail address: mlange@bgc-jena.mpg.de

## Contents

| | |
|---|---:|
| 1. Introduction | 186 |
| 2. Plant diversity effects on the soil microbial community and soil processes and functions | 187 |
|    2.1 Microbial community composition and diversity | 187 |
|    2.2 Soil water balance | 190 |
|    2.3 Nutrient cycling | 195 |
|    2.4 Plant carbon allocation to soil, microbial net assimilation and microbial carbon storage | 202 |
| 3. Consequences of the element and water cycles and their coupling for the BEF relationships | 207 |
| Acknowledgements | 211 |
| References | 211 |

## Abstract

Soils are important for ecosystem functions and services. However, soil processes are complex and changes of solid phase soil properties, such as soil organic matter contents are slow. As a consequence, a comprehensive understanding of the role of soil in the biodiversity-ecosystem functioning (BEF) relationship is still lacking. Thus, long-term observations and experiments are needed in biodiversity research in order to better understand how biodiversity influences soil properties and thus the BEF relationships. To elucidate the integrated response of soil-related functions and processes to plant diversity, we reviewed literature on the water, nutrient and carbon cycles in biodiversity research with specific focus on the Jena Experiment. Furthermore, we took advantage of the long-term observations of water, nutrient and carbon dynamics gathered in the

Jena Experiment to investigate changes of the plant diversity effect over time on theses cycles and the accompanying plant-microbial interactions. We found that soil organic carbon and soil nitrogen stocks in the top 15 cm constantly increased over time and that this increase was positively related to plant species richness. In contrast, the concentrations of the quantitatively most important nutrient ions nitrate and phosphate in soil solution decreased with time, likely because of the ongoing removal of nutrients by plant biomass harvest. We furthermore observed a shift in the microbial community composition, which was triggered by an increased availability of plant-derived carbon at higher plant species richness over time, suggesting that plant communities compensated for nutrient losses by stimulating the microbial nutrient cycling. In addition, water including dissolved nutrients and carbon percolated deeper in plots of higher plant diversity. Thereby, higher plant diversity spatially extended the nutrient cycling through the microbial communities to deeper soil layers from which nutrients are transferred to the topsoil by deep-rooting plants. Although microbial nutrient cycling cannot fully compensate for negative plant diversity effects on nutrient availability in soil solution, this suggests that over time the role of plant-derived inputs becomes increasingly important for ecosystem functioning. It furthermore implies that plant species richness tightens plant-microbial interactions, which in the long-term feed back on other ecosystem functions, such as productivity.

## 1. Introduction

The rapid and extensive loss of biodiversity (Barnosky et al., 2011) has marked consequences for multiple ecosystem functions and services, for example, reduced biomass production, carbon storage and stability (e.g., Balvanera et al., 2006; Hooper et al., 2012). The soil is of particular importance for ecosystem functions and services (Wall and Six, 2015), since it controls, for example, the water and nutrient supply for plant growth, soil carbon storage and water cleaning and retention. To pinpoint mechanisms underlying the effects of biodiversity loss on ecosystem functions and services experiments are essential (Eisenhauer et al., 2016, 2019). However, while many studies demonstrated a strong impact of biodiversity loss on ecosystem functions like productivity (e.g., Hector et al., 1999; Marquard et al., 2009; Roscher et al., 2005; Tilman et al., 2001), the biodiversity effects on soil functions, such as nutrient cycling and provisioning, carbon storage and water dynamics remain underrated. This might be due to the fact that soil organisms, like the microbial community respond to plant diversity with a time lag (Eisenhauer et al., 2010), and that solid soil properties, such as the organic matter concentration, generally react slowly (Poeplau et al., 2011). Moreover, soil-related functions and processes are intimately coupled and feed back on each other. This high complexity is difficult to disentangle. Therefore, plant diversity effects on nutrient and carbon cycling and water dynamics need to be considered together in order to gain a deeper understanding of the underlying mechanisms.

The cycles of nutrients, carbon and water are interlinked by plants and soil microorganisms. The above- and belowground components of terrestrial ecosystems essentially depend on each other (Wardle et al., 2004), since plants provide carbon sources for soil fauna and microorganisms, while microorganisms and detritivore animals decompose and recycle organic matter and thereby increase the availability of nutrients for plants (Porazinska et al., 2003; van der Heijden et al., 2008). Moreover, plants and soil, and with it nutrient and carbon cycles, are linked by the availability of water. The availability of water strongly impacts plant growth, which directly influences carbon allocation to soils. Besides, microbial activity is strongly driven by soil moisture (Moyano et al., 2013). Thereby, plants probably indirectly influence the activity of the microbial community. Water uptake by plants drys the soil and their shading reduces evaporation losses and keeps the soil moist (Fischer et al., 2019). However, so far it is not clear how biodiversity is related with the complex interplay between nutrient, carbon and water cycles. Here, we review the current literature on different soil functions and processes and report long-term effects of plant diversity on nutrient, carbon and water dynamics to elucidate how plant diversity impacts the element and water cycling and how plant diversity changed their coupling over time.

## 2. Plant diversity effects on the soil microbial community and soil processes and functions

### 2.1 Microbial community composition and diversity

The structure and activity of the soil microbial community is closely linked to plant communities through biomass production, litter quality, seasonal variability of litter production, root-shoot carbon allocation, and root exudates (e.g., Hooper et al., 2000; Porazinska et al., 2003; Potthoff et al., 2006). Plant diversity influences a wide range of these litter input properties and thus the microbial community structure and activity (Eisenhauer et al., 2010; Habekost et al., 2008; Strecker et al., 2016; Zak et al., 2003). In the frame of the Jena Experiment, Lange et al. (2014) investigated how plant diversity affected soil microbial biomass and the microbial community structure based on the phospholipid fatty acid method (refer to Section 2.4). This study found that species richness was the main driver of soil microbial biomass, while the fungal-to-bacterial biomass ratio was positively affected by plant functional group richness and the presence of legumes. The effect of plant species richness on soil microbial biomass was mediated via root inputs and their nitrogen concentration. The detrimental effect of root nitrogen concentration on microbial biomass was closely related to increased root biomass; with increasing plant species richness root biomass increased, while at the same time nitrogen concentration decreased, which confirms earlier findings (Bessler et al., 2012; Eisenhauer et al., 2010). Specifying the role of plant diversity on the biomass and structure of the soil microbial community, Eisenhauer et al. (2017) set up a microcosm experiment. The authors reported that plant diversity increased shoot and root biomass, the amount of root exudates as well as the bacterial and fungal biomass. Furthermore, Eisenhauer et al. (2017) found that fungal biomass increased most with increasing plant diversity resulting in a significant shift in the fungal-to-bacterial biomass ratio at high plant diversity. This was, however, not confirmed by the field study of Lange et al. (2014), in which the fungal-to-bacterial biomass ratio was highest at intermediate diversity levels. The discrepancy between these studies might be caused by different ranges of the plant diversity gradient in both studies: the microcosm experiment comprised six plant species at the highest diversity level, which was similar to the intermediate diversity level in the Jena Experiment with eight plant species.

The nonlinear relation of plant diversity and fungal-to-bacterial biomass ratio in the Jena Experiment furthermore points to other mechanisms shaping soil microbial community structure. Beside plant inputs, the structure and activity of soil microbial communities depend also on soil properties, for instance pH, soil temperature, texture, and moisture (Griffiths et al., 2011; Papatheodorou et al., 2004; Steenwerth et al., 2008; Thoms et al., 2010). Lange et al. (2014) found that the positive plant diversity effect was strongly mediated by a higher leaf area index resulting in higher soil moisture in the top soil layer. Higher plant diversity increased canopy density of the plant stands, measured as leaf area index, which presumably reduced evaporation from the soil surface (Rosenkranz et al., 2012) and increased soil moisture in the uppermost layer of the soil, at least in the first years of the Jena Experiment (Fischer et al., 2019; Leimer et al., 2014). Thus, the positive plant diversity effect on the microbial biomass, and in particular on bacterial biomass was mainly driven by improved microclimatic conditions.

Another important factor controlling the soil moisture is soil texture: with smaller particle size the water holding capacity of the soil increases. However, the results of Lange et al. (2014) suggest that the positive effect of plant diversity on soil microbial biomass may even exceed that of soil texture via changes in soil moisture. Nonetheless, many studies investigating the effects of plant diversity on soil microbial communities focus on plant inputs, thereby neglecting the influence of plant diversity and composition on microclimatic conditions, such as soil moisture. The study by Lange et al. (2014) revealed that bacteria and fungi were similarly affected by plant diversity, however, bacteria were more controlled by changes in abiotic properties, while fungi were more affected by the input of organic materials.

In addition to the phospholipid fatty acid analyses, terminal-restriction fragment length polymorphism (TRFLP) revealed that genetic diversity of the bacterial and fungal community also increased with plant species richness (Lange et al., 2015). The increased diversity, in turn, has been reported to increase microbial activity (Bell et al., 2005). However, the plant diversity effect on genetic diversity of the microbial community was not consistent among different studies. While the earlier method of gene recognition (TRFLP) showed positive responses of bacterial and fungal diversity to plant species richness, the current and more advanced method, 454-pyrosequencing, showed differential responses of soil bacteria, fungi, archaea, and protists to plant species richness (Dassen et al., 2017). These differences may be due to different sampling times, but also due to the use of the different methods, varying in depth and phylogenetic resolution (Bent et al., 2007). These inconsistent effects of plant diversity highlight the need for deeper and long-term investigations of the genetic diversity and composition of the soil microbial community.

Generally, increasing plant species and plant functional group richness positively affect the diversity of soil biota, probably because plant species differ in root morphology, root chemical composition, and temporal variability of resource inputs (Hooper et al., 2000). The increased morphological, chemical, and temporal variability of belowground structures and resources likely results in increased diversity of niches supporting more diverse assemblages of soil biota (Eisenhauer et al., 2010; Hooper et al., 2000; Lavelle et al., 1995). Thus, the impact of plant diversity on soil biota may alter ecosystem processes mediated by microorganisms, such as the decomposition rates of organic matter (Orwin et al., 2006) and soil organic matter storage (Fornara and Tilman, 2008; Steinbeiss et al., 2008). As a consequence, plant diversity-mediated shifts in the microbial community biomass and composition are likely to exert feedback effects on other soil processes and plant growth (van der Heijden et al., 2008; Wardle et al., 2004; Zak et al., 2003).

## 2.2 Soil water balance

Soil moisture, and thus water availability play a crucial role for ecosystem functioning as a partitioned resource (for transpiration see Harpole and Tilman (2007)) as well as potential limiting factor (oxygen deficiency). Silvertown et al. (1999) and Araya et al. (2011) showed that niches generated by soil hydrology are crucial for plant community assembly. In contrast, much of the work on the relationship between biodiversity and ecosystem function has been concerned with drought stress (Kahmen et al., 2005; Tilman and Downing, 1994) and community water use, with differing results (De Boeck et al., 2006; Stocker et al., 1999). This is because soil water is not only a resource, but at the same time a property emerging from ecosystem processes and carbon uptake.

The relationships between soil moisture patterns and plant diversity also vary. Leimer et al. (2014) investigated the effects of plant species and functional richness and functional composition on soil water contents and derived water fluxes in the Jena Experiment during the establishment period between 2003 and 2007. In this period, there were no overall effects of species or functional group richness on water contents, evapotranspiration, and downward flux of the soil layers 0–0.3 and 0.3–0.7 m depth. This is in contrast to other studies, in which an increasing shading effect of vegetation on the water content of the topsoil was reported (e.g., Rosenkranz et al., 2012) and a water depletion effect in the subsoil. The reason for the discrepancies is likely the different depth and time considered. The shading effect seems to be restricted to the uppermost few cm of the soil and was observed in the Jena Experiment in studies targeting topsoil (Chen et al., 2017; Fischer et al., 2019; Lange et al., 2014; Rosenkranz et al., 2012). The depletion effect by deep roots is constrained to the deeper subsoil and also appeared only 8 years after establishment (Fischer et al., 2019).

In addition to plant diversity effects, Leimer et al. (2014) showed for the early phase of the Jena Experiment that presence of specific functional groups significantly changed water contents and fluxes with partly opposing effects in the upper versus the lower soil layer. The presence of tall herbs increased soil water contents in the topsoil during dry conditions and decreased soil water contents in the subsoil during wet conditions. The presence of grasses generally decreased water contents in the topsoil, particularly during dry phases; increased evapotranspiration and decreased modelled downward flux from topsoil; and decreased evapotranspiration from subsoil. The presence of legumes, in contrast, decreased evapotranspiration and increased downward flux from topsoil and evapotranspiration from subsoil.

Specific functional groups likely affected the water balance via specific root traits (e.g., shallow dense roots of grasses and deep taproots of tall herbs) or specific shading intensity caused by functional group effects on vegetation cover.

Fischer et al. (2019) revisited the analysis of soil water contents in the Jena Experiments 10 years after establishment. This study revealed that the plant community effects on the soil water conditions changed with advancing time of grassland establishment on former arable soil. While in the first years, soil moisture distribution was dominated by canopy processes and effects of functional groups on root uptake depth, in the later phase, after approximately 8 years, increasing plant diversity had affected soil hydraulic properties. Both soil porosity and infiltration capacity increased with species diversity (Fischer et al., 2015). This was attributed to improved aggregation of the soil in response to increasing soil organic matter accumulation and favoured microbial activity with increasing plant species richness (Gould et al., 2016; Peres et al., 2013). Thus, the biodiversity effect on soil moisture depends on the degree of ecosystem establishment after conversion from another land use and on the considered soil depth, which may explain partly contrasting results in the literature such as findings that soil moisture decreased with diversity in a natural grassland, probably due to enhanced water uptake and faster percolation as a consequence of a stable soil structure (Mokany et al., 2008).

At the same time, a study in the Montpellier European Ecotron confirmed deeper root water uptake, below the main rooting horizon, in more diverse plots (Guderle et al., 2018), although root distribution was unaffected by species diversity. The latter observation is in accordance with other findings at the Jena Experiment site, showing that species diversity increases root biomass (Ravenek et al., 2014) and rooting depth (Oram et al., 2018), but not the overall ecosystem root distribution (Oram et al., 2018). However, root water uptake locations can differ strongly from root biomass distributions, and can also shift quickly depending on soil water availability, especially during times of strong and sustained transpiration (Garrigues et al., 2006). Moreover, besides biomass, root morphology affects the depth of root water uptake in soil (Barkaoui et al., 2016).

In fact, the results by Guderle et al. (2018) derived from the soil water balance are in contrast to results from tracer studies (Bachmann et al., 2015; Jesch et al., 2018), which did not show any evidence of water uptake from deeper rooting horizons of diverse plots. The detection of complementarity effects might be related to the timing of the measurements and soil depth which is considered: on the one hand, Guderle et al. (2018) measured at peak plant biomass, at high water demand, while Bachmann et al. (2015) obtained samples at low leaf cover (spring and shortly after mowing), when the low water demand could be covered from shallow soil. Thus, it could be speculated that there is only complementarity during water stress periods, while there is none when the soil contains enough plant-available soil water. Moreover, Guderle et al. (2018) considered soil depth down to 60 cm and showed largest effects in 60 cm depth, whereas Bachmann et al. (2015) and Jesch et al. (2018) mainly considered the upper 30 cm. These contradictory results highlight the need for further research in this field. Nevertheless, the combination of higher infiltration capacity in the top soil and deeper uptake in diverse mixtures may have additionally contributed to enhance water use efficiency (Milcu et al., 2014), by decreasing soil evaporation (Milcu et al., 2016). Notably, while also functional groups affected soil properties indirectly via promoting or suppressing earthworms (Fischer et al., 2014) and had a strong effect on vertical soil water distribution (Fischer et al., 2019; Leimer et al., 2014), functional diversity, as opposed to species diversity, did not increase overall soil water depletion (Fischer et al., 2019).

In addition to soil water conditions, we assessed the ecosystem water budget of the grassland by measuring incident rainfall (water input above the vegetation canopy) and throughfall (water input to the soil after the passage through the vegetation canopy). The throughfall was measured with collectors placed 0.3 m above the soil surface. The difference between incident rainfall and throughfall represents either the loss of water ("interception loss"; difference > 0) or an input of water ("interception gain," difference < 0) during the passage through the vegetation. Overall, we found higher throughfall than incident rainfall illustrating that there was an interception gain of the grassland rather than a loss (Fig. 1), which could be caused by dew input. The interception gain increased with time (Fig. 1), possibly because of the establishment of the plant community associated with a denser and more structured canopy. Interestingly, this effect was more pronounced in diverse mixtures than in mixtures of low plant species richness, and could be an additional microclimatic explanation, besides shading, for the higher soil water contents in the topsoil (Chen et al., 2017; Lange et al., 2014; Rosenkranz et al., 2012).

In the Montpellier European Ecotron, Milcu et al. (2014) investigated whether and how plant diversity affects ecosystem properties related to carbon uptake efficiency, such as water use efficiency (WUE). The study

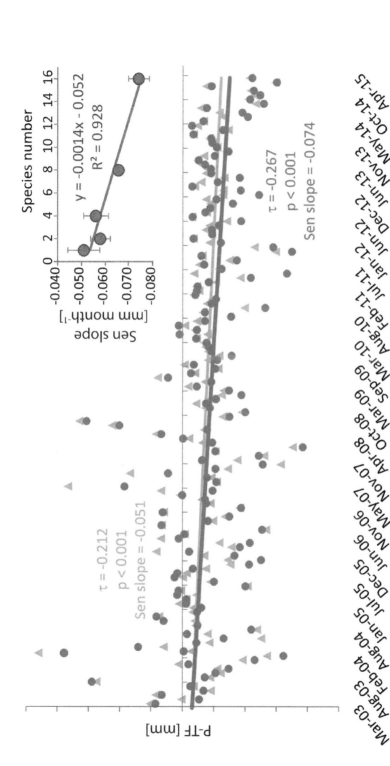

**Fig. 1** Temporal course of the mean interception gain (volume of incident rainfall [P]—throughfall [TF]) in the 2-species and 16-species mixtures of Block 2 (the first fully instrumented Block) between 2003 and 2015. The temporal trend was tested with the seasonal Mann-Kendall test after aggregation of the fortnightly measurements to monthly values (Hirsch et al., 1982). The τ value describes the strength of the trend. The Sen slope refers to the nonparametric slope of the temporal trend. The inlet represents the regression of the Sen slope on the plant species number. The regression line was added for illustration purposes.

found that higher diversity led to an increase in WUE by 30–50%. This finding was based on increased ecosystem evapotranspiration with simultaneously higher gross primary production in high plant diversity plots relative to low plant diversity plots. The increased WUE with plant diversity was explained by more diverse nitrogen distribution in the canopy, which is likely to allow more efficient use of light. In a follow-up modelling approach, the impact of species richness and functional diversity of plants on ecosystem water vapour fluxes were investigated (Milcu et al., 2016). The model results suggest that, at low plant species richness, a higher proportion of the available energy was diverted to evaporation (a nonproductive flux), while at higher species richness the proportion of ecosystem transpiration (a productivity-related water flux) increased. These results provide evidence that, at the peak of the growing season, higher leaf area index and lower percentage of bare ground at high plant diversity diverts more of the available water to transpiration, a flux closely coupled with photosynthesis and productivity. Higher rates of transpiration presumably contribute to the positive effect of diversity on productivity. However, so far the long-term effects of plant diversity and community composition on the soil moisture dynamics and how soil moisture dynamics are related to element cycling and storage are poorly understood.

The finding of Fischer et al. (2019), demonstrating that the effects of plant diversity on soil water contents changed with time, is an example for the interaction and couplings between the water and soil element cycles mediated by plants and microorganisms. Notably, the relation between plant diversity and soil water content moved away from being determined by transpiration (extraction-related) to being influenced by soil properties (storage-related, Table 1). The increasing plant biomass with increasing plant species richness provided shading at the beginning

**Table 1** Relation of soil water contents to plant species richness in the Jena Experiment using separate mixed effects models for each year.

| Depth | 2003 | 2004 | 2005 | 2008 | 2009 | 2010 | 2011 | 2012 | 2013 |
|---|---|---|---|---|---|---|---|---|---|
| 0–20 | Plant ↑ | Plant ↑ | – | Plant ↓ | – | Plant ↓ | Plant ↓ | Soil ** ↓ | – |
| 20–40 | – | – | Plant ↓ | Plant ↓ | – | – | Soil *** ↓ | Soil *** ↓ | Soil ↓ |
| 40–60 | – | – | – | – | – | – | Soil ** ↓ | Plant *↓ | – |

Direction of the arrow indicates whether soil water content increased (↑) or decreased (↓) with plant diversity; significance level: >0.1 (–, not significant, no trend), <0.05 (*), <0.01 (**) or <0.001 (***). The variables explained the soil water pattern better when replacing species richness by them were either plant related (e.g., leaf area, root biomass, cover) or soil related (e.g., soil organic matter, porosity, bulk density). Note the change of direction of the diversity-soil moisture relation over time and thus a favourable soil moisture. In addition, higher carbon inputs into the soil probably stimulated the mainly carbon-limited microorganisms to grow (Eisenhauer et al., 2010; Lange et al., 2014) and accelerated the organic matter turnover, thereby improving overall nutrient supply. In the long run, this may have further reaching consequences: favoured by the increased infiltration capacities in diverse plots, deep percolation of water with its nutrient and carbon concentrations is enhanced, thus likely allowing for the advancement of soil development to greater depth and at a faster rate. The latter might be even further pronounced by the increasing interception gain enhancing percolation in the species-rich mixtures (Fig. 1).

Thus, besides direct effects on water fluxes (evaporation and transpiration), plant community composition, including species richness and functional group assembly, influences soil functions, such as soil microbial biomass and activity (Drenovsky et al., 2004; Eisenhauer et al., 2010; Lange et al., 2014; Zak et al., 2003), soil fauna (Eisenhauer et al., 2011), and soil organic matter dynamics (Lange et al., 2015; Steinbeiss et al., 2008; refer to Section 2.4), all of which affect soil properties related to soil water percolation, e.g., soil aggregation (Gould et al., 2016; Peres et al., 2013), soil pore volume, and water infiltration capacity (Fischer et al., 2015). The close relationship between the biogeochemical and water cycles can incur further indirect effects of biodiversity on the availability of other resources that are mediated by soil water (Allan et al., 2013; Tilman et al., 2001). For example, diversity effects on in situ net ammonification rates were attributed to higher water content in the topsoil of species-rich grassland communities (Rosenkranz et al., 2012). In addition, the presence of certain plant functional groups, like legumes and grasses affected the abundance and activity of soil organisms (Eisenhauer et al., 2011; Milcu et al., 2008), which also modified soil water infiltration rates (Fischer et al., 2014).

## 2.3 Nutrient cycling

The effect of a biodiversity loss on ecosystem processes, such as soil nitrogen and phosphorus cycling was investigated in several manipulative field experiments (Hooper and Vitousek, 1998; Spehn et al., 2005). In soil, increased nitrogen uptake by the plant community was shown to decrease plant-available nitrogen concentrations or nitrate leaching in highly diverse mixtures (Leimer et al., 2016; Oelmann et al., 2007b; Tilman et al., 1996).

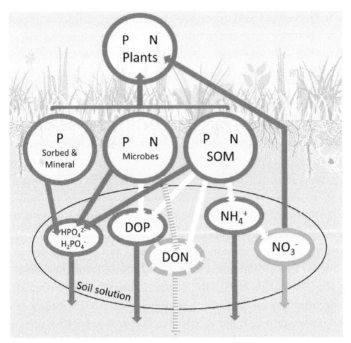

**Fig. 2** Plant diversity effects on the nitrogen and phosphorus cycle. Circles and ellipses represent the solid soil and dissolved nutrient pools, respectively. Arrows indicate fluxes between pools. Colours depict the direction of the slope of the relationship between plant diversity and the pool/flux considered: green = positive slope, dark yellow = negative slope, grey = nonsignificant, white = not studied yet. Dashed colours indicate effects observed in single years. *The figure illustrates results from studies conducted in the Jena Experiment, refer to section 2.3.*

Four years after establishment of the Jena Experiment, Leimer et al. (2016) could show that increasing plant species richness decreased dissolved organic nitrogen leaching (Fig. 2). This finding was attributed to enhanced use of dissolved organic nitrogen as a carbon and nitrogen source and enhanced mineralization of dissolved organic nitrogen by soil microorganisms. An increase of species richness decreased total dissolved nitrogen leaching likely driven by the complementary use of nitrate by diverse mixtures. Legumes increased dissolved organic nitrogen and total dissolved nitrogen leaching likely because of their $N_2$-fixing ability and higher litter production. Grasses decreased total dissolved nitrogen leaching because of more exhaustive use of nitrate and water. The results of this study demonstrate that leaching of dissolved organic nitrogen and total dissolved nitrogen is coupled to microbial and plant carbon cycling.

The effects of plant species richness and of grasses on dissolved nitrogen concentrations can be explained by a more exhaustive uptake by plants (Hooper and Vitousek, 1998; Niklaus et al., 2001; Scherer-Lorenzen et al., 2003; Tilman et al., 1996). Resource use complementarity has frequently been suggested as explanation for the increased plant uptake of nitrogen, but could not be directly confirmed in a tracer studies by (Jesch et al., 2018). If plants take up more nitrogen because of increased biomass production in diverse systems (Oelmann et al., 2007b), the uptake of all other essential elements, e.g., phosphorus must also increase. Karanika et al. (2007) presented support for this hypothesis based on a pot experiment. Similarly, Oelmann et al. (2011b) found that the phosphorus stock in aboveground biomass and the phosphorus exploitation increased with increasing plant species richness in the Jena Experiment (Fig. 2). However, similar to results of Hooper and Vitousek (1998), Oelmann et al. (2011b) did neither find effects of plant species richness on readily plant-available phosphorus concentrations in soil solution nor on the more strongly bound phosphorus fractions in the Jena Experiment (Fig. 2). A key to explain these patterns is to investigate the release of phosphorus during decomposition of organic matter and of subsequent retention by plants and microorganisms.

The microbial mineralization of nutrients was directly approached by assessing the microbial phosphatase activities (Hacker et al., 2015). This study reported that phosphatase activities increased with plant species richness, which was mediated by the availability of organic carbon and microbial biomass in the rhizosphere. Therefore, plant species richness effects on phosphorus mobilization in soil were driven by interactions between microorganisms in soil, organic carbon as a substrate for microorganisms, and the roots (rhizosphere) as a habitat for microorganisms. The control of phosphorus mobilization by soil microorganisms rather than by plant demand suggests that phosphorus mobilized by microorganisms is directly taken up by plants without entering plant-available phosphorus pools in soil (Fig. 2). These results indicate that the coupling between carbon and phosphorus cycling in soil becomes tighter with increasing plant diversity.

Furthermore, the studies by Hacker et al. (2015, 2017) revealed that microbial biomass phosphorus concentrations in soil increased with increasing plant species richness (Fig. 2). Functional group richness as well as the presence of specific plant functional groups showed no significant effect on microbial biomass phosphorus. The significant relationship between enzyme activities involved in the phosphorus cycle/microbial biomass phosphorus and plant species richness suggests that phosphorus nutrition of plant communities is governed by tightly coupled biological processes in the studied grassland (Hacker et al., 2015, 2017). The latter conclusion was corroborated by the results by Hacker et al. (2019) that one-third of the oxygen atoms in the plant-available phosphate pool were microbially exchanged within only six days after the addition of $^{18}$O-labeled water. Moreover, plant species richness was related to the $\delta^{18}$O values of soil water via evaporation. Because of the complete exchange of oxygen atoms in a phosphate molecule with ambient water during microbial phosphorus turnover in soil, plant species richness effects were also visible in $\delta^{18}$O values of bioavailable phosphate. These results suggest that microbial phosphorus demand and associated microbial phosphorus released by dying cells are important controls of net phosphorus release rates in soil (Fig. 2). In a study specifically addressing the phosphorus mobilization in the rhizosphere by root exudation of protons, which mainly increases the dissolution of hardly soluble Ca phosphates in the soil of the Jena Experiment, Hacker et al. (2017) determined the fast and slow reacting phosphorus pools by fitting a biexponential function to the release of phosphorus in response to a constant $H^+$ pressure. This study found that only the presence of legumes influenced the phosphorus release from mineral pools. Legumes decreased both, the fast and slow reacting phosphorus pools. Hacker et al. (2017) attributed this to the increased phosphorus demand and associated ability to access hardly available phosphorus fractions of legumes. Plant species richness-induced shifts of these processes did not affect phosphorus fractions in soil (Oelmann et al., 2011b), likely because of the rhizosphere acting as a tight link between soil organic matter, microorganisms and plant roots (Fig. 2).

Here, we used a linear mixed effect model to investigate the temporal course of plant species richness effects on plant-available nutrients. Specifically, we considered the soil solution concentrations (sampled at 0.3 m soil depth) of nitrogen in nitrate ($NO_3$-N) and ammonium ($NH_4$-N) and phosphorus in phosphate ($PO_4$-P), providing information on the interplay between microbial nutrient cycling and plant and microbial uptake of the mineralized nutrients in the rhizosphere. As reported for the early years of the experiment, nitrate concentrations decreased with increasing plant species richness and this effect was consistent over time (Table 2). Therefore, potentially positive plant species richness effects on nitrogen release associated with organic matter turnover (Oelmann et al., 2007b, 2011a) cannot compensate for negative plant diversity effects on nitrogen availability in soil (Oelmann et al., 2007b, 2011a) driven by increased nitrogen uptake of aboveground plant biomass. The continuous removal of nitrogen with

**Table 2** Results of linear mixed effects models (LMM), testing the impact of plant species richness (PSR), functional group composition and their changing effects over time on yearly averages of the soil solution concentrations (sampled at 0.3 m soil depth) of nitrogen in nitrate ($NO_3$-N) and ammonium ($NH_4$-N) and phosphorus in phosphate ($PO_4$-P) (2003–2015).

| | $NO_3$-N | | | $NH_4$-N | | | $PO_4$-P | | |
|---|---|---|---|---|---|---|---|---|---|
| | L-ratio | P value | | L-ratio | P value | | L-ratio | P value | |
| Year | 3.85 | **0.049** | ↓ | **9.36** | **0.002** | ↓ | **45.23** | **<0.001** | ↓ |
| PSR (log-scale) | **8.78** | **0.003** | ↓ | 0.15 | 0.69 | | 0.69 | 0.40 | |
| Grasses (y/n) | **18.05** | **<0.001** | ↓ | 0.23 | 0.63 | | 0.09 | 0.76 | |
| Legumes (y/n) | **30.49** | **<0.001** | ↑ | 1.15 | 0.28 | | 0.62 | 0.43 | |
| Tall herbs (y/n) | 1.32 | 0.25 | | 0.20 | 0.65 | | **6.90** | **0.008** | ↓ |
| Small herbs (y/n) | **7.45** | **0.006** | ↓ | 1.30 | 0.25 | | 0.04 | 0.84 | |
| Year × PSR (log-scale) | 0.30 | 0.58 | | 0.00 | 0.99 | | 1.18 | 0.28 | |
| Year × grasses (y/n) | 1.50 | 0.22 | | 0.23 | 0.63 | | 2.05 | 0.15 | |
| Year × legumes (y/n) | 1.71 | 0.19 | | 0.00 | 0.99 | | 0.07 | 0.78 | |
| Year × tall herbs (y/n) | 0.99 | 0.32 | | 0.45 | 0.50 | | 3.72 | 0.05 | |
| Year × small herbs (y/n) | 0.01 | 0.93 | | 0.32 | 0.57 | | 3.22 | 0.07 | |

Given are results of likelihood ratio tests (L-ratios and corresponding P values). Arrows indicate significant increase (↑) or decrease (↓) of the response variables (nitrate, ammonium, phosphate) to the variables of plant diversity and composition. LMM were fitted with "year" as random slope and "block" and "plot" as random intercepts. Bold text indicates significant effects ($\alpha = 0.05$).

the harvest decreased the nitrogen availability in soil (significant negative effect of "year" in Table 2) indicating that increased nitrogen uptake became more important over time. Deposition of nitrogen from the atmosphere accounts for less than one-fifth of the annual nitrogen removal by the harvest and thus, cannot compensate harvest nitrogen losses in the long term (Oelmann et al., 2007a). Similar to plant diversity effects, the effects of plant functional groups on nitrogen availability in soil reported in the early years of the experiment persisted over time (grasses, legumes, small herbs; Table 2). Therefore, although there was a succession of processes in soil (increase of carbon storage, increase of microbial biomass and activity) influencing the nitrogen cycle, the role of plants for the nitrogen cycle dominated irrespective of the duration of the experiment. In contrast, phosphorus availability in soil seemed to be insensitive to the effect of plant diversity or plant functional groups (Table 2), as was also reported in the early years of the experiment (Oelmann et al., 2011b). The only exception is the decrease of phosphate concentrations if tall herbs were present (Table 2), likely due to their increased exploitation of phosphorus in soil (Oelmann et al., 2011b). Similar to nitrogen, there was a depletion of the dissolved phosphate concentrations in soil solution with time, because of phosphorus removal with the harvest (significant negative effect of "year" in Table 2). Therefore, the plant diversity-dependent coupling among carbon, nitrogen, phosphorus, and water belowground maintain plant aboveground biomass production over time.

Beside mineral nutrients, the soil water also holds dissolved organic nitrogen and carbon. These organic, nonmineralized molecules can be utilized by the soil microbial community, thereby gaining energy and potentially providing plant available nitrogen (Leimer et al., 2016; Schimel and Bennett, 2004). Furthermore, the transport of these dissolved molecules into deeper soil layers can support up to 30% of the microbial activity below 40 cm (Neff and Asner, 2001). In addition to the total fluxes of dissolved organic nitrogen and carbon that are transported to deeper soil layers, the concentrations alone are important for their microbial utilization (Don et al., 2013). Here, we used linear mixed effect models to study how plant diversity affected the concentrations of dissolved organic nitrogen and dissolved organic carbon over time, below the rooting depth of most plants, at 0.3 m soil depth. The analyses revealed three major patterns in the Jena Experiment. First, mean annual concentrations of dissolved organic nitrogen varied largely among years with overall decreasing values with time, while mean annual concentrations of dissolved organic carbon have strongly decreased since the establishment of the experimental field site without large variations among years (Fig. 3). Second, plant species richness increased dissolved carbon concentrations (Table 3). Moreover, the dissolved organic nitrogen concentration was increased in the presence of legumes and decreased in the presence of grasses and tall herbs. And third, while there was no significant effect of plant species richness on total dissolved carbon concentration in the early years, a strong positive effect on total dissolved carbon concentrations was observed in the late years, as indicated by the significant interaction term year × PSR (Table 3, Fig. 3). Similarly, total dissolved nitrogen concentrations were inconsistently affected by plant species richness, too. More specifically, plant species richness decreased total dissolved nitrogen leaching in the 4th year after the establishment of the Jena

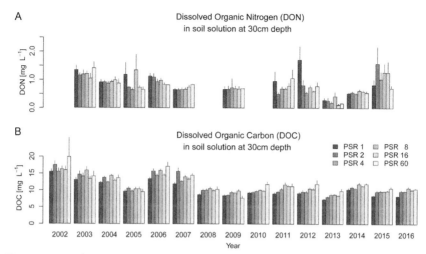

**Fig. 3** Impact of plant species richness (PSR) on concentrations of (A) dissolved organic nitrogen (DON) and (B) dissolved organic carbon (DOC) in different experiment years. Shown are mean annual concentrations measured along the PSR gradient. Soil solution was sampled in 30 cm depth.

**Table 3** Results of linear mixed effects models (LMM), testing the impact of plant species richness (PSR), functional group composition and their changing effects over time on yearly averages of dissolved organic carbon (DOC) and dissolved organic nitrogen (DON) concentrations in soil solution (2003–2015).

|  | DOC | | | DON | | |
|---|---|---|---|---|---|---|
|  | **L-ratio** | **P value** |  | **L-ratio** | **P value** |  |
| Year | **322.28** | **<0.001** | ↓ | **13.90** | **<0.001** | ↓ |
| PSR (log-scale) | **4.84** | **0.028** | ↑ | 1.65 | 0.20 | |
| Grasses (y/n) | 1.98 | 0.16 | ↑ | **9.11** | **0.002** | ↓ |
| Legumes (y/n) | 1.62 | 0.20 | ↑ | **7.38** | **0.007** | ↑ |
| Tall herbs (y/n) | 0.05 | 0.82 | | **4.40** | **0.040** | ↓ |
| Small herbs (y/n) | 2.88 | 0.09 | ↑ | 2.96 | 0.08 | |
| Year × SR (log-scale) | **6.47** | **0.011** | | 0.00 | 0.98 | |
| Year × grasses (y/n) | 0.12 | 0.73 | | 0.31 | 0.57 | |
| Year × legumes (y/n) | 0.21 | 0.64 | | 0.00 | 0.96 | |
| Year × tall herbs (y/n) | 1.57 | 0.21 | | 0.03 | 0.87 | |

Given are results of likelihood ratio tests (L-ratios and corresponding $P$ values). Arrows indicate significant increase (↑) or decrease (↓) of the response variables (nitrate, ammonium, phosphate) to the variables of plant diversity and composition. LMM were fitted with "year" as random slope and "block" and "plot" as random intercepts. Bold text indicates significant effects ($\alpha = 0.05$).

Experiment (Leimer et al., 2016). However, this effect did not persist over time. Together, this indicates lower nitrogen and carbon export with continued establishment of the field site, which is most likely related to the shift from a fertilized arable field to an unfertilized meadow. This shift is likely to deplete the bioavailable nutrient pools in soil over time, resulting in an overall decrease of plant productivity (Ravenek et al., 2014), which in turn may result in the decrease of dissolved organic carbon concentrations over time.

## 2.4 Plant carbon allocation to soil, microbial net assimilation and microbial carbon storage

Besides abiotic properties (Voroney and Heck, 2015), soil microbial activity is largely controlled by plant inputs (Wardle, 2002; but see Lange et al., 2014). Plants exude easily decomposable carbon forms into the soil (Jacoby et al., 2017; Kuzyakov and Xu, 2013) in order to enhance microbial nutrient cycling (Macdonald et al., 2018). It is estimated that 15% of photosynthetically fixed carbon is allocated into the soil via plant roots (Farrar et al., 2003). However, not all soil microorganisms equally depend directly on plant-derived carbon resources. Soil microorganisms can be classified in terms of their preferentially decomposed carbon sources: copiotrophs are able to decompose labile carbon sources, e.g., root exudates (Wardle et al., 2004), and oligotrophs have higher nutrient affinity and are capable of decomposing soil organic carbon and plant-derived litter (Fierer et al., 2007). This classification coincides largely with grouping the soil microorganisms by their abundance in different compartments of the soil (Bahn et al., 2013; Bird et al., 2011; Denef et al., 2009; Esperschutz et al., 2009; Kramer and Gleixner, 2006) using the analyses of phospholipid fatty acids (Frostegård et al., 2011; Zelles, 1997). In particular, Gram negative (G−) bacteria are recognized as rhizospheric microorganisms (Denef et al., 2009), preferentially decomposing labile substrates, such as recently fixed plant-derived carbon from root exudates and plant cell debris. In contrast, Gram-positive (G+) bacteria dominate the bacterial community in the bulk soil, and they are able to utilize more complex substrates (Bahn et al., 2013; Kramer and Gleixner, 2008).

To investigate how plant diversity influences the microbial assimilation of plant-derived carbon, the phospholipid fatty acid method in combination with stable isotopes is a highly useful tool (Mellado-Vázquez et al., 2019; Pett-Ridge and Firestone, 2017). Mellado-Vázquez et al. (2016) investigated how plant diversity drives the soil microbial uptake of plant-derived carbon. In this study a continuous $^{13}CO_2$ label was applied in a controlled environment (Montpellier European Ecotron) for 3 weeks to 12 soil-vegetation monoliths originating from the Jena Experiment. Mellado-Vázquez et al. (2016) reported an increased uptake of plant-derived carbon by G− bacteria and arbuscular mycorrhizal fungi with higher plant diversity. Root biomass but not the amount and $\delta^{13}C$ value of root sugars mediated the positive plant diversity effect observed on G− bacteria, whereas the specific interaction between plant and arbuscular mycorrhizal fungi was independent from any plant trait. This extended the view that changes in plant diversity only indirectly induce shifts in the carbon assimilation of soil microorganisms via changes in the microbial community structure (De Deyn et al., 2011a); and it provides evidence that the carbon assimilation of rhizosphere bacteria increases if the root provides more access to decomposable products. In contrast to the microbial groups living in the rhizosphere, the $\delta^{13}C$ values of microorganisms dominating the bulk soil (G+ bacteria, actinobacteria and saprotrophic fungi) were not significantly affected by plant diversity. Instead, their $\delta^{13}C$ values were positively correlated to $\delta^{13}C$ values of soil, indicating that soil organic carbon is their major carbon source (Bahn et al., 2013; Elfstrand et al., 2008; Kramer and Gleixner, 2008). These findings were in contrast to earlier studies (Chung et al., 2007, 2009) reporting that plant diversity increases microbial carbon uptake at higher plant diversity by enhanced root exudation. The results rather indicate that higher plant diversity increased the access to plant-derived carbon for root-associated microorganisms, which are more capable of utilizing easily decomposable carbon inputs (i.e., root exudates and plant cell debris) in the rhizosphere through more root biomass. They further indicate that plant diversity facilitated the accessibility of plant-derived carbon but not the above-belowground transfer rates. This facilitating effect enabled more diverse plant communities to use carbon dioxide complementarily (Gubsch et al., 2010) and most likely nutrient resources both from soil organic matter mineralization for better growth.

To investigate how plant diversity impacted the carbon allocation to the soil microbial community in the long term, we considered the ratio of G+ to G− as a proxy for the availability of easily decomposable carbon for the microbial community (Fanin et al., 2019), namely, root exudates (Mellado-Vázquez et al., 2016). Using linear mixed effect models, we here show that the G+ to G− ratio decreased with time, indicating that more easily degradable carbon is available for the decomposing bacteria in later years of the assessment (Fig. 4). Furthermore, no plant species richness effect was observed in the first year of this measurement (2007), while in the later years (2012, 2016) the carbon availability increased with plant species

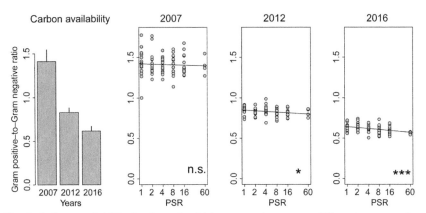

**Fig. 4** Carbon availability for the microbial community among different years and as affected by plant species richness (PSR). The ratio between Gram-positive (G+) to Gram-negative (G−) bacteria was considered as a proxy for the availability of easily decomposable carbon for the microbial community (Fanin et al., 2019). The assessments of the G+ and G− bacteria are based on PLFA markers (G+: i15:0, a15:0, i16:0, i17:0 and a17:0, G−: 16:1ω7, 17:1, 18:1ω7 (Zelles, 1997)). Significant effects of PSR are indicated with *$P < 0.05$, ***$P < 0.001$. n.s. is not significant.

richness as indicated by a negative relation of the G+ to G− ratio to plant species richness (Fig. 4). This result is in line with a recent global meta-analysis (Chen et al., 2019) and suggests that over time the role of plant-derived inputs becomes increasingly important for microbial functioning.

The enhanced belowground inputs associated with increased biomass production at higher diversity levels lead to increased soil organic carbon accumulation (De Deyn et al., 2011b; Fornara and Tilman, 2008; Steinbeiss et al., 2008). Generally, the amount of carbon stored in soil represents the balance between plant shoot and root litter production, root exudates and their microbial decomposition (Jastrow et al., 2007). Thus, the increase in carbon storage with plant diversity therefore either reflects higher primary production (Ravenek et al., 2014; Tilman et al., 2001) and allocation of plant-derived carbon to the soil (De Deyn et al., 2012) or longer persistence of plant-derived organic materials due to slower decomposition (Jastrow et al., 2007). Lange et al. (2015) aimed at elucidating the underlying mechanisms of this plant-soil relationship. The study reported that higher plant diversity increased rhizosphere carbon inputs into the microbial community resulting in increased microbial activity, which in turn unexpectedly increased soil carbon storage. In the 11 years of the experiment, the carbon storage in highly diverse plots exceeded the carbon storage in monocultures by 150% (390 ± 114 g kg$^{-1}$ (mean ± s.d.) in communities with 60 plant species; $155 \pm 77\,\mathrm{g\,kg^{-1}}$ in communities with 1 plant species). Increases in soil carbon were related to the enhanced accumulation of recently fixed carbon in high-diversity plots, while plant diversity had less pronounced effects on the decomposition rate of carbon already present before the establishment of the experiment (Lange et al. 2015). The microbial carbon storage was driven by higher levels of carbon input into the soil and more favourable microclimatic conditions caused by more diverse plant communities, that result in more active, more abundant (Lange et al., 2014) and more diverse soil microbial communities (Bell et al., 2005). Microbial activity increased the turnover rates of root litter and exudates as indicated by increased microbial respiration (Eisenhauer et al., 2010; Strecker et al., 2016). Thus, microbial products associated with increased microbial respiration, for example, microbial necromass, end up in slow-cycling carbon pools in the form of reduced organic material (Gleixner et al., 2002; Miltner et al., 2012; Schmidt et al., 2011). The study by Lange et al. (2015) demonstrates that elevated carbon storage at high plant diversity is a direct function of the metabolic activity of soil microorganisms (Liang et al., 2017). It further indicates that the increase in carbon storage is mainly limited by the integration of new carbon into soil and less by the decomposition of existing soil carbon. Therefore, in line with the recent debate (Liang and Balser, 2011; Schimel and Schaeffer, 2012), the findings of this study challenge previous views underestimating or even negating the influence of the abundance and activity of soil microorganisms on soil carbon storage. Instead, it suggests to reconsider the role of soil microorganisms as sources rather than sinks for slow-cycling organic matter. Thus, the activity and composition of soil microbial communities can serve as a proxy for carbon transfer into sustainable slow-cycling forms of soil carbon, and plant diversity and associated soil microbial communities can significantly contribute to sequestration of atmospheric carbon dioxide.

Investigating the shifts over time of biodiversity effects on soil organic matter, we here showed that soil organic carbon and total nitrogen contents continuously increased in the Jena Experiment since its establishment (Table 4, Fig. 5). These increases were positively related to plant species richness (Table 4, Fig. 5), indicating a general positive effect of plant species richness on soil organic carbon and nitrogen storage. This result confirms the increase in carbon storage in the topsoil (0–5 cm) reported by Lange et al. (2015) at greater soil depth (0–15 cm) and in the longer term (14 years). The accumulation continuously increased over time, and thereby the relationships between plant species richness and soil organic carbon and

**Table 4** Results of linear mixed effects models (LMM), testing the impact of plant species richness (PSR), functional group composition and their changing effects over time on soil organic carbon ($C_{org}$) and total soil nitrogen ($N_{tot}$) contents.

|  | $C_{org}$ | | | $N_{tot}$ | | |
|---|---|---|---|---|---|---|
|  | L-ratio | P value | | L-ratio | P value | |
| Year | **576.25** | **<0.001** | ↑ | **31.46** | **<0.001** | ↑ |
| PSR (log-scale) | **6.11** | **0.013** | ↑ | **8.70** | **0.003** | ↑ |
| Grasses (y/n) | 0.30 | 0.58 | | 1.46 | 0.23 | |
| Legumes (y/n) | 0.01 | 0.93 | | 0.04 | 0.85 | |
| Tall herbs (y/n) | 1.42 | 0.23 | | 0.20 | 0.66 | |
| Small herbs (y/n) | 2.67 | 0.10 | | 2.25 | 0.13 | |
| Year × PSR (log-scale) | **98.53** | **<0.001** | | **39.96** | **<0.001** | |
| Year × grasses (y/n) | 0.39 | 0.53 | | 0.16 | 0.69 | |
| Year × legumes (y/n) | 0.33 | 0.57 | | **5.24** | **0.022** | |
| Year × tall herbs (y/n) | 0.15 | 0.70 | | 0.82 | 0.37 | |
| Year × small herbs (y/n) | 3.02 | 0.08 | | 1.05 | 0.31 | |
| Year × small herbs (y/n) | 0.25 | 0.62 | | 0.04 | 0.85 | |

Given are results of likelihood ratio tests (L-ratios and corresponding P values). Arrows indicate significant increase (↑) or decrease (↓) of the response variables ($C_{org}$, $N_{tot}$) to the variables of plant diversity and composition. LMM were fitted with "year" as random slope and "block" and "plot" as random intercept. Bold text indicates significant effects ($\alpha = 0.05$).

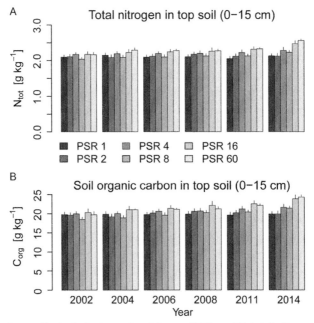

**Fig. 5** Long-term effect of plant species richness (PSR) on (A) total nitrogen ($N_{tot}$) and (B) soil organic carbon ($C_{org}$) contents.

nitrogen became more positive as indicated by the significant interaction term year × PSR (Table 4). Besides the general positive effect of plant species richness on soil organic matter storage, the significant interaction term year × legumes on total nitrogen contents showed that over time more nitrogen was stored in the presence of legumes. Compared to the species richness effect, however, the legume effect was less pronounced, as the nonsignificant overall effect shows (Table 4).

In summary, the increased access to root exudates and plant cell debris as well as larger soil organic carbon stocks found at high plant diversity results in strong effects on the soil microbial community, since more carbon resources exist for both rhizosphere and soil-related microorganisms (Eisenhauer et al., 2010; Mellado-Vázquez et al., 2016). The more abundant and active microbial community in turn has positive effects on soil carbon storage (Lange et al., 2015) because more microbial necromass is produced and finally stored as soil organic matter (Gleixner et al., 2002; Miltner et al., 2012). This relation highlights the dual role of the soil microorganisms in decomposing and producing soil organic matter at the same time (Gleixner, 2013; Lange et al., 2015; Schmidt et al., 2011). In turn, the more active and abundant microbial community due to larger carbon availability at high plant diversity alters nutrient cycling (Orwin et al., 2006). and thus nutrient availability for plants and consequently plant growth. Moreover, the increased organic carbon content at high diversity likely affects the soil water dynamics, because soil aggregation (Peres et al., 2013) and soil porosity (Fischer et al., 2015) increase with higher soil carbon storage, leading to faster drainage of rainwater to deeper soil layers (Fischer et al., 2019). This means that overall biodiversity improves the interlinkage of plants with soil microorganisms and carbon, nutrient and water cycles.

## 3. Consequences of the element and water cycles and their coupling for the BEF relationships

Long-term biodiversity experiments have revealed that the positive biodiversity-ecosystem functioning (BEF) relationship strengthens with time (Cardinale et al., 2007; Eisenhauer et al., 2019; Reich et al., 2012; Vogel et al., 2019). A strengthening control of biodiversity was, e.g., reported for aboveground plant biomass (Cardinale et al., 2007; Marquard et al., 2009; Reich et al., 2012), belowground plant biomass (Ravenek et al., 2014) and soil organisms (Eisenhauer et al., 2012). In contrast, no consistency in the strengthening of the BEF relationship with time was found for water, nutrient and carbon cycles in the soil (Meyer et al., 2016). The strengthening of the BEF relationship over time is assumed to be related to plant diversity-dependent shifts in the linkage between the plant community and soil microbial community over time (Eisenhauer et al., 2012; Vogel et al., 2019). Bardgett et al. (2005) proposed that over short timescales (hours to seasons) interactions among plant roots and the soil microbial community are highly dynamic and affect the cycling of nutrients, thereby influencing plant nutrient supply and growth. At intermediate timescales (decades to millennia), the changes in resource supply to soil, and feedback mechanisms between individual plants and their soil biological communities determine the nutrient cycling and plant growth.

The Jena Experiment (Roscher et al., 2004) has been running for >15 years. This is, of course, close to the lower limit of the intermediate timescale, but we expect to observe effects that have been proposed to play a role at intermediate timescales. Our expectations are based on the fundamental changes that probably occur in the plant-microbial interaction at a fast pace due to the land-use change from an arable field to a grassland, which took place during the establishment of the experiment. Because this land-use change is accompanied by changes in resource supply to soil (e.g., carbon), we expect a more dominant role of processes that play at intermediate timescales. In the case of the Jena Experiment, the initial conditions were that of a fertilized arable field. Because of the subsequent management of the experiment, namely, no fertilization and removal of the harvested shoot biomass, the soil conditions developed towards a seminatural meadow. Diversity effects observed on the short timescale were mainly related to plant growth, like for example, shoot biomass (Marquard et al., 2009) or the exploitation of plant-available nutrient pools (Fig. 6; Oelmann et al., 2007a,b, 2011b). In contrast, other soil related processes such as export of dissolved organic carbon by leaching or microbial activity and biomass (Eisenhauer et al., 2010; Strecker et al., 2016) responded to the plant diversity gradient with a delay of several years after the experiment had been established. These delays are probably due to the complexity of the involved processes, as for instance the dual role of the soil microorganisms: decomposing and producing soil organic matter and dissolved organic carbon at the same time (Gleixner, 2013; Lange et al., 2015; Schmidt et al., 2011). Thus, at the intermediate timescales, more abundant and more active microbial communities in turn lead to higher soil organic matter storage over time with higher plant diversity (Fig. 6; Eisenhauer et al., 2010; Lange et al., 2015; Strecker et al., 2016). Further, the increased soil carbon storage with plant diversity resulted in

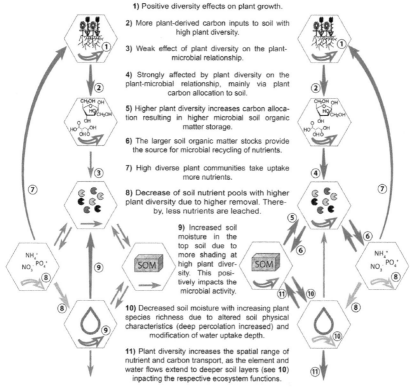

**Fig. 6** Proposed mechanisms how plant diversity drives water and soil element cycling (A) on short (hours to seasons) and (B) on intermediate timescales (decades to millennia). Pools are depicted by hexagons and element fluxes by arrows. Changes of the pools are reflected by the curved arrows inside the hexagons. Colours show the direction of the plant diversity effect: green=positive, dark yellow=negative, grey=nonsignificant or not consistent.

larger soil organic matter stocks and ultimately lead to a strong feedback loop with soil microorganisms, as the larger soil organic matter stocks support a more abundant and diverse soil microbial (Fig. 6) and animal community (Eisenhauer et al., 2011). Taken together, this contributes to maintain soil fertility at the intermediate timescale in spite of the regular nutrient removal via biomass harvest. Thus, higher plant diversity probably fosters a microbial life that increases soil biodiversity and contributes to maintain soil fertility (Klopf et al., 2017; Nielsen et al., 2015). The fact that aboveground phosphorus stocks, which are increased by increasing plant diversity, cannot be linked to depletion in soil phosphorus pools (Fig. 2) indicates that there are shortcuts between microorganisms and plants that are not accessible with the traditional methods used to determine nutrient availability in soil (Fig. 6). These shortcuts, promoted by increased microbial recycling of organic matter, appear to be accelerated by increased root exudation rates or facilitated uptake of exudates by the microbial community (Mellado-Vázquez et al., 2016). However, plant diversity not only increases the strength of the plant-microbial linkage in the rhizosphere, it also likely extends the positive plant diversity effects to deeper soil layers at the intermediate timescale (Fig. 6). With higher plant diversity, the concentrations of dissolved organic carbon in the soil solution are increased. This continuous transport of small amounts of organic material to deeper soil layers is likely to contribute to the formation of soil structure (Neff and Asner, 2001). In addition, the vertical transport of organic material increases, not only because of higher total dissolved carbon concentrations, but also because of enhanced deep percolation of water with higher plant diversity (Fischer et al., 2019).

We conclude that the carbon cycle represents the crucial cog in the wheel of plant diversity effects on coupled water, nutrient and carbon cycles in grassland mixtures (Fig. 6). On the one hand, plant diversity-dependent organic carbon accumulation in soil contributes to diversity-dependent nutrient release and subsequent nutrient uptake by plants and soil microorganisms. On the other hand, diversity-dependent organic carbon storage by plants and in soil affects soil water in different ways depending on the time since the establishment of the experiment. These effects then again translate into diversity-dependent nutrient release and leaching. Yet, it remains to be seen whether the diversity-dependent organic matter accumulation in soil continues once steady-state conditions, i.e., an organic matter content typical for grasslands, will have been reached.

Our results rely on simultaneous studies of the effect of plant diversity on the nutrient, carbon and water cycles in the Jena Experiment. Multitracer isotope studies linking plants, soil organisms and soil can help to prove direct couplings and feedbacks between the water, nutrient and carbon cycles.

We identified organic matter accumulation in soil as a crucial component in the coupling of nutrient, carbon and water cycles, and we acknowledge that such an accumulation may saturate in the longer term. However, we hypothesize that the steady state contents of soil organic matter are positively related to diversity. Therefore, future studies should aim at (i) investigating these relationships in biodiversity experiments for several decades and (ii) disentangling the complex interplay of factors involved in carbon cycling including plant diversity in established seminatural grasslands.

## Acknowledgements

We thank Uta Gerighausen and Steffen Rühlow for their help during sampling of soil and soil water, the RoMA at the MPI for Biogeochemistry provided the data on carbon and nitrogen. We gratefully acknowledge Nico Eisenhauer, Wolfgang W. Weisser, Ernst-Detlef Schulze, Christiane Roscher, Alex Weigelt and Anne Ebeling and all the people who were involved in planning, set up and maintaining of the experiment. We would also like to thank two anonymous reviewers and Nico Eisenhauer as serial volume editor for their reviews that helped to improve the manuscript. The Jena Experiment is funded by the Deutsche Forschungsgemeinschaft (DFG FOR 456, FOR 1451, DFG Oe516/3-1, -2; DFG Wi1601/20-1; DFG Gl262/14 and 19), with additional support from the Friedrich Schiller University of Jena, the Max Planck Society and the Swiss National Science Foundation (SNF). Markus Lange is funded by the Zwillenberg-Tietz foundation. Financial support through the ProExzellenz Initiative from the German federal state of Thuringia to the Friedrich Schiller University Jena within the research project AquaDiva@Jena for conducting the research is gratefully acknowledged.

## References

Allan, E., Weisser, W.W., Fischer, M., Schulze, E.-D., Weigelt, A., Roscher, C., Baade, J., Barnard, R.L., Bessler, H., Buchmann, N., Ebeling, A., Eisenhauer, N., Engels, C., Fergus, A.J.F., Gleixner, G., Gubsch, M., Halle, S., Klein, A.M., Kertscher, I., Kuu, A., Lange, M., Le Roux, X., Meyer, S.T., Migunova, V.D., Milcu, A., Niklaus, P.A., Oelmann, Y., Pasalic, E., Petermann, J.S., Poly, F., Rottstock, T., Sabais, A.C.W., Scherber, C., Scherer-Lorenzen, M., Scheu, S., Steinbeiss, S., Schwichtenberg, G., Temperton, V., Tscharntke, T., Voigt, W., Wilcke, W., Wirth, C., Schmid, B., 2013. A comparison of the strength of biodiversity effects across multiple functions. Oecologia 173, 223–237.

Araya, Y.N., Silvertown, J., Gowing, D.J., McConway, K.J., Linder, H.P., Midgley, G., 2011. A fundamental, eco-hydrological basis for niche segregation in plant communities. New Phytol. 189, 253–258.

Bachmann, D., Gockele, A., Ravenek, J.M., Roscher, C., Strecker, T., Weigelt, A., Buchmann, N., 2015. No evidence of complementary water use along a plant species richness gradient in temperate experimental grasslands. PLoS One 10 e0116367.

Bahn, M., Lattanzi, F.A., Hasibeder, R., Wild, B., Koranda, M., Danese, V., Bruggemann, N., Schmitt, M., Siegwolf, R., Richter, A., 2013. Responses of belowground carbon allocation dynamics to extended shading in mountain grassland. New Phytol. 198, 116–126.

Balvanera, P., Pfisterer, A.B., Buchmann, N., He, J.S., Nakashizuka, T., Raffaelli, D., Schmid, B., 2006. Quantifying the evidence for biodiversity effects on ecosystem functioning and services. Ecol. Lett. 9, 1146–1156.

Bardgett, R.D., Bowman, W.D., Kaufmann, R., Schmidt, S.K., 2005. A temporal approach to linking aboveground and belowground ecology. Trends Ecol. Evol. 20, 634–641.

Barkaoui, K., Roumet, C., Volaire, F., 2016. Mean root trait more than root trait diversity determines drought resilience in native and cultivated Mediterranean grass mixtures. Agr. Ecosyst. Environ. 231, 122–132.

Barnosky, A.D., Matzke, N., Tomiya, S., Wogan, G.O.U., Swartz, B., Quental, T.B., Marshall, C., McGuire, J.L., Lindsey, E.L., Maguire, K.C., Mersey, B., Ferrer, E.A., 2011. Has the Earth's sixth mass extinction already arrived? Nature 471, 51–57.

Bell, T., Newman, J.A., Silverman, B.W., Turner, S.L., Lilley, A.K., 2005. The contribution of species richness and composition to bacterial services. Nature 436, 1157–1160.

Bent, S.J., Pierson, J.D., Forney, L.J., 2007. Measuring species richness based on microbial community fingerprints: the emperor has no clothes. Appl. Environ. Microbiol. 73, 2399.

Bessler, H., Oelmann, Y., Roscher, C., Buchmann, N., Scherer-Lorenzen, M., Schulze, E.-D., Temperton, V.M., Wilcke, W., Engels, C., 2012. Nitrogen uptake by grassland communities: contribution of N-2 fixation, facilitation, complementarity, and species dominance. Plant Soil 358, 301–322.

Bird, J.A., Herman, D.J., Firestone, M.K., 2011. Rhizosphere priming of soil organic matter by bacterial groups in a grassland soil. Soil Biol. Biochem. 43, 718–725.

Cardinale, B.J., Wright, J.P., Cadotte, M.W., Carroll, I.T., Hector, A., Srivastava, D.S., Loreau, M., Weis, J.J., 2007. Impacts of plant diversity on biomass production increase through time because of species complementarity. Proc. Natl. Acad. Sci. U. S. A. 104, 18123–18128.

Chen, H.M., Mommer, L., van Ruijven, J., de Kroon, H., Fischer, C., Gessler, A., Hildebrandt, A., Scherer-Lorenzen, M., Wirth, C., Weigelt, A., 2017. Plant species richness negatively affects root decomposition in grasslands. J. Ecol. 105, 209–218.

Chen, C., Chen, H.Y.H., Chen, X., Huang, Z., 2019. Meta-analysis shows positive effects of plant diversity on microbial biomass and respiration. Nat. Commun. 10, 1332.

Chung, H., Zak, D.R., Reich, P.B., Ellsworth, D.S., 2007. Plant species richness, elevated $CO_2$, and atmospheric nitrogen deposition alter soil microbial community composition and function. Glob. Chang. Biol. 13, 980–989.

Chung, H.G., Zak, D.R., Reich, P.B., 2009. Microbial assimilation of new photosynthate is altered by plant species richness and nitrogen deposition. Biogeochemistry 94, 233–242.

Dassen, S., Cortois, R., Martens, H., de Hollander, M., Kowalchuk, G.A., van der Putten, W.H., de Deyn, G.B., 2017. Differential responses of soil bacteria, fungi, archaea and protists to plant species richness and plant functional group identity. Mol. Ecol. 26, 4085–4098.

De Boeck, H.J., Lemmens, C., Bossuyt, H., Malchair, S., Carnol, M., Merckx, R., Nijs, I., Ceulemans, R., 2006. How do climate warming and plant species richness affect water use in experimental grasslands? Plant Soil 288, 249–261.

De Deyn, G.B., Quirk, H., Oakley, S., Ostle, N., Bardgett, R.D., 2011a. Rapid transfer of photosynthetic carbon through the plant-soil system in differently managed species-rich grasslands. Biogeosciences 8, 1131–1139.

De Deyn, G.B., Shiel, R.S., Ostle, N.J., McNamara, N.P., Oakley, S., Young, I., Freeman, C., Fenner, N., Quirk, H., Bardgett, R.D., 2011b. Additional carbon sequestration benefits of grassland diversity restoration. J. Appl. Ecol. 48, 600–608.

De Deyn, G.B., Quirk, H., Oakley, S., Ostle, N.J., Bardgett, R.D., 2012. Increased plant carbon translocation linked to overyielding in grassland species mixtures. PLoS One 7, e45926.

Denef, K., Roobroeck, D., Wadu, M.C.W.M., Lootens, P., Boeckx, P., 2009. Microbial community composition and rhizodeposit-carbon assimilation in differently managed temperate grassland soils. Soil Biol. Biochem. 41, 144–153.

Don, A., Roedenbeck, C., Gleixner, G., 2013. Unexpected control of soil carbon turnover by soil carbon concentration. Environ. Chem. Lett. 11, 407–413.

Drenovsky, R.E., Vo, D., Graham, K.J., Scow, K.M., 2004. Soil water content and organic carbon availability are major determinants of soil microbial community composition. Microb. Ecol. 48, 424–430.

Eisenhauer, N., Bessler, H., Engels, C., Gleixner, G., Habekost, M., Milcu, A., Partsch, S., Sabais, A.C.W., Scherber, C., Steinbeiss, S., Weigelt, A., Weisser, W.W., Scheu, S., 2010. Plant diversity effects on soil microorganisms support the singular hypothesis. Ecology 91, 485–496.

Eisenhauer, N., Milcu, A., Sabais, A.C.W., Bessler, H., Brenner, J., Engels, C., Klarner, B., Maraun, M., Partsch, S., Roscher, C., Schonert, F., Temperton, V.M., Thomisch, K., Weigelt, A., Weisser, W.W., Scheu, S., 2011. Plant diversity surpasses plant functional groups and plant productivity as driver of soil biota in the long term. PLoS One 6, e16055.

Eisenhauer, N., Reich, P.B., Scheu, S., 2012. Increasing plant diversity effects on productivity with time due to delayed soil biota effects on plants. Basic Appl. Ecol. 13, 571–578.

Eisenhauer, N., Barnes, A.D., Cesarz, S., Craven, D., Ferlian, O., Gottschall, F., Hines, J., Sendek, A., Siebert, J., Thakur, M.P., Türke, M., 2016. Biodiversity–ecosystem function experiments reveal the mechanisms underlying the consequences of biodiversity change in real world ecosystems. J. Veg. Sci. 27, 1061–1070.

Eisenhauer, N., Lanoue, A., Strecker, T., Scheu, S., Steinauer, K., Thakur, M.P., Mommer, L., 2017. Root biomass and exudates link plant diversity with soil bacterial and fungal biomass. Sci. Rep. 7, 44641.

Eisenhauer, N., Schielzeth, H., Barnes, A.D., Barry, K.E., Bonn, A., Brose, U., Bruelheide, H., Buchmann, N., Buscot, F., Ebeling, A., Ferlian, O., Freschet, G.T., Giling, D.P., Hättenschwiler, S., Hillebrand, H., Hines, J., Isbell, F., Koller-France, E., König-Ries, B., de Kroon, H., Meyer, S.T., Milcu, A., Müller, J., Nock, C.A., Petermann, J.S., Roscher, C., Scherber, C., Scherer-Lorenzen, M., Schmid, B., Schnitzer, S.A., Schuldt, A., Tscharntke, T., Türke, M., van Dam, N.M., van der Plas, F., Vogel, A., Wagg, C., Wardle, D.A., Weigelt, A., Weisser, W.W., Wirth, C., Jochum, M., 2019. A multitrophic perspective on biodiversity–ecosystem functioning research. Adv. Ecol. Res. 61, 1–54.

Elfstrand, S., Lagerlöf, J., Hedlund, K., Mårtensson, A., 2008. Carbon routes from decomposing plant residues and living roots into soil food webs assessed with $^{13}C$ labelling. Soil Biol. Biochem. 40, 2530–2539.

Esperschutz, J., Buegger, F., Winkler, J.B., Munch, J.C., Schloter, M., Gattinger, A., 2009. Microbial response to exudates in the rhizosphere of young beech trees (*Fagus sylvatica* L.) after dormancy. Soil Biol. Biochem. 41, 1976–1985.

Fanin, N., Kardol, P., Farrell, M., Nilsson, M.-C., Gundale, M.J., Wardle, D.A., 2019. The ratio of gram-positive to gram-negative bacterial PLFA markers as an indicator of carbon availability in organic soils. Soil Biol. Biochem. 128, 111–114.

Farrar, J., Hawes, M., Jones, D., Lindow, S., 2003. How roots control the flux of carbon to the rhizosphere. Ecology 84, 827–837.

Fierer, N., Bradford, M.A., Jackson, R.B., 2007. Toward an ecological classification of soil bacteria. Ecology 88, 1354–1364.

Fischer, C., Roscher, C., Jensen, B., Eisenhauer, N., Baade, J., Attinger, S., Scheu, S., Weisser, W.W., Schumacher, J., Hildebrandt, A., 2014. How do earthworms, soil texture and plant composition affect infiltration along an experimental plant diversity gradient in grassland? PLoS One 9, e98987.

Fischer, C., Tischer, J., Roscher, C., Eisenhauer, N., Ravenek, J., Gleixner, G., Attinger, S., Jensen, B., de Kroon, H., Mommer, L., Scheu, S., Hildebrandt, A., 2015. Plant species diversity affects infiltration capacity in an experimental grassland through changes in soil properties. Plant Soil 397, 1–16.

Fischer, C., Leimer, S., Roscher, C., Ravenek, J., de Kroon, H., Kreutziger, Y., Baade, J., Beßler, H., Eisenhauer, N., Weigelt, A., Mommer, L., Lange, M., Gleixner, G., Wilcke, W., Schröder, B., Hildebrandt, A., 2019. Plant species richness and functional groups have different effects on soil water content in a decade-long grassland experiment. J. Ecol. 107, 127–141.

Fornara, D.A., Tilman, D., 2008. Plant functional composition influences rates of soil carbon and nitrogen accumulation. J. Ecol. 96, 314–322.
Frostegård, Å., Tunlid, A., Bååth, E., 2011. Use and misuse of PLFA measurements in soils. Soil Biol. Biochem. 43, 1621–1625.
Garrigues, E., Doussan, C., Pierret, A., 2006. Water uptake by plant roots: I—formation and propagation of a water extraction front in mature root systems as evidenced by 2D light transmission imaging. Plant Soil 283, 83–98.
Gleixner, G., 2013. Soil organic matter dynamics: a biological perspective derived from the use of compound-specific isotopes studies. Ecol. Res. 28, 683–695.
Gleixner, G., Poirier, N., Bol, R., Balesdent, J., 2002. Molecular dynamics of organic matter in a cultivated soil. Org. Geochem. 33, 357–366.
Gould, I.J., Quinton, J.N., Weigelt, A., De Deyn, G.B., Bardgett, R.D., 2016. Plant diversity and root traits benefit physical properties key to soil function in grasslands. Ecol. Lett. 19, 1140–1149.
Griffiths, R.I., Thomson, B.C., James, P., Bell, T., Bailey, M., Whiteley, A.S., 2011. The bacterial biogeography of British soils. Environ. Microbiol. 13, 1642–1654.
Gubsch, M., Buchmann, N., Schmid, B., Schulze, E.-D., Lipowsky, A., Roscher, C., 2010. Differential effects of plant diversity on functional trait variation of grass species. Ann. Bot. 107, 157–169.
Guderle, M., Bachmann, D., Milcu, A., Gockele, A., Bechmann, M., Fischer, C., Roscher, C., Landais, D., Ravel, O., Devidal, S., Roy, J., Gessler, A., Buchmann, N., Weigelt, A., Hildebrandt, A., 2018. Dynamic niche partitioning in root water uptake facilitates efficient water use in more diverse grassland plant communities. Funct. Ecol. 32, 214–227.
Habekost, M., Eisenhauer, N., Scheu, S., Steinbeiss, S., Weigelt, A., Gleixner, G., 2008. Seasonal changes in the soil microbial community in a grassland plant diversity gradient four years after establishment. Soil Biol. Biochem. 40, 2588–2595.
Hacker, N., Ebeling, A., Gessler, A., Gleixner, G., Mace, O.G., de Kroon, H., Lange, M., Mommer, L., Eisenhauer, N., Ravenek, J., Scheu, S., Weigelt, A., Wagg, C., Wilcke, W., Oelmann, Y., 2015. Plant diversity shapes microbe-rhizosphere effects on P mobilisation from organic matter in soil. Ecol. Lett. 18, 1356–1365.
Hacker, N., Gleixner, G., Lange, M., Wilcke, W., Oelmann, Y., 2017. Phosphorus release from mineral soil by acid hydrolysis: method development, kinetics, and plant community composition effects. Soil Sci. Soc. Am. J. 81, 1389–1400.
Hacker, N., Wilcke, W., Oelmann, Y., 2019. The oxygen isotope composition of bioavailable phosphate in soil reflects the oxygen isotope composition in soil water driven by plant diversity effects on evaporation. Geochim. Cosmochim. Acta 248, 387–399.
Harpole, W.S., Tilman, D., 2007. Grassland species loss resulting from reduced niche dimension. Nature 446, 791–793.
Hector, A., Schmid, B., Beierkuhnlein, C., Caldeira, M.C., Diemer, M., Dimitrakopoulos, P.G., Finn, J.A., Freitas, H., Giller, P.S., Good, J., Harris, R., Hogberg, P., Huss-Danell, K., Joshi, J., Jumpponen, A., Korner, C., Leadley, P.W., Loreau, M., Minns, A., Mulder, C.P.H., O'Donovan, G., Otway, S.J., Pereira, J.S., Prinz, A., Read, D.J., Scherer-Lorenzen, M., Schulze, E.D., Siamantziouras, A.S.D., Spehn, E.M., Terry, A.C., Troumbis, A.Y., Woodward, F.I., Yachi, S., Lawton, J.H., 1999. Plant diversity and productivity experiments in European grasslands. Science 286, 1123–1127.
Hirsch, R.M., Slack, J.R., Smith, R.A., 1982. Techniques of trend analysis for monthly water-quality data. Water Resour. Res. 18, 107–121.
Hooper, D.U., Vitousek, P.M., 1998. Effects of plant composition and diversity on nutrient cycling. Ecol. Monogr. 68, 121–149.
Hooper, D.U., Bignell, D.E., Brown, V.K., Brussaard, L., Dangerfield, J.M., Wall, D.H., Wardle, D.A., Coleman, D.C., Giller, K.E., Lavelle, P., Van der Putten, W.H., De Ruiter, P.C., Rusek, J., Silver, W.L., Tiedje, J.M., Wolters, V., 2000. Interactions

between aboveground and belowground biodiversity in terrestrial ecosystems: patterns, mechanisms, and feedbacks. Bioscience 50, 1049–1061.

Hooper, D.U., Adair, E.C., Cardinale, B.J., Byrnes, J.E.K., Hungate, B.A., Matulich, K.L., Gonzalez, A., Duffy, J.E., Gamfeldt, L., O'Connor, M.I., 2012. A global synthesis reveals biodiversity loss as a major driver of ecosystem change. Nature 486, 105–U129.

Jacoby, R., Peukert, M., Succurro, A., Koprivova, A., Kopriva, S., 2017. The role of soil microorganisms in plant mineral nutrition-current knowledge and future directions. Front. Plant Sci. 8, 1617.

Jastrow, J.D., Amonette, J.E., Bailey, V.L., 2007. Mechanisms controlling soil carbon turnover and their potential application for enhancing carbon sequestration. Clim. Change 80, 5–23.

Jesch, A., Barry, K.E., Ravenek, J.M., Bachmann, D., Strecker, T., Weigelt, A., Buchmann, N., de Kroon, H., Gessler, A., Mommer, L., Roscher, C., Scherer-Lorenzen, M., 2018. Below-ground resource partitioning alone cannot explain the biodiversity-ecosystem function relationship: a field test using multiple tracers. J. Ecol. 106, 2002–2018.

Kahmen, A., Perner, J., Buchmann, N., 2005. Diversity-dependent productivity in semi-natural grasslands following climate perturbations. Funct. Ecol. 19, 594–601.

Karanika, E.D., Alifragis, D.A., Mamolos, A.P., Veresoglou, D.S., 2007. Differentiation between responses of primary productivity and phosphorus exploitation to species richness. Plant Soil 297, 69–81.

Klopf, R.P., Baer, S.G., Bach, E.M., Six, J., 2017. Restoration and management for plant diversity enhances the rate of belowground ecosystem recovery. Ecol. Appl. 27, 355–362.

Kramer, C., Gleixner, G., 2006. Variable use of plant- and soil-derived carbon by microorganisms in agricultural soils. Soil Biol. Biochem. 38, 3267–3278.

Kramer, C., Gleixner, G., 2008. Soil organic matter in soil depth profiles: distinct carbon preferences of microbial groups during carbon transformation. Soil Biol. Biochem. 40, 425–433.

Kuzyakov, Y., Xu, X.L., 2013. Competition between roots and microorganisms for nitrogen: mechanisms and ecological relevance. New Phytol. 198, 656–669.

Lange, M., Habekost, M., Eisenhauer, N., Roscher, C., Bessler, H., Engels, C., Oelmann, Y., Scheu, S., Wilcke, W., Schulze, E.-D., Gleixner, G., 2014. Biotic and abiotic properties mediating plant diversity effects on soil microbial communities in an experimental grassland. PLoS One 9 (5), e96182.

Lange, M., Eisenhauer, N., Sierra, C.A., Bessler, H., Engels, C., Griffiths, R.I., Mellado-Vazquez, P.G., Malik, A.A., Roy, J., Scheu, S., Steinbeiss, S., Thomson, B.C., Trumbore, S.E., Gleixner, G., 2015. Plant diversity increases soil microbial activity and soil carbon storage. Nat. Commun. 6, 6707.

Lavelle, P., Lattaud, C., Trigo, D., Barois, I., 1995. Mutualism and biodiversity in soils. Plant Soil 170, 23–33.

Leimer, S., Kreutziger, Y., Rosenkranz, S., Bessler, H., Engels, C., Hildebrandt, A., Oelmann, Y., Weisser, W.W., Wirth, C., Wilcke, W., 2014. Plant diversity effects on the water balance of an experimental grassland. Ecohydrology 7, 1378–1391.

Leimer, S., Oelmann, Y., Eisenhauer, N., Milcu, A., Roscher, C., Scheu, S., Weigelt, A., Wirth, C., Wilcke, W., 2016. Mechanisms behind plant diversity effects on inorganic and organic N leaching from temperate grassland. Biogeochemistry 131, 339–353.

Liang, C., Balser, T.C., 2011. Microbial production of recalcitrant organic matter in global soils: implications for productivity and climate policy. Nat. Rev. Microbiol. 9, 75.

Liang, C., Schimel, J.P., Jastrow, J.D., 2017. The importance of anabolism in microbial control over soil carbon storage. Nat. Microbiol. 2, 17105.

Macdonald, C.A., Delgado-Baquerizo, M., Reay, D.S., Hicks, L.C., Singh, B.K., 2018. Chapter 6—soil nutrients and soil carbon storage: modulators and mechanisms. In: Singh, B.K. (Ed.), Soil Carbon Storage. Academic Press, pp. 167–205.

Marquard, E., Weigelt, A., Temperton, V.M., Roscher, C., Schumacher, J., Buchmann, N., Fischer, M., Weisser, W.W., Schmid, B., 2009. Plant species richness and functional composition drive overyielding in a six-year grassland experiment. Ecology 90, 3290–3302.

Mellado-Vázquez, P.G., Lange, M., Bachmann, D., Gockele, A., Karlowsky, S., Milcu, A., Piel, C., Roscher, C., Roy, J., Gleixner, G., 2016. Plant diversity generates enhanced soil microbial access to recently photosynthesized carbon in the rhizosphere. Soil Biol. Biochem. 94, 122–132.

Mellado-Vázquez, P.G., Lange, M., Gleixner, G., 2019. Soil microbial communities and their carbon assimilation are affected by soil properties and season but not by plants differing in their photosynthetic pathways (C3 vs. C4). Biogeochemistry 142, 175–187.

Meyer, S.T., Ebeling, A., Eisenhauer, N., Hertzog, L., Hillebrand, H., Milcu, A., Pompe, S., Abbas, M., Bessler, H., Buchmann, N., De Luca, E., Engels, C., Fischer, M., Gleixner, G., Hudewenz, A., Klein, A.M., de Kroon, H., Leimer, S., Loranger, H., Mommer, L., Oelmann, Y., Ravenek, J.M., Roscher, C., Rottstock, T., Scherber, C., Scherer-Lorenzen, M., Scheu, S., Schmid, B., Schulze, E.D., Staudler, A., Strecker, T., Temperton, V., Tscharntke, T., Vogel, A., Voigt, W., Weigelt, A., Wilcke, W., Weisser, W.W., 2016. Effects of biodiversity strengthen over time as ecosystem functioning declines at low and increases at high biodiversity. Ecosphere 7, 14.

Milcu, A., Partsch, S., Scherber, C., Weisser, W.W., Scheu, S., 2008. Earthworms and legumes control litter decomposition in a plant diversity gradient. Ecology 89, 1872–1882.

Milcu, A., Roscher, C., Gessler, A., Bachmann, D., Gockele, A., Guderle, M., Landais, D., Piel, C., Escape, C., Devidal, S., Ravel, O., Buchmann, N., Gleixner, G., Hildebrandt, A., Roy, J., 2014. Functional diversity of leaf nitrogen concentrations drives grassland carbon fluxes. Ecol. Lett. 17, 435–444.

Milcu, A., Eugster, W., Bachmann, D., Guderle, M., Roscher, C., Gockele, A., Landais, D., Ravel, O., Gessler, A., Lange, M., Ebeling, A., Weisser, W.W., Roy, J., Hildebrandt, A., Buchmann, N., 2016. Plant functional diversity increases grassland productivity-related water vapor fluxes: an ecotron and modeling approach. Ecology 97, 2044–2054.

Miltner, A., Bombach, P., Schmidt-Bruecken, B., Kaestner, M., 2012. SOM genesis: microbial biomass as a significant source. Biogeochemistry 111, 41–55.

Mokany, K., Ash, J., Roxburgh, S., 2008. Functional identity is more important than diversity in influencing ecosystem processes in a temperate native grassland. J. Ecol. 96, 884–893.

Moyano, F.E., Manzoni, S., Chenu, C., 2013. Responses of soil heterotrophic respiration to moisture availability: an exploration of processes and models. Soil Biol. Biochem. 59, 72–85.

Neff, J.C., Asner, G.P., 2001. Dissolved organic carbon in terrestrial ecosystems: synthesis and a model. Ecosystems 4, 29–48.

Nielsen, U.N., Wall, D.H., Six, J., 2015. Soil biodiversity and the environment. Annu. Rev. Environ. Resour. 40, 63–90.

Niklaus, P.A., Kandeler, E., Leadley, P.W., Schmid, B., Tscherko, D., Korner, C., 2001. A link between plant diversity, elevated $CO_2$ and soil nitrate. Oecologia 127, 540–548.

Oelmann, Y., Kreutziger, Y., Temperton, V.M., Buchmann, N., Roscher, C., Schumacher, J., Schulze, E.D., Weisser, W.W., Wilcke, W., 2007a. Nitrogen and phosphorus budgets in experimental grasslands of variable diversity. J. Environ. Qual. 36, 396–407.

Oelmann, Y., Wilcke, W., Temperton, V.M., Buchmann, N., Roscher, C., Schumacher, J., Schulze, E.D., Weisser, W.W., 2007b. Soil and plant nitrogen pools as related to plant diversity in an experimental grassland. Soil Sci. Soc. Am. J. 71, 720–729.

Oelmann, Y., Buchmann, N., Gleixner, G., Habekost, M., Roscher, C., Rosenkranz, S., Schulze, E.-D., Steinbeiss, S., Temperton, V.M., Weigelt, A., Weisser, W.W., Wilcke, W., 2011a. Plant diversity effects on aboveground and belowground N pools in temperate grassland ecosystems: development in the first 5 years after establishment. Global Biogeochem. Cycles 25 GB2014.

Oelmann, Y., Richter, A.K., Roscher, C., Rosenkranz, S., Temperton, V.M., Weisser, W.W., Wilcke, W., 2011b. Does plant diversity influence phosphorus cycling in experimental grasslands? Geoderma 167-68, 178–187.

Oram, N.J., Ravenek, J.M., Barry, K.E., Weigelt, A., Chen, H.M., Gessler, A., Gockele, A., de Kroon, H., van der Paauw, J.W., Scherer-Lorenzen, M., Smit-Tiekstra, A., van Ruijven, J., Mommer, L., 2018. Below-ground complementarity effects in a grassland biodiversity experiment are related to deep-rooting species. J. Ecol. 106, 265–277.

Orwin, K.H., Wardle, D.A., Greenfield, L.G., 2006. Ecological consequences of carbon substrate identity and diversity in a laboratory study. Ecology 87, 580–593.

Papatheodorou, E.M., Argyropoulou, M.D., Stamou, G.P., 2004. The effects of large- and small-scale differences in soil temperature and moisture on bacterial functional diversity and the community of bacterivorous nematodes. Appl. Soil Ecol. 25, 37–49.

Peres, G., Cluzeau, D., Menasseri, S., Soussana, J., Bessler, H., Engels, C., Habekost, M., Gleixner, G., Weigelt, A., Weisser, W., Scheu, S., Eisenhauer, N., 2013. Mechanisms linking plant community properties to soil aggregate stability in an experimental grassland plant diversity gradient. Plant Soil 373, 285–299.

Pett-Ridge, J., Firestone, M.K., 2017. Using stable isotopes to explore root-microbe-mineral interactions in soil. Rhizosphere 3, 244–253.

Poeplau, C., Don, A., Vesterdal, L., Leifeld, J., Van Wesemael, B., Schumacher, J., Gensior, A., 2011. Temporal dynamics of soil organic carbon after land-use change in the temperate zone—carbon response functions as a model approach. Glob. Chang. Biol. 17, 2415–2427.

Porazinska, D.L., Bardgett, R.D., Blaauw, M.B., Hunt, H.W., Parsons, A.N., Seastedt, T.R., Wall, D.H., 2003. Relationships at the aboveground-belowground interface: plants, soil biota, and soil processes. Ecol. Monogr. 73, 377–395.

Potthoff, M., Steenwerth, K.L., Jackson, L.E., Drenovsky, R.E., Scow, K.M., Joergensen, R.G., 2006. Soil microbial community composition as affected by restoration practices in California grassland. Soil Biol. Biochem. 38, 1851–1860.

Ravenek, J.M., Bessler, H., Engels, C., Scherer-Lorenzen, M., Gessler, A., Gockele, A., De Luca, E., Temperton, V.M., Ebeling, A., Roscher, C., Schmid, B., Weisser, W.W., Wirth, C., de Kroon, H., Weigelt, A., Mommer, L., 2014. Long-term study of root biomass in a biodiversity experiment reveals shifts in diversity effects over time. Oikos 123, 1528–1536.

Reich, P.B., Tilman, D., Isbell, F., Mueller, K., Hobbie, S.E., Flynn, D.F.B., Eisenhauer, N., 2012. Impacts of biodiversity loss escalate through time as redundancy fades. Science 336, 589–592.

Roscher, C., Schumacher, J., Baade, J., Wilcke, W., Gleixner, G., Weisser, W.W., Schmid, B., Schulze, E.D., 2004. The role of biodiversity for element cycling and trophic interactions: an experimental approach in a grassland community. Basic Appl. Ecol. 5, 107–121.

Roscher, C., Temperton, V.M., Scherer-Lorenzen, M., Schmitz, M., Schumacher, J., Schmid, B., Buchmann, N., Weisser, W.W., Schulze, E.D., 2005. Overyielding in experimental grassland communities—irrespective of species pool or spatial scale. Ecol. Lett. 8, 419–429.

Rosenkranz, S., Wilcke, W., Eisenhauer, N., Oelmann, Y., 2012. Net ammonification as influenced by plant diversity in experimental grasslands. Soil Biol. Biochem. 48, 78–87.
Scherer-Lorenzen, M., Palmborg, C., Prinz, A., Schulze, E.D., 2003. The role of plant diversity and composition for nitrate leaching in grasslands. Ecology 84, 1539–1552.
Schimel, J.P., Bennett, J., 2004. Nitrogen mineralization: challenges of a changing paradigm. Ecology 85, 591–602.
Schimel, J.P., Schaeffer, S.M., 2012. Microbial control over carbon cycling in soil. Front. Microbiol. 3, 348.
Schmidt, M.W.I., Torn, M.S., Abiven, S., Dittmar, T., Guggenberger, G., Janssens, I.A., Kleber, M., Koegel-Knabner, I., Lehmann, J., Manning, D.A.C., Nannipieri, P., Rasse, D.P., Weiner, S., Trumbore, S.E., 2011. Persistence of soil organic matter as an ecosystem property. Nature 478, 49–56.
Silvertown, J., Dodd, M.E., Gowing, D.J.G., Mountford, J.O., 1999. Hydrologically defined niches reveal a basis for species richness in plant communities. Nature 400, 61–63.
Spehn, E.M., Hector, A., Joshi, J., Scherer-Lorenzen, M., Schmid, B., Bazeley-White, E., Beierkuhnlein, C., Caldeira, M.C., Diemer, M., Dimitrakopoulos, P.G., Finn, J.A., Freitas, H., Giller, P.S., Good, J., Harris, R., Hogberg, P., Huss-Danell, K., Jumpponen, A., Koricheva, J., Leadley, P.W., Loreau, M., Minns, A., Mulder, C.P.H., O'Donovan, G., Otway, S.J., Palmborg, C., Pereira, J.S., Pfisterer, A.B., Prinz, A., Read, D.J., Schulze, E.D., Siamantziouras, A.S.D., Terry, A.C., Troumbis, A.Y., Woodward, F.I., Yachi, S., Lawton, J.H., 2005. Ecosystem effects of biodiversity manipulations in European grasslands. Ecol. Monogr. 75, 37–63.
Steenwerth, K.L., Drenovsky, R.E., Lambert, J.J., Kluepfel, D.A., Scow, K.M., Smart, D.R., 2008. Soil morphology, depth and grapevine root frequency influence microbial communities in a pinot noir vineyard. Soil Biol. Biochem. 40, 1330–1340.
Steinbeiss, S., Bessler, H., Engels, C., Temperton, V.M., Buchmann, N., Roscher, C., Kreutziger, Y., Baade, J., Habekost, M., Gleixner, G., 2008. Plant diversity positively affects short-term soil carbon storage in experimental grasslands. Glob. Chang. Biol. 14, 2937–2949.
Stocker, R., Korner, C., Schmid, B., Niklaus, P.A., Leadley, P.W., 1999. A field study of the effects of elevated $CO_2$ and plant species diversity on ecosystem-level gas exchange in a planted calcareous grassland. Glob. Chang. Biol. 5, 95–105.
Strecker, T., Mace, O.G., Scheu, S., Eisenhauer, N., 2016. Functional composition of plant communities determines the spatial and temporal stability of soil microbial properties in a long-term plant diversity experiment. Oikos 125, 1743–1754.
Thoms, C., Gattinger, A., Jacob, M., Thomas, F.M., Gleixner, G., 2010. Direct and indirect effects of tree diversity drive soil microbial diversity in temperate deciduous forest. Soil Biol. Biochem. 42, 1558–1565.
Tilman, D., Downing, J.A., 1994. Biodiversity and stability in grasslands. Nature 367, 363–365.
Tilman, D., Wedin, D., Knops, J., 1996. Productivity and sustainability influenced by biodiversity in grassland ecosystems. Nature 379, 718–720.
Tilman, D., Reich, P.B., Knops, J., Wedin, D., Mielke, T., Lehman, C., 2001. Diversity and productivity in a long-term grassland experiment. Science 294, 843–845.
van der Heijden, M.G.A., Bardgett, R.D., van Straalen, N.M., 2008. The unseen majority: soil microbes as drivers of plant diversity and productivity in terrestrial ecosystems. Ecol. Lett. 11, 296–310.
Vogel, A., Ebeling, A., Gleixner, G., Roscher, C., Scheu, S., Ciobanu, M., Koller-France, E., Lange, M., Lochner, A., Meyer, S.T., Oelmann, Y., Wilcke, W., Schmid, B., Eisenhauer, N., 2019. A new experimental approach to test why biodiversity effects strengthen as ecosystems age. Adv. Ecol. Res. 61, 221–264.

Voroney, R.P., Heck, R.J., 2015. Chapter 2—the soil habitat. In: Paul, E.A. (Ed.), Soil Microbiology, Ecology and Biochemistry, fourth ed. Academic Press, Boston, pp. 15–39.

Wall, D.H., Six, J., 2015. Give soils their due. Science 347, 695.

Wardle, D.A., 2002. Linking the aboveground and belowground components. In: Levin, S.A., Horn, H.S. (Eds.), Monographs in Population Biology. Princeton University Press, New Jersey, p. 392.

Wardle, D.A., Bardgett, R.D., Klironomos, J.N., Setala, H., van der Putten, W.H., Wall, D.H., 2004. Ecological linkages between aboveground and belowground biota. Science 304, 1629–1633.

Zak, D.R., Holmes, W.E., White, D.C., Aaron, D.P., Tilman, D., 2003. Plant diversity, soil microbial communities, and ecosystem function: are there any links? Ecology 84, 2042–2050.

Zelles, L., 1997. Phospholipid fatty acid profiles in selected members of soil microbial communities. Chemosphere 35, 275–294.

# CHAPTER SEVEN

# A new experimental approach to test why biodiversity effects strengthen as ecosystems age

Anja Vogel[a,b,c,*], Anne Ebeling[c], Gerd Gleixner[d], Christiane Roscher[a,e], Stefan Scheu[f], Marcel Ciobanu[g], Eva Koller-France[h,i], Markus Lange[d], Alfred Lochner[a,b], Sebastian T. Meyer[j], Yvonne Oelmann[h], Wolfgang Wilcke[i], Bernhard Schmid[k,l,†], Nico Eisenhauer[a,b,†]

[a]German Centre for Integrative Biodiversity Research (iDiv) Halle-Jena-Leipzig, Leipzig, Germany
[b]Institute of Biology, Leipzig University, Leipzig, Germany
[c]Institute of Ecology and Evolution, Friedrich Schiller University Jena, Jena, Germany
[d]Department of Biogeochemical Processes, Max Planck Institute for Biogeochemistry, Jena, Germany
[e]UFZ, Helmholtz Centre for Environmental Research, Physiological Diversity, Leipzig, Germany
[f]J.F. Blumenbach Institute of Zoology and Anthropology, University of Göettingen, Göettingen, Germany
[g]Institute of Biological Research, Branch of the National Institute of Research and Development for Biological Sciences, Cluj-Napoca, Romania
[h]Geoecology, University of Tübingen, Tübingen, Germany
[i]Institute of Geography and Geoecology, Karlsruhe Institute of Technology (KIT), Karlsruhe, Germany
[j]Department of Ecology and Ecosystem Management, Technische Universität München, Freising-Weihenstephan, Germany
[k]Department of Geography, University of Zürich, Zürich, Switzerland
[l]Institute of Ecology, College of Urban and Environmental Sciences, Peking University, Beijing, China
*Corresponding author: e-mail address: anja.vogel@uni-jena.de

## Contents

1. Introduction — 222
2. Methods — 227
   - 2.1 Study site — 227
   - 2.2 The ΔBEF experiment — 227
   - 2.3 Measurements — 230
   - 2.4 Data analysis — 233
   - 2.5 The soil barrier experiment — 235
3. Results — 236
   - 3.1 Establishment of the treatments — 236
4. Discussion — 250
   - 4.1 Establishment of the treatments — 250
   - 4.2 Treatment effects on the plant diversity–productivity relationship — 254
5. Conclusions — 256
Acknowledgements — 256
References — 257

---

† Authors jointly supervised this work.

## Abstract

Previous experimental studies found strengthening relationships between biodiversity and ecosystem functioning (BEF) over time. Simultaneous temporal changes of abiotic and biotic conditions, such as in the composition of soil communities, soil carbon and nutrient concentrations, plant community assembly or selection processes, are currently discussed as potential drivers for strengthening BEF relationships. Despite the popularity of these explanations, experimental tests of underlying mechanisms of strengthening BEF relationships over time are scarce, and confounding influences of calendar year cannot be ruled out unless ecosystems of different age are compared in the same calendar years. To address this critical gap of knowledge, we reestablished the plant communities of a long-term biodiversity experiment that had started in 2002 (the Jena Experiment) with new seeds and old or new soil again in 2016. Comparing these treatments with the original communities set up in 2002, we tested whether old communities had stronger plant diversity effects on plant productivity than young ones and if this depended on soil- or plant-related processes. Our first results show that in old communities, the effect of plant diversity on productivity was indeed stronger than in young communities and that this could not be explained by the age of the soil only. However, we found significant effects of soil on the composition of soil organisms, which might be relevant for other ecosystem functions and may have stronger effects over time. Our new experimental approach enables us to test which mechanisms cause strengthening BEF relationships for many different ecosystem functions independent of the study year.

## 1. Introduction

There is growing consensus that biodiversity increases many provisioning and regulating ecosystem services (Cardinale et al., 2012; Duffy et al., 2017; Eisenhauer et al., 2019; Hooper et al., 2012; Naeem et al., 2012), of which the best-studied function is aboveground plant biomass production (Hector et al., 1999; Marquard et al., 2009; Tilman et al., 1997). Moreover, experimental studies manipulating plant diversity found that the magnitude of biodiversity effects on ecosystem functions (BEF) increases over time, i.e., as ecosystems 'age' (Guerrero-Ramírez et al., 2017; Huang et al., 2018; Meyer et al., 2016; Reich et al., 2012). This result has important implications because it suggests that experimental results obtained from newly-assembled, young communities probably underestimate the consequences of biodiversity loss. Studying the mechanisms explaining why biodiversity effects strengthen over time as communities age is crucial in order to understand how important local species loss is for ecosystem functions in real-world ecosystems. Either increases in function at high plant diversity, decreases of function at low plant diversity, or a combination of both have been shown to cause strengthening BEF relationships over time (Guerrero-Ramírez et al., 2017; Meyer et al., 2016). These dissimilar trajectories of plant community performance illustrate context-dependent underlying mechanisms for increasing biodiversity effects (Guerrero-Ramírez et al., 2017). However, in all these previous studies, community age was confounded with calendar year, meaning that uncontrolled environmental conditions of particular years within the study periods may have contributed to biodiversity effects on particular ecosystem functions. Here, we describe a new experimental approach overcoming these shortcomings to get a deeper understanding of the mechanisms driving strengthening biodiversity effects as communities age.

Previous BEF experiments revealed changes in both plant- and soil-related processes over time, leading to hypotheses about the mechanistic links of both groups of processes in driving ecosystem functioning. The plant community level itself underlies temporal changes with potential consequences for the soil system (Reich et al., 2012). Overall, it has been shown that the contribution of multiple plant species to diversity effects on productivity increases over time (Allan et al., 2011; Cardinale et al., 2007, 2011; Fargione et al., 2007; Huang et al., 2018; Reich et al., 2012), which is commonly referred to as complementarity effects. The notion is that diverse communities use the resource pool more efficiently than species-poor communities do (Barry et al., 2019; Loreau and Hector, 2001; Mueller et al., 2013; Roscher et al., 2008). By contrast, the dominance of specific productive species in mixtures also has been shown to contribute to plant diversity effects, but such selection effects seem to play a minor role in strengthening biodiversity effects as communities age (Huang et al., 2018; Reich et al., 2012).

Higher complementarity effects are associated with higher fluctuations of plant species over time in diverse communities (Allan et al., 2011). This means that the identity of those plant species, which contribute most to ecosystem function, changes over time (Allan et al., 2011; Isbell et al., 2011). Shifts in plant species composition can be associated with changes in the representation of functional traits in communities which are important for community productivity (Diaz and Cabido, 2001; HilleRisLambers et al., 2004) or with their maintenance due to compensatory dynamics (Bai et al., 2004; Yachi and Loreau, 1999). The diversity and dominance of community traits indeed can explain a significant proportion of productivity changes over time and across diversity gradients (Reich et al., 2012; Roscher et al., 2012, 2013). However, it is not only the abundance and composition of plant species, which determine the functional diversity of plant communities and finally ecosystem functioning. Analysis in a long-term biodiversity experiment revealed that besides phenotypic variation in trait expression along the plant diversity gradient (Gubsch et al., 2011; Lipowsky et al., 2015; Roscher et al., 2011a), selection for specific plant phenotypes also increases biodiversity effects on productivity (Lipowsky et al., 2011; van Moorsel et al., 2018; Zuppinger-Dingley et al., 2014). Such evolutionary effects point to the increase of complementarity among plant species due to niche differentiation (Tilman and Snell-Rood, 2014).

Beside temporal dynamics in the plant communities themselves, changes in the soil nutrient content and soil biota over time give rise to further potential mechanisms of strengthening biodiversity effects. Long-term experiments show that the plant diversity effect on soil microbial biomass (Eisenhauer et al., 2010; Habekost et al., 2008; Lange et al., 2014, 2015; Strecker et al., 2016; Thakur et al., 2015) as well as the abundance and diversity of soil micro-, meso- and macrofauna (Eisenhauer et al., 2011a, b) developed only after a time lag of 4 years in the Jena Experiment. As these different soil organisms play key roles in many ecosystem functions related to decomposition, nutrient cycling, and plant growth, observed shifts in soil communities were suggested to have feedback effects on plant community composition and biomass production (Eisenhauer, 2012; Eisenhauer et al., 2012). Negative plant-soil feedback effects at low plant diversity (Latz et al., 2012, 2016; Maron et al., 2011; Schnitzer et al., 2011) or facilitation at high plant diversity via biotic interactions with mutualistic partners (Latz et al., 2012; Wagg et al., 2015; Wright et al., 2017) are alternative mechanisms for positive and increasing BEF relationships (Eisenhauer et al., 2012). Furthermore, there is evidence of increased soil-C and -N storage by increased root biomass and increased microbial activity in diverse compared with species-poor plant communities (Fornara and Tilman, 2008; Lange et al., 2015; Mellado-Vázquez et al., 2016; Steinbeiss et al., 2008). High resource availability is one potential mechanism leading to increased function at high plant diversity, as indicated by the significance of soil-C content in a meta-analysis of grassland and forest biodiversity experiments (Guerrero-Ramírez et al., 2017) and by increased N content in a manipulative study (Fridley, 2002).

It is, however, difficult, to separate effects of community age from effects caused by particular calendar years, especially for ecosystem functions that have been measured with few temporal replicates. General temporal trends, such as globally increasing N deposition (Galloway et al., 2004), increasing greenhouse gas emissions (IPCC, 2014), warming climate or specific weather conditions during the observation period of biodiversity experiments, also might have contributed to the observed effects. Elevated atmospheric $CO_2$ concentrations and N enrichment increase productivity at least in experiments of short duration (Fridley, 2002; He et al., 2002; Isbell et al., 2013; Reich et al., 2001). Warming can increase aboveground productivity at high plant diversity, as has been shown in short-term experiments and also after several years (Cowles et al., 2016, but see De Boeck et al., 2008). Moreover, extreme climate events, such as droughts and floods, have been shown to influence plant productivity (Isbell et al., 2015) and BEF relationships (Wright et al., 2015). Consequently, gradual and abrupt changes in environmental conditions can contribute to strengthening effects of plant diversity on productivity.

To account for potential confounding effects of calendar years within the study period, we reestablished plant communities of a long-term biodiversity experiment (the Jena Experiment) in a split-plot experiment with new seeds, and old or new soil in 2016 and compared measurements in these young plant communities with measurements taken in the same calendar year in the old communities set up in 2002 (Roscher et al., 2004). Together, our setup comprises three treatments varying in the age of the plant and soil communities and forms the so-called $\Delta$BEF Experiment (DELTa-BEF; short for DEterminants of Long-Term Biodiversity Effects on Ecosystem Functioning Experiment). One treatment comprises young communities without plot-specific soil and plant history, because soil was replaced by arable soil from an adjacent crop field similar to the conditions at the beginning of the Jena Experiment in 2002 (Roscher et al., 2004), and new plant seeds were derived from commercial suppliers. In a second treatment, we kept the old soil with its plot-specific history but reseeded new plant communities. The old communities of the Jena Experiment form the third treatment and have a history of plot-specific changes in biotic and abiotic soil properties as well as in plant community composition over time as outlined above. With this experimental setup, we thus could study the roles of soil age and plant-community age independent of confounding effects of particular calendar years by comparing the different treatments within calendar years and by comparing the newly established plot (second treatment) with the original Jena Experiment of the same age but established and measured 14 years earlier. We hypothesized that young communities without a shared history of soil and plant communities would show the weakest BEF relationship similar to results of short-term BEF experiments (Fig. 1A; Reich et al., 2012). Moreover, we hypothesized that old communities with a shared soil as well

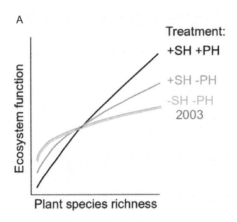

**Fig. 1** Hypothetical relationship of plant species richness and ecosystem functions in the different treatments of the ΔBEF-experiment (A). The soil of the −SH −PH treatment was excavated and replaced by arable soil (B), while the soil in +SH −PH was extricated from large roots (C). (D and E) show the location of the treatments on one plot as an example. The location of treatments was assigned randomly on five potential positions (E).

as plant history would show the steepest BEF relationship, while the treatment with old plot-specific soil history but young plant communities would have an intermediate relationship. Here, we describe the establishment of the experiment and present first treatment effects on the BEF relationship by focusing on plant biomass production as ecosystem function.

## 2. Methods
### 2.1 Study site

We established the ΔBEF Experiment on the main plots of the Jena Experiment. This large biodiversity experiment was established in 2002 in an alluvial plain of the Saale river in Jena (latitude 50.95, longitude 11.62, altitude 130 m a.s.l., MAP 610 mm, MAT 9.9 °C; Hoffmann et al., 2014) and is described in detail in Roscher et al. (2004). The soil is an Eutric Fluvisol (IUSS Working Group WRB, 2015), which had been managed as a highly fertilized arable field for growing vegetables and cereals for at least four decades prior to the establishment of the Jena Experiment. Sixty grassland plant species typical for semi-natural mesophilic grassland communities of this region served as species pool to establish the experimental plant communities. Seeds were ordered by commercial seed suppliers. Plant communities varying in species richness (1, 2, 4, 8, 16, and 60 species) and functional group richness (1, 2, 3, 4 functional groups: legumes, grasses, small herbs, and tall herbs) were randomly assembled from the respective functional groups and sown in May 2002 in four blocks (according to a gradient in edaphic conditions across the field site; Roscher et al., 2004). The seed density of 1000 viable seeds per $m^2$ was equally divided among the species in each mixture and adjusted to germination rates from standard laboratory tests. Since 2002, the colonization of non-target plant species in the experimental plots has been precluded by hand weeding two to three times per year. Weeding has been shown not to compromise observed plant diversity effects on ecosystem functions (Weisser et al., 2017). Further management includes mowing and removal of the mown biomass twice per year (at peak biomass in late spring and late summer) according to the traditional management of such grasslands in the study region.

### 2.2 The ΔBEF experiment

In May 2016, i.e., 14 years after the establishment of the Jena Experiment, we set up 3 subplots on all 80 plots of the main plant diversity experiment (two monoculture plots had been given up in the meantime up due to very low cover of <10%; Weisser et al., 2017). These subplots have a size of $1.5 \times 3 \, m^2$, and the location of each subplot was a random choice out of five potential locations at the edge of each long-term plot (Fig. 1D and E). This design maintained the original plots and allowed the continuation of long-term time series measurements. The three subplots forming the ΔBEF Experiment vary in soil history (new soil without soil history, −SH vs. old soil of the single experimental plots since 2002 with soil history, +SH; Table 1) and in plant community history (newly established plant communities seeded in 2016 without plant history, −PH vs. old plant communities sown in 2002 with plant history, +PH). We completely excavated the soil and plant layer of the treatment without soil history and without plant history (−SH −PH; Fig. 1B) to a depth of 30 cm which represents the main rooting zone in the Jena Experiment (Bessler et al., 2009, Ravenek et al., 2014) and replaced it by soil of an adjacent arable field of the same soil layer. Similar to the conditions of the field site of the Jena Experiment, this new soil also had been highly fertilized and used for growing of vegetables and cereals for decades. Accordingly, the arable field soil had properties that were in the middle of the range of the soil properties determined for the plots of the Jena Experiment in 2002 (pH 7.3; $C_{org}$ 20.5 g kg$^{-1}$; $N_{tot}$ 2.3 g kg$^{-1}$ compared with pH 7.1–8.4; $C_{org}$ 5–33 g kg$^{-1}$; $N_{tot}$ 1–2.7 g kg$^{-1}$ in 2002; Roscher et al., 2004).

For the second treatment with soil history but without plant history (+SH −PH), we removed the plant sod by using a digger, while keeping the soil of the respective plots (Fig. 1C). To remove large roots in the soil in this treatment, we mixed and homogenized the soil to a depth of 30 cm with the digger. A lateral mixture of the soil types with their surroundings was prevented by the installation of plastic sheets (PE-HD, 1 mm thick) as soil barriers of the top 0–30 cm of the soil in the −SH −PH and +SH −PH treatments. The plastic sheets were flexible, which facilitated the installation, and were insusceptible to erosion. We recompacted the soil of both treatments using a vibrating plate before the seeding of plant communities.

As done in 2002, we sowed the same plot-specific plant species mixtures on 23 May 2016 by hand in the two treatments with young plant communities (−SH −PH and +SH −PH) with a density of 1000 germinable seeds per m$^2$, equally divided among the target species. Therefore, we ordered seed material from commercial suppliers in 2016, conducted germination tests under standardized conditions, and adjusted the sown seed numbers for each species according to their respective germination rates closely replicating the protocols used during the establishment of the original Jena Experiment (Roscher et al., 2004). The seed material had no plot-specific coexistence and interaction history (−PH). Seeds were mixed with a handful of site-specific sieved and dried soil to guarantee equal seed distributions across the subplot.

**Table 1** Name and description of the experimental treatments of the ΔBEF experiment.

| Name of treatment | Soil history | Age of soil history (year) | Soil history contrast for mixed models | Plant history | Age of plant history (year) | Plant history contrast for mixed models | Year contrast for mixed models |
|---|---|---|---|---|---|---|---|
| −SH −PH | Conventional (crop production) | 1 | −SH | Unspecific (commercial supplier) | 1 | −PH | 2017 |
| +SH −PH | Plotspecific (experimental grassland) | 15 | +SH | Unspecific (commercial supplier) | 1 | −PH | 2017 |
| +SH +PH | Plotspecific (experimental grassland) | 15 | +SH | Plotspecific (experimental grassland) | 15 | +PH | 2017 |
| 2003 | Conventional (crop production) | 1 | −SH | Unspecific (commercial supplier) | 1 | −PH | 2003 |

Here, no soil or plant history (−SH, −PH) refers to a legacy of non-experimental conditions, while soil or plant history (+SH, +PH) refers to a legacy of 15 years of experimental grassland diversity treatments. The levels of the soil and plant histories were combined to the treatments of the ΔBEF experiment.

Our third treatment (+SH +PH) represents an undisturbed subplot of each experimental plot of the Jena Experiment (Fig. 1D). The respective old communities have a shared soil and plant history since 2002. No new seeds were sown into the plots of this third treatment.

## 2.3 Measurements

Before sowing of the ΔBEF Experiment, we determined the seed bank to estimate the number of viable seeds of the target communities per area. For each experimental block, we took three composite samples each consisting of four soil cores (diameter: 5.7 cm, depth: 5 cm) at different positions of the homogenized arable field soil prior to refilling the subplots (to assess the viable seed bank of (−SH −PH). In the (+SH +PH) treatment, we collected a composite sample of four soil cores to 5 cm depth (diameter: 5.7 cm) along a transect in each plot. In the (+SH −PH) treatment, we took a composite sample consisting of five soil cores (diameter: 1.8 cm, depth: 30 cm). This was done because the soil of this treatment was homogenized down to this depth to remove large roots, which might have redistributed seeds across soil layers. Furthermore, we removed for this treatment the upper ∼2 cm of the soil cores from our samples, because this soil layer was also removed together with the grass sod during the setup of this treatment. To assess the number of viable seeds per $m^2$, we accounted for this difference in soil sampling by multiplying the sampled area (five times a core with a diameter of 1.8 cm) by factor 6 (six times 5 cm depth increments = 30 cm sampling depth). Overall, the sampling volume for determining the soil seed bank was low, but could not be increased to avoid too much disturbance on the subplots. Soil samples were transported to the laboratory and stored at 4 °C until processing. In order to reduce the soil volume and establish better conditions for seed germination, we applied the bulk reduction method (ter Heerdt et al., 1996). Each soil sample was washed through a cascade of two sieves (2 mm mesh size to eliminate roots and rhizomes, and 0.2 mm mesh size to remove fine-soil without seeds). Afterward, samples were spread out in thin layers (<5 mm) in trays filled with a heat-sterilized sand-soil mixture (1:1). The trays were placed in an open greenhouse with a roof, which automatically closed during rain, and cultivated at ambient temperatures to promote germination by natural daily temperature fluctuations. They were covered with a fine gauze to prevent contamination by wind-borne seeds. In addition, control trays containing only the sterilized sand/soil mixture were installed to control for contamination originating from the substrate. Trays were watered regularly and checked for emerging seedlings. Seedlings were determined to species level and then removed. Although most seedlings emerged in the first weeks, the experiment was continued until next spring to break the dormancy of some species with chilling. The community-level seed bank was summarized from all emerging seedlings of species belonging to the target community of the respective plot.

After sowing of the plant communities in the field, we counted the number of emerging seedlings in a randomly placed quadrat of $50 \times 50\,cm^2$. We did the first count between 22 and 24 June 2016, and a second count between 4 and 5 April 2017, i.e., 1 and 11 months after the sowing, respectively. The second count mainly served to check for seedling emergence of species with seeds that require a chilling period for dormancy breaking and do not appear at all or at very low numbers in the first growing season when sowing in spring (Heisse et al., 2007).

We counted individuals of each species in a randomly placed quadrat of $50 \times 50\,cm^2$ from 27 September to 11 October 2016 to determine plant density. In contrast to −SH −PH and +SH −PH, it was challenging to visually separate single individuals in the +SH +PH treatment. The numbers, therefore, represent an estimate of individuals in this treatment.

Aboveground biomass was harvested from 29 May to 6 June 2017 (12 months after sowing of the new plant communities) and 29–31 August 2017 (15 months after sowing the new plant communities) before mowing the entire experimental plots. The vegetation was clipped 3 cm above soil surface within two randomly placed frames ($20 \times 50\,cm^2$) per subplot and sorted by sown species, while non-target weeds, dead and not-identifiable plant material were removed. After drying at 70 °C for 48 h, samples were weighed and dry mass of target species was summed to community-level values for each subplot. Annual biomass was calculated as the sum of the two individual harvests. We measured leaf area index in every subplot before each biomass harvest by one above measurement as reference and five measurements below the canopy along a transect in the middle of each subplot (LAI-2200C, LI-COR Biosciences, Lincoln, Nebraska, USA). Because of the small plot size, we did not account for the most vertical sensor readings.

We used a decimal scale (based on Londo, 1976) to estimate plant species-specific cover on the whole subplot area prior to every biomass harvest. Cover data were supplemented by similar data collected in 2003 for the original Jena Experiment. This represents an initial treatment without soil history and without plant history comparable to the new −SH −PH treatment but differing in calendar year (2003 vs. 2017; see van Moorsel et al. (2018) for further justification of this approach). We used cover data to calculate realized target species richness, species evenness, and the dissimilarity between the treatments based on differences in species abundances (see below).

Two soil sampling campaigns were conducted to determine abiotic and biotic soil properties. Soil organic C concentration was determined in all treatments of the experiment in spring of 2016, so immediately after the soil treatments had been established but prior to sowing. Soil organic C in the old +SH +PH treatment was determined using a split-tube sampler (4.8 cm diameter), by taking three soil cores per plot to a depth of 30 cm. Soil cores were segmented into 5 cm-depth sections and pooled per depth sections and plot (Lange et al., 2015; Steinbeiss et al., 2008). Soil of the young treatments (−SH −PH, +SH −PH) was sampled using a soil corer (2.0 cm diameter), by taking 10 soil cores per plot to a depth of 5 cm. Soil cores were pooled per plot, and the soil was then dried, sieved, and milled. Subsequently, total C was determined by combustion with an elemental analyser at 1150 °C (Elementaranalysator vario Max CN, Elementar Analysensysteme GmbH, Hanau, Germany). Inorganic C concentration was measured after oxidative removal of organic C for 16 h at 450 °C in a muffle furnace. Finally, organic C concentration was calculated as the difference between total and inorganic C (Lange et al., 2015; Steinbeiss et al., 2008) for the top 5 cm soil layer.

Soil sampling for the other soil properties took place in June 2017 (13 months after setting up of the treatments). We pooled six soil cores (4 cm in diameter, 5 cm depth) per treatment, homogenized them in the field, and stored them at 4 °C until further processing. We sieved one subsample of the soil at <2 mm and stored it 8 months at −20 °C for microbial analysis (see below). The other subsample was used to determine concentrations of soil mineral nitrogen concentrations ($NH_4^+$-N and $NO_3^-$-N) and gravimetric soil moisture content. We removed visible roots and stones in this subsample and extracted field-fresh soil samples with 1 M KCl solution within 24 h of sampling and filtrated the samples after 30 min of shaking. The concentrations of $NH_4^+$-N and $NO_3^-$-N were determined colorimetrically by continuous flow analysis (SanPlus, Skalar, NL) using KCl as a blank. The concentration of soil mineral nitrogen was calculated as the sum of the $NH_4^+$-N and $NO_3^-$-N concentrations.

Soil microbial-biomass carbon of approximately 5 g soil (fresh weight) was measured using an $O_2$-microcompensation apparatus (Scheu, 1992).

We measured the microbial respiratory response after incubation of ~24 h at hourly intervals for at least 8 h at 20 °C. Substrate-induced respiration was calculated from the respiratory response to D-glucose for 10 h at 20 °C (Eisenhauer et al., 2013). We added glucose according to preliminary studies to saturate the catabolic enzymes of microorganisms (4 mg g$^{-1}$ dry weight dissolved in 400 μL deionized water; Strecker et al., 2016). The mean of the lowest three readings within the first 10 h (between the initial peak caused by disturbing the soil and the peak caused by microbial growth) was taken as maximum initial respiratory response (MIRR; μL O$_2$ g$^{-1}$ soil dry weight h$^{-1}$) and microbial biomass (μg C g$^{-1}$ soil dry weight) was calculated as 38 × MIRR (Beck et al., 1997). The values represent the biomass of all active bacteria and fungi in the soil.

Another subsample of the soil was analysed for the soil nematode community as an indicator group of soil food web structure. Soil nematodes were extracted from 25 g of fresh soil using a modified Baermann method (Ruess, 1995) and fixed with 4% formaldehyde solution at 65 °C. After counting, nematodes were determined to genus level (adults) or family level (juveniles) and assigned to herbivorous, bacterivorous, fungivorous, omnivorous, and predatory trophic groups (Bongers and Bongers, 1998; Yeates et al., 1993).

## 2.4 Data analysis

We used cover estimates to calculate Simpson evenness $D$ for each subplot by $D_k = \sum_{i=1}^{n} p_{ik}^2$, where $p_{ik}$ is the relative abundance of species $i$ in a subplot $k$ and $n$ is the number of sown species in the subplot.

To analyse differences in species compositions of the plant and the soil nematode communities, we calculated Bray-Curtis dissimilarities (BC) for all possible treatment combinations per plot. $BC$ varies between zero (=identical composition regarding species presences and abundances in both treatments) and one (=no species in common) and is calculated as $BC_{ik} = \frac{\sum_{i=1}^{n}|p_{ij} - p_{ik}|}{\sum_{i=1}^{n}|p_{ij} + p_{ik}|}$, where $p_{ik}$ is the relative abundance of species $i$ in the two subplots $j$ and $k$. We used species-specific cover estimates as abundance measures for the plant communities. In case of the soil nematode communities, we used nematode counts per soil weight as abundance measures and calculated dissimilarities including the whole nematode community and single trophic groups, respectively. Note that the interpretation of $BC$ varies between plant and nematode communities in our experiment. In case of the plant communities, only abundance shifts and extinctions can cause deviations of $BC$ from 0, while for the nematode communities it is also species turnover or species gains and losses that can contribute to differences among communities. We therefore additionally calculated the Jaccard index based on presence-absence data according to $J = \frac{S_{jk}}{S_j + S_k - S_{jk}}$, with $S_j$ as the number of taxa in subplot $j$, $S_k$ as the number of taxa in subplot $k$, and $S_{jk}$ as the number of shared taxa in both subplots. This index addresses the turnover of communities between two treatment pairs (Anderson et al., 2011). We computed $BC$ and $J$ with the function *vegdist* in the package *vegan* (Oksanen et al., 2017). We performed mixed models to analyse whether dissimilarities varied among treatment pairs. Additionally, we used nonmetric multidimensional scaling (NMDS) for the ordination of the subplots based on Bray-Curtis distances of the nematode communities using the function *metaMDS* in the package *vegan*.

Data of aboveground biomass and cover estimates in 2003 supplemented our three ΔBEF treatments by the initial treatment (−SH −PH in 2003). This enabled us to separate the treatment effect into three contrasts for our statistical analysis. One contrast (year) separates the effect of the different calendar years (2003 vs. 2017, respectively), one contrasts the histories of plant communities (−PH vs. +PH), and one contrasts the soil histories (−SH vs. +SH, see Table 1). We fitted a series of mixed effects models for the annual aboveground biomass in 2017. Random intercept terms accounted for the nested structure of the ΔBEF Experiment, i.e., they consisted of block, plot nested in block, and treatment nested in plot and block. As fixed effects, we fitted species richness (as a log-linear term), followed by treatment and the interaction of both (overall model A). We separated the term treatment into contrasts for year, soil history, and plant history to test for their particular effects in additional models (models B–D). In model B, we ordered the treatment contrasts in the sequence year—soil history—plant history (Table S1 in the online version at https://doi.org/10.1016/bs.aecr.2019.06.006). This model, therefore, is testing for the *pure effect of plant history*, since we accounted for the variance due to different calendar years and soil types first. The sequence of the treatment contrast in model C was year—plant history—soil history, and it therefore tests for the *pure soil history effect*. In model D, we tested for the *pure effect of different calendar years* by fitting the treatment contrasts as plant history—soil history—year. To achieve the assumption of normality of residuals, we log-transformed plant and microbial biomass, soil mineral N concentration, plant density, realized richness and evenness and root-transformed nematode abundance,

abundance of viable seeds and number of seedlings. We analysed the response variables of soil and plant community properties in the same way, with the only exception that we skipped the contrast for different calendar years, when we had only data of 2017 but not of 2003 available. We performed all analyses using R 3.3.3 (R Development Core Team, 2011).

## 2.5 The soil barrier experiment

We were not able to install plastic sheets in the (+SH +PH) treatment without disturbing the communities and interfering with existing installations. To quantify this potentially confounding effect (e.g., higher water runoff in subplots with than without sheet installation), we set up an additional experiment to test for the effects of soil barriers on different ecosystem properties. We used eight abandoned plots of earlier experiments at the field site, which had not been weeded since 2010, and hence had well-established plant communities. In each of these plots, we established two subplots of the same size as in the $\Delta$BEF Experiment and installed plastic sheets by using a soil trencher in one treatment. The other treatment was left without a soil barrier and served as control. We installed three transects of 20 cm width in parallel to the long edges of the subplots to determine potential edge effects. We measured soil moisture, leaf area index and aboveground biomass in parallel to the measurements in the $\Delta$BEF Experiment to test for the effects of soil barriers and edge effects. A portable soil moisture probe (ML3 ThetaProbe, Delta-T Devices, Cambridge, UK) was used to determine the volumetric soil water content indirectly by time-domain-reflectometry in 6 cm depth at five replicate points along each transect. In two replicate frames of $20 \times 50\,cm^2$ per transect, aboveground biomass was clipped at 3 cm height. We measured leaf area index exactly as described for the $\Delta$BEF Experiment in only one transect in the middle of the subplot as doen for the $\Delta$BEF Experiment. Single transects of only 20 cm width would have been too small for this measure. We performed linear mixed models to test for the effect of treatment (with or without soil barrier), transect and the interaction of both. None of the measured variables showed any significant effect of the soil barriers (soil moisture: $F_{1,7}=2.79$, $P=0.139$; leaf area index: $F_{1,7}=0.369$, $P=0.563$; aboveground biomass: $F_{1,7}=0.38$, $P=0.559$) in any transect (soil moisture: $F_{2,220}=0.02$, $P=0.975$; aboveground biomass: $F_{1,16}=1.18$, $P=0.293$). Therefore, we assume that the results of the $\Delta$BEF Experiment are not biased by soil barrier effects.

## 3. Results
### 3.1 Establishment of the treatments
#### 3.1.1 Treatment effects on plant communities

The arable soil from the neighbouring agricultural field used for the establishment of the −SH −PH treatment did not contain any viable seeds from the experimental grassland species pool. The soil of the +SH −PH treatment with plot-specific history contained a viable seed bank comprising some of the target species of the respective plots. The removal of the plant communities and soil homogenization to 30 cm depth, however, reduced the number of viable seeds in the seed bank of this treatment compared with the undisturbed +SH +PH treatment (PH in Table 2, Fig. 2A). The number of viable seeds in the seed bank increased with plant species richness in both treatments (SR significant, but PH × SR not significant in Table 2 and Fig. 2A). The number of target species seedlings, which emerged after sowing of the new experimental communities, was independent of the soil history (SH in model C not significant in Table 2 and Fig. 2B) and higher in the newly sown compared to the old plant communities (PH in model C in Table 2). This was not surprising considering that there was no sowing into the old communities. Furthermore, plant species richness increased the number of emerging seedlings, but this effect was stronger in old compared to the newly sown plant communities (PH × SR in model C in Table 2 and Fig. 2B), also when we compared the plant communities independently of the soil type (PH × SR in model B in Table 2 and Fig. 2B). The soil itself did not change the diversity effect in the newly sown plant communities (SH × SR in model C not significant, Fig. 2B). Mature plants had a higher density in the old compared to the newly sown plant communities (PH in model C in Table 2 and Fig. 2C) and this result was independent of the soil type (PH in model B in Table 2 and Fig. 2C). The density of plant individuals increased with plant species richness and this effect was stronger in the old compared to the new communities (PH × SR in models B and C in Table 2). The soil itself did neither alter the density nor the plant species richness effect on the density of mature plant individuals (SH and SH × SR in model C not significant, Table 2 and Fig. 2C).

Leaf area index was higher in the −SH −PH treatment without any specific soil or plant history compared to +SH −PH and +SH +PH treatments (SH in model B and C in Table 2 and Fig. 2D). This effect of soil history was independent of community species richness (SH × SR interaction in model

**Table 2** Effects of experimental treatments on plant community properties.

| | Model | Seedbank (# viable seeds) | | | | Number of seedlings | | | | Density | | | LAI spring | | |
|---|---|---|---|---|---|---|---|---|---|---|---|---|---|---|---|
| | | numDf | denDf | F | P | numDf | denDf | F | P | denDf | F | P | denDf | F | P |
| (Intercept) | A, B, C, D | 1 | 74 | 107.39 | <0.000 *** | 1 | 148 | 1553.03 | 0.000 *** | 138 | 2956.42 | 0.000 *** | 144 | 206.03 | 0.000 *** |
| Plant species richness (SR), log-linear | A, B, C, D | 1 | 71 | 22.79 | <0.000 *** | 1 | 71 | 9.43 | 0.003 ** | 71 | 22.50 | 0.000 *** | 71 | 28.09 | 0.000 *** |
| **Treatments (T)** | **A** | | | | | **2** | **148** | **39.29** | **0.000 *** ** | **138** | **22.02** | **0.000 *** ** | **144** | **35.80** | **0.000 *** ** |
| Year (Y) | B | | | | | | | | | | | | | | |
| Soil history (SH) | B | | | | | 1 | 148 | 16.93 | 0.000 *** | 138 | 14.60 | 0.000 *** | 144 | 70.47 | 0.000 *** |
| **Plant history (PH)** | **B** | **1** | **74** | **53.75** | **<0.000 *** ** | **1** | **148** | **61.64** | **0.000 *** ** | **138** | **29.43** | **0.000 *** ** | **144** | **1.14** | **0.288** |
| Y | C | | | | | | | | | | | | | | |
| PH | C | | | | | 1 | 148 | 78.44 | 0.000 *** | 138 | 43.60 | 0.000 *** | 144 | 10.59 | 0.001 ** |
| **SH** | **C** | | | | | **1** | **148** | **0.13** | **0.717** | **138** | **0.44** | **0.510** | **144** | **61.01** | **0.000 *** ** |
| PH | D | | | | | | | | | | | | | | |
| SH | D | | | | | | | | | | | | | | |
| **Y** | **D** | | | | | | | | | | | | | | |
| T×SR | A | | | | | 2 | 148 | 4.09 | 0.019 * | 138 | 8.46 | 0.000 *** | 144 | 6.78 | 0.002 ** |
| Y: SR | B | | | | | | | | | | | | | | |

*Continued*

**Table 2** Effects of experimental treatments on plant community properties.—cont'd

| | | Seedbank (# viable seeds) | | | | Number of seedlings | | | | Density | | | | LAI spring | | |
|---|---|---|---|---|---|---|---|---|---|---|---|---|---|---|---|---|
| | Model | numDf | denDf | F | p | numDf | denDf | F | p | denDf | F | p | | denDf | F | p |
| SH×SR | B | | | | | 1 | 148 | 3.52 | 0.063 . | 138 | 9.38 | 0.003 ** | | 144 | 7.58 | 0.007 ** |
| **PH×SR** | **B** | 1 | 74 | 0.14 | 0.712 | **1** | **148** | **4.66** | **0.033 *** | **138** | **7.54** | **0.007 *** | | **144** | **5.97** | **0.016 *** |
| Y: SR | C | | | | | | | | | | | | | | | |
| PH×SR | C | | | | | 1 | 148 | 7.88 | 0.006 ** | 138 | 15.17 | 0.000 *** | | 144 | 12.11 | 0.001 *** |
| **SH×SR** | **C** | | | | | **1** | **148** | **0.30** | **0.586** | **138** | **1.75** | **0.188** | | **144** | **1.45** | **0.231** |
| PH×SR | D | | | | | | | | | | | | | | | |
| SH×SR | D | | | | | | | | | | | | | | | |
| **Y: SR** | **D** | | | | | | | | | | | | | | | |

| | | Realized plant species richness | | | | Simson evenness spring | | |
|---|---|---|---|---|---|---|---|---|
| | Model | numDf | denDf | F | p | denDf | F | p |
| (Intercept) | A, B, C, D | 1 | 220 | 8155.10 | <0.001 *** | 234 | 625.88 | <0.001 *** |
| Plant species richness (SR), log-linear | A, B, C, D | 1 | 71 | 4859.88 | <0.001 *** | 75 | 319.18 | <0.001 *** |
| **Treatments (T)** | **A** | **3** | **220** | **17.68** | **<0.001 *** | **234** | **1.51** | **0.212** |
| Year (Y) | B | 1 | 220 | 14.26 | <0.001 *** | 234 | 2.29 | 0.131 |
| Soil history (SH) | B | 1 | 220 | 9.15 | 0.003 ** | 234 | 0.26 | 0.609 |
| **Plant history (PH)** | **B** | **1** | **220** | **29.64** | **<0.001 *** | **234** | **1.99** | **0.160** |
| Y | C | 1 | 220 | 14.26 | <0.001 *** | 234 | 2.29 | 0.131 |

| Term | Model | Df | Df | F | P |  | Df | F | P |  |
|---|---|---|---|---|---|---|---|---|---|---|
| PH | C | 1 | 220 | 38.78 | <0.001 | *** | 234 | 0.93 | 0.335 |  |
| **SH** | **C** | **1** | 220 | **0.01** | **0.941** |  | **234** | **1.32** | **0.252** |  |
| PH | D | 1 | 220 | 50.75 | <0.001 | *** | 234 | 2.00 | 0.159 |  |
| SH | D | 1 | 220 | 0.48 | 0.490 |  | 234 | 0.19 | 0.659 |  |
| **Y** | **D** | **1** | **220** | **1.81** | **0.180** |  | **234** | **2.35** | **0.127** |  |
| T × SR | A | 3 | 220 | 16.24 | <0.001 | *** | 234 | 2.94 | 0.034 | * |
| Y × SR | B | 1 | 220 | 11.53 | 0.001 | *** | 234 | 0.33 | 0.566 |  |
| SH × SR | B | 1 | 220 | 9.22 | 0.003 | ** | 234 | 5.82 | 0.017 | * |
| **PH × SR** | **B** | **1** | **220** | **27.99** | **<0.001** | *** | **234** | **2.68** | **0.103** |  |
| Y × SR | C | 1 | 220 | 11.53 | 0.001 | *** | 234 | 0.33 | 0.566 |  |
| PH × SR | C | 1 | 220 | 37.20 | <0.001 | *** | 234 | 6.88 | 0.009 | ** |
| **SH × SR** | **C** | **1** | **220** | **0.00** | **0.962** |  | **234** | **1.61** | **0.205** |  |
| PHSR | D | 1 | 220 | 47.23 | <0.001 | *** | 234 | 7.10 | 0.008 | ** |
| SH × SR | D | 1 | 220 | 0.43 | 0.514 |  | 234 | 0.87 | 0.351 |  |
| **Y × SR** | **D** | **1** | **220** | **1.08** | **0.301** |  | **234** | **0.85** | **0.357** |  |

Df = degrees of freedom.

We obtained results from different linear mixed-effects model analyses: treatment was fitted as a factor with three to four levels in model A and depicted into a year (2003 vs. 2017, according to Table 1), a soil history (−SH vs. +SH, according to Table 1) and a plant history contrast (−PH vs. +PH, according to Table 1) in models B, C and D with varying order of the contrasts. In this way, model B tests for the pure plant history effect by contrasting treatment +SH −PH against treatment +SH +PH, model C tests for the pure soil history effect by contrasting treatment −SH −PH against treatment +SH −PH, and model D tests for the effect of the calendar year by contrasting treatment −SH −PH against 2003. ***$P \leq 0.001$, **$P \leq 0.01$, *$P \leq 0.05$, ⋅$P \leq 0.1$. Plant species richness was modelled as a log-linear term.

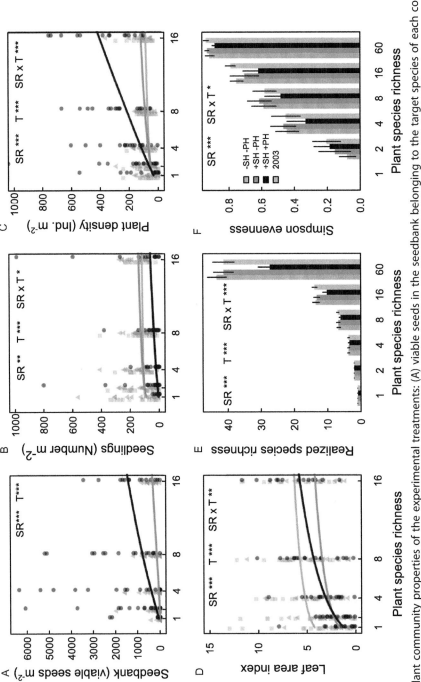

Fig. 2 Plant community properties of the experimental treatments: (A) viable seeds in the seedbank belonging to the target species of each community. The treatment −SH −PH did not contain any viable seeds from the experimental species pool. (B) summed number of seedlings emerged in summer 2016 and spring 2017, (C) number of plant individuals counted in established communities in summer 2017, and (D) leaf area index during peak biomass in spring 2017. (E) realized plant species richness and (F) Simpson evenness of the plant communities based on cover estimates in spring 2017 and 2003. Lines represent the predicted means and were derived from a mixed model with a log-linear term for plant species richness (SR), treatment (T) and the interaction of both (SR × T) as fixed effects. Significant effects are indicated on top ***$P \leq 0.001$, **$P \leq 0.01$, *$P \leq 0.05$, $P \leq 0.1$.

C not significant, Table 2). However, old communities had a much higher leaf area index at high plant species richness compared to young communities on old soil (PH × SR in model B significant, Table 2).

Plant species numbers in the newly sown treatments established well after sowing and showed no significant difference between the two soil treatments (SH in model C not significant, Table 2 and Fig. 2E). Old communities (+SH +PH) had a lower realized plant species richness compared to young communities (PH in model C in Table 2 and Fig. 2E), although in 2003 they established with a similar realized species richness compared to the new −SH −PH treatment (year in model D not significant, Table 2). The lower realized number of plant species in old than in new communities was more pronounced at high sown species richness (significant PH × SR in models B and C in Table 2 and Fig. 2E) and independent of the soil history (SH × SR in model C not significant, Table 2). The calendar year did not affect the plant species richness effect on the realized number of species (Y × SR of model D in Table 2 not significant).

The evenness of the communities overall increased with plant species richness, but the strength of this relationship varied among treatments (T × SR of model A in Table 2 and Fig. 2F). Old plant communities had a lower evenness at high species richness than young communities (significant PH × SR in model C, Table 2 and Fig. 2F). Pure soil or plant history did not change the plant diversity effect on community evenness (PH × SR in model B and SH × SR in model C not significant, Table 2). Furthermore, the evenness of newly sown communities was independent of the calendar year of establishment (Y and Y × SR in model D in Table 2 and Fig. 2F).

Although we sowed the subplots originally with the same plant species composition and proportions, the realized composition of the plant communities differed among the treatments (Treatment pair: F5,300 = 29.25, $P<0.001$), especially in diversity treatments of more than two species (Fig. 3, SR: F1,57 = 66.83, $P<0.001$; Treatment pair × SR: F5,300 = 4.04, $P=0.002$). The difference can be explained by the age of the plant communities (Bray-Curtis-distance of treatment pairs with same vs. different age of plant communities: $F_{1,304} = 137.36$, $P<0.001$; two-way interaction with SR: $F_{1,304} = 15.66$, $P<0.001$). This is indicated by the higher Bray-Curtis-dissimilarity in treatment pairs of different plant history (between +SH +PH and +SH −PH, green in Fig. 3A; between +SH +PH and −SH −PH, grey in Fig. 3A; between +SH +PH and 2003, grey in Fig. 3B) than in treatment pairs with similar plant history (between −SH −PH and +SH −PH, brown in Fig. 3A; −SH −PH and 2003, orange in Fig. 3B; +SH −PH and 2003,

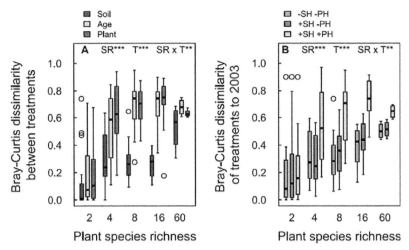

**Fig. 3** Pairwise Bray-Curtis dissimilarities in plant community composition (A) between two treatment pairs differing in soil history only (−SH −PH and +SH −PH, brown), plant history (+SH −PH and +SH +PH, green) as well as plant and soil history (−SH −PH and +SH +PH, grey). Bray-Curtis dissimilarities in plant community composition (B) between the single treatments and 2003 (orange: −SH −PH vs 2003; blue: +SH −PH vs 2003; grey: +SH +PH vs 2003) for the single plant species richness levels. Significant effects are indicated on top ***$P \leq 0.001$, **$P \leq 0.01$, *$P \leq 0.05$, $\cdot P \leq 0.1$.

blue in Fig. 3B). Contrary to this, the history of the soil caused minor compositional shifts of plant communities (treatment pairs with same vs. different soil history: $F_{1,304} = 0.48$, $P = 0.489$; two-way interaction with SR: $F_{1,304} = 0.001$, $P = 0.975$). Accordingly, Bray-Curtis dissimilarity in treatment pairs of different soil history (between −SH −PH and +SH −PH, brown in Fig. 3A) was low. Furthermore, compositional shifts due to the calendar year (Bray-Curtis dissimilarity between −SH −PH and 2003, orange in Fig. 3B) were lower compared to the compositional shifts due to the age of the plant communities (Bray-Curtis dissimilarity between +SH −PH and +SH +PH as well as between +SH +PH and 2003, grey in Fig. 3A and B).

### 3.1.2 Treatment effects on soil properties

Soil abiotic and biotic properties were measured from the same soil samples to test for treatment effects and to get baseline data on resource- and biotic interaction-based hypotheses underlying BEF relationships. Soil moisture differed significantly among treatments and along the plant diversity gradient (T in model A in Table 3 and Fig. 4A). Soil moisture was not affected by soil

**Table 3** Effects of experimental treatments on abiotic and biotic soil properties.

| | | | Soil moisture | | | Soil mineral N | | | Soil organic carbon | | | Soil microbial biomass | | | Soil nematode abundance | | |
|---|---|---|---|---|---|---|---|---|---|---|---|---|---|---|---|---|---|
| | Model | numDf | denDf | F | P | denDf | F | P | denDf | F | P | denDf | F | P | denDf | F | P |
| (Intercept) | A, B, C | 1 | 148 | 1567.58 | <0.001 *** | 116 | 8.40 | 0.005 ** | 148 | 410.99 | <0.001 *** | 148 | 5379.88 | <0.001 *** | 79 | 1303.21 | <0.001 *** |
| Plant species richness (SR), log-linear | A, B, C | 1 | 71 | 37.16 | <0.001 *** | 56 | 0.16 | 0.693 | 71 | 38.982 | <0.001 *** | 71 | 17.08 | <0.001 *** | 39 | 26.23 | <0.001 *** |
| Treatments (T) | A | 2 | 148 | 40.80 | <0.001 *** | 116 | 22.07 | <0.001 *** | 148 | 175.69 | <0.001 *** | 148 | 107.79 | <0.001 *** | 79 | 17.36 | <0.001 *** |
| Soil history (SH) | B | 1 | 148 | 31.98 | <0.001 *** | 116 | 12.54 | 0.001 *** | 148 | 21.833 | <0.001 *** | 148 | 5.93 | 0.016 * | 79 | 21.24 | <0.001 *** |
| Plant history (PH) | B | 1 | 148 | 49.63 | <0.001 *** | 116 | 31.60 | <0.001 *** | 148 | 329.55 | <0.001 *** | 148 | 209.65 | <0.001 *** | 79 | 13.48 | <0.001 *** |
| Plant history (PH) | C | 1 | 148 | 79.72 | <0.001 *** | 116 | 44.08 | <0.001 *** | 148 | 179.16 | <0.001 *** | 148 | 189.24 | <0.001 *** | 79 | 30.98 | <0.001 *** |
| Soil history (SH) | C | 1 | 148 | 1.89 | 0.171 | 116 | 0.06 | 0.805 | 148 | 172.22 | <0.001 *** | 148 | 26.33 | <0.001 *** | 79 | 3.74 | 0.057 . |
| T×SR | A | 2 | 148 | 15.50 | <0.001 *** | 116 | 0.23 | 0.792 | 148 | 21.195 | <0.001 *** | 148 | 2.91 | 0.058 . | 79 | 2.15 | 0.124 |
| SH×SR | B | 1 | 148 | 13.48 | <0.001 *** | 116 | 0.10 | 0.750 | 148 | 19.805 | <0.001 *** | 148 | 3.54 | 0.062 . | 79 | 4.26 | 0.042 * |
| PH×SR | B | 1 | 148 | 17.52 | <0.001 *** | 116 | 0.37 | 0.547 | 148 | 22.585 | <0.001 *** | 148 | 2.28 | 0.133 | 79 | 0.03 | 0.862 |
| PH×SR | C | 1 | 148 | 29.81 | <0.001 *** | 116 | 0.47 | 0.496 | 148 | 40.206 | <0.001 *** | 148 | 5.06 | 0.026 * | 79 | 0.91 | 0.343 |
| SH×SR | C | 1 | 148 | 1.18 | 0.279 | 116 | 0.00 | 0.980 | 148 | 2.1841 | 0.142 | 148 | 0.76 | 0.384 | 79 | 3.38 | 0.070 . |

Df= degrees of freedom.
We obtained results from different linear mixed effects model analyses: treatment was fitted as an overall factor with three levels in model A and depicted into a soil history (−SH vs. +SH, according to Table 1) and a plant history contrast (−PH vs. +PH, according to Table 1) in models B and C with varying order of the contrasts. In this way, model B tests for the pure plant history effect by contrasting treatment +SH −PH against treatment +SH +PH, and model C tests for the pure soil history effect by contrasting treatment −SH −PH against treatment +SH −PH. ***$P \leq 0.001$, **$P \leq 0.01$, *$P \leq 0.05$, $P \leq 0.1$. Plant species richness was modelled as a log-linear term.

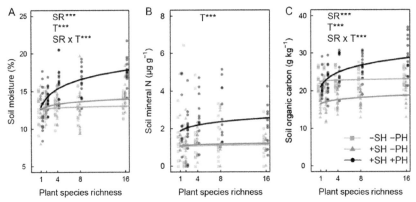

**Fig. 4** Soil abiotic properties of the experimental treatments: soil moisture (A), soil mineral nitrogen concentration (sum of nitrate- and ammonium-N, B), and soil organic carbon content at 0–5 cm depth. Lines represent the predicted means and were derived from a mixed model with a log-linear term for plant species richness (SR), treatment (T) and the interaction of both (T × SR) as fixed effects. Significant effects of the models are indicated on top ***$P \leq 0.001$, **$P \leq 0.01$, *$P \leq 0.05$, ˙$P \leq 0.1$.

history but by plant history as shown by the differences between the +SH −PH and +SH +PH treatments (PH in model B in Table 3). Additionally, the increase in soil moisture across the plant diversity gradient was much higher in old (+SH +PH treatment) than in new communities (PH × SR in model B and C, Table 3 and Fig. 4A). Plant-available N (nitrate and ammonium) concentrations in the soil were much higher in the +SH +PH treatment than in the newly seeded treatments without plant history (PH in model C in Table 3 and Fig. 4B). Beyond this, we did not find any significant differences in soil mineral N concentrations between the +SH −PH and −SH −PH treatments (SH in model C not significant, Table 3), indicating the absence of a pure soil history effect (Table 3). We did not find a significant plant species richness effect on soil mineral N concentration (SR and interactions with SR not significant in all models, Table 3 and Fig. 4B). Soil organic C content differed in all treatments and across the plant diversity gradient (T in model A in Table 3 and Fig. 4C). As expected, we found a lower soil organic C content in the new compared to the old communities (PH in model C, Table 3 and Fig. 4C). The organic C content of the +SH −PH treatment with old soil was even lower than that of the −SH −PH treatment with new soil (SH in model C, Table 3 and Fig. 4C). The positive effect of plant species richness on soil organic C content only occurred in the old communities (PH × SR

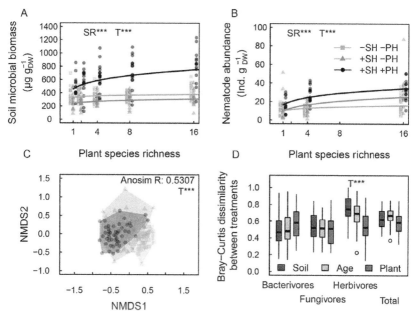

**Fig. 5** Soil biotic properties of the experimental treatments: microbial biomass (A) and soil nematode abundance (B). Lines represent the predicted means and were derived from a mixed model with a log-linear term for plant species richness (SR), treatment (T) and the interaction of both (T × SR) as fixed effects. (C) shows the NMDS ordination of the nematode communities of all subplots based on pairwise Bray-Curtis distances. The Bray-Curtis dissimilarities between all possible treatment pairs (D) for the single trophic groups separately and for the nematode community as a whole. Dissimilarities were based on soil (between −SH −PH and +SH −PH, brown), age (between −SH −PH and +SH +PH, grey), and plant history (between +SH −PH and +SH +PH, green). Significant effects of the models are indicated on top ***$P \leq 0.001$, **$P \leq 0.01$, *$P \leq 0.05$, ·$P \leq 0.1$.

in model C, Table 3 and Fig. 4C) and therefore it was not affected by the soil history (SH × SR in model C not significant, Table 3).

Soil microbial biomass and nematode abundance increased with plant species richness and were affected by the treatments (Table 3, Fig. 5A and B). Soil microbial biomass was highest in the +SH +PH treatment with shared soil and plant history (PH in model C significant, Table 3). Also independent of the soil history, plant history significantly increased soil microbial biomass (PH in model B in Table 3). Interestingly, soil microbial biomass was lower in +SH −PH compared to the −SH −PH treatment, resulting in a negative effect of pure soil history (SH in model C in Table 3 and Fig. 5A). The treatments only marginally changed the plant diversity effect (Table 3).

Similar to microbial biomass, nematode abundance was highest in the +SH +PH treatment, indicating a positive plant history effect (PH in model C in Table 3 and Fig. 5B). The lowest nematode abundance was found in the −SH −PH treatment (SH in model B in Table 3). Plant species richness had a stronger positive effect on nematode communities in old than in new soil (SH and SH × SR in model B in Table 3). Non-metric multidimensional scaling of the nematode communities revealed differences of nematode communities among treatments (Fig. 5C). Differences in the composition of the nematode communities, measured as Bray-Curtis dissimilarities among all treatment pairs, were not significantly affected by treatment or plant species richness (Fig. 5D, SR: $F_{1,39}=0.09$, $P>0.2$; Treatment pair: $F_{2,68}=1.33$, $P>0.2$). However, the treatment effect on the nematode community composition depended on the trophic groups of the nematodes (Fig. 5D, trophic group × treatment pair interaction: $F_{6,336}=7.53$, $P<0.001$). The dissimilarity between the soil history treatments was most pronounced within the community of plant-feeding nematodes, whereas dissimilarities due to plant history did not change between the other trophic groups. Ignoring the abundances of the nematode taxa by calculating the Jaccard index, we found similar results (Fig. S1 in the online version at https://doi.org/10.1016/bs.aecr.2019.06.006, Treatment pair: $F_{2,68}=4.83$, $P=0.01$). The composition of plant-feeding nematodes was more dissimilar between the two soil treatments (−SH −PH and +SH −PH treatment) compared with the treatment pairs including different plant history (Fig. S1 in the online version at https://doi.org/10.1016/bs.aecr.2019.06.006, Treatment pair: $F_{2,68}=3.40$, $P=0.04$). The dissimilarity of the total nematode community depended on the treatment pair (Fig. S1 in the online version at https://doi.org/10.1016/bs.aecr.2019.06.006, Treatment pair: $F_{2,68}=6.35$, $P=0.003$) and was highest between the −SH −PH and +SH −PH treatments. Our results indicate that the composition of nematode taxa changes more strongly with soil history than with plant history, especially within the trophic group of plant-feeding nematodes.

### 3.1.3 Treatment effects on the plant diversity–productivity relationship

Aboveground biomass production differed significantly among treatments (Fig. 6A, model A in Table 4). The treatments with soil history (+SH −PH and +SH +PH treatments) had lower aboveground biomass compared to the −SH −PH treatment without soil history (SH in model B in Table 4 and Fig. 6A). Comparing the young plant communities only (−SH −PH

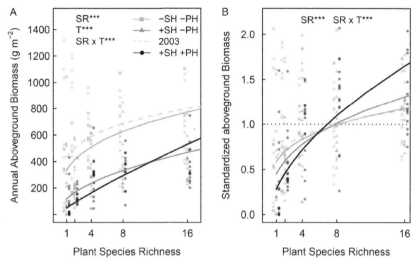

**Fig. 6** Annual aboveground biomass production (A) and standardized aboveground biomass (B) of the experimental treatments across the plant species richness gradient. Lines represent the predicted means and were derived from a mixed model with a log-linear term for plant species richness (SR), treatment (T) and the interaction of both (T × SR) as fixed effects. Significant effects are indicated on bottom ***$P \leq 0.001$, **$P \leq 0.01$, *$P \leq 0.05$, ·$P \leq 0.1$.

and +SH −PH treatment) revealed that soil history clearly decreased aboveground biomass productivity (SH in model C in Table 4 and Fig. 6A). Community biomass production of the −SH −PH treatment in 2017 was as high as the community biomass production in 2003 (Y in model D in Table 4 not significant, Fig. 6A). In general, aboveground biomass production increased with plant species richness in all treatments (SR in all models in Table 4 and Fig. 6A). However, the strength of species richness effect differed among treatments (T × SR in model A in Table 4, Fig. 6A). As hypothesized, we found the steepest slope of the plant species richness-productivity relationship in the old communities with shared plant and soil history (+SH +PH; PH in model D in Table 4 and Fig. 6B). Separating the treatment effect revealed that pure soil history did not significantly alter the species richness effect on aboveground biomass production (SH × SR in model C not significant, Table 1 and Fig. 5B), and that the calendar year in young communities also did not change the plant diversity effect on aboveground productivity (Y × SR in model D not significant, Table 4 and Fig. 6B). This was also true when we standardized the aboveground biomass by mean biomass of the respective treatment to compare the slopes on the same scale (Table 4 and Fig. 6B).

**Table 4** Effects of experimental treatments on annual aboveground biomass and standardized annual aboveground biomass.

| | Model | numDf | denDf | Annual biomass | | | | Standardized biomass | | |
|---|---|---|---|---|---|---|---|---|---|---|
| | | | | F | P | | | F | P | |
| (Intercept) | A, B, C, D | 1 | 222 | 7750.20 | <0.001 | *** | | 4.89 | 0.028 | * |
| Plant species richness (SR), log-linear | A, B, C, D | 1 | 71 | 50.33 | <0.001 | *** | | 44.54 | <0.001 | *** |
| **Treatments (T)** | **A** | **3** | **222** | **49.84** | **<0.001** | *** | | **2.08** | **0.104** | |
| Year (Y) | B | 1 | 222 | 60.30 | <0.001 | *** | | 2.80 | 0.096 | . |
| Soil history (SH) | B | 1 | 222 | 79.97 | <0.001 | *** | | 2.41 | 0.122 | |
| **Plant history (PH)** | **B** | **1** | **222** | **9.25** | **0.003** | ** | | **1.03** | **0.312** | |
| Y | C | 1 | 222 | 60.30 | <0.001 | *** | | 2.80 | 0.096 | . |
| PH | C | 1 | 222 | 50.48 | <0.001 | *** | | 2.73 | 0.100 | . |
| **SH** | **C** | **1** | **222** | **38.74** | **<0.001** | *** | | **0.70** | **0.402** | |
| PH | D | 1 | 222 | 86.25 | <0.001 | *** | | 4.48 | 0.035 | * |
| SH | D | 1 | 222 | 61.88 | <0.001 | *** | | 1.54 | 0.217 | |
| **Y** | **D** | **1** | **222** | **1.39** | **0.240** | | | **0.22** | **0.640** | |
| **T × SR** | **A** | **3** | **222** | **9.40** | **<0.001** | *** | | **8.91** | **<0.001** | *** |
| Y × SR | B | 1 | 222 | 9.70 | 0.002 | ** | | 8.44 | 0.004 | ** |
| SH × SR | B | 1 | 222 | 10.94 | 0.001 | ** | | 10.82 | 0.001 | ** |

| | | Df | | F | p | | F | p | |
|---|---|---|---|---|---|---|---|---|---|
| PH×SR | B | 1 | 222 | 7.55 | 0.007 | ** | 7.48 | 0.007 | ** |
| Y×SR | C | 1 | 222 | 9.70 | 0.002 | ** | 8.44 | 0.004 | ** |
| PH×SR | C | 1 | 222 | 16.27 | <0.001 | *** | 16.11 | <0.001 | *** |
| SH×SR | C | 1 | 222 | 2.22 | 0.137 | | 2.19 | 0.140 | |
| PH×SR | D | 1 | 222 | 23.43 | <0.001 | *** | 22.58 | <0.001 | *** |
| SH×SR | D | 1 | 222 | 4.36 | 0.038 | * | 3.94 | 0.049 | * |
| Y×SR | D | 1 | 222 | 0.40 | 0.528 | | 0.22 | 0.637 | |

Df= degrees of freedom.
Standardization was done by division of aboveground biomass by the mean biomass per treatment. We obtained results from different linear mixed effects model analyses: treatment was fitted as a factor with four levels in model A and depicted into a year (2003 vs. 2017, according to Table 1), a soil history (−SH vs. +SH, according to Table 1) and a plant history contrast (−PH vs. +PH, according to Table 1) in models B, C and D with varying order of the contrasts. In this way, model B tests for the pure plant history effect by contrasting treatment +SH −PH against treatment +SH +PH, model C tests for the pure soil history effect by contrasting treatment −SH −PH against treatment +SH −PH, and model D tests for the effect of the calendar year by contrasting treatment −SH −PH against 2003. ***: $P \leq 0.001$, **: $P \leq 0.01$, *: $P \leq 0.05$, .: $P \leq 0.1$. Plant species richness was modelled as a log-linear term

## 4. Discussion

Here, we present the hypotheses, setup, and first results of a new experiment testing the mechanistic underpinnings of strengthening plant diversity effects over time in a long-term biodiversity experiment. The new ΔBEF Experiment is a cross-sectional approach and complements the long-term grassland biodiversity platform of the Jena Experiment by the opportunity to separate community age effects from effects of the respective calendar year with its specific environmental conditions. Moreover, effects of plant community-specific soil history can be tested, and long-term plots of the Jena Experiment can serve as a reference for conditions of communities with soil and plant history.

### 4.1 Establishment of the treatments

Our analysis of the seed bank revealed that the plant communities in the −SH −PH treatment can be considered as naïve, because no seeds of the experimental species pool were found in samples of the arable soil. This was expected, since the management of arable fields aims to exclude grassland species by tilling, pesticide application, and fertilization for agricultural production. Moreover, the plant communities of the +SH −PH treatment can also be considered as nearly naïve. Although we found viable seeds in the seed bank of this treatment (on average 304 viable seeds per $m^2$), the number was only 25% as high as in the +SH +PH treatment (on average 1284 viable seeds per $m^2$). Furthermore, the density of emerging seedlings observed in the field in the +SH −PH treatment did not differ from the completely naïve communities of the −SH −PH treatment. The treatments without plant history (−SH −PH and +SH −PH) were also more similar to each other than to the old communities of the +SH +PH treatment; they had a lower density of mature individuals, higher realized plant species, higher evenness, and differed little in their plant species composition. Our comparisons with the plant communities in 2003 (Figs 2E, F, and 3B) confirmed that these differences between the new and old plant communities reflect the compositional shifts over time in the old communities as observed previously (Allan et al., 2011; Roscher et al., 2011b). Moreover, the positive effects of plant species richness on the number of emerging seedlings, plant density, leaf area index, and on compositional shifts were more pronounced in old than in new plant communities. Notably, in this initial assessment of the ΔBEF Experiment, we did not find any significant differences between new plant communities established on old or new (arable) soil. Future studies will explore if soil history effects emerge over time.

Differences between the new (−SH −PH and +SH −PH treatments) and the old communities (+SH +PH treatment) were not only present at the level of plant-community properties but also in the soil. Soil moisture of the upper soil layer was significantly lower in the new than in the old communities (Fig. 4A). Similarly, Fischer et al. (2019) reported lower topsoil water content in the first years of the Jena Experiment than in later years. They explained increases in the water content of the topsoil during the first years by higher leaf area index of the plant canopy. However, in the present study, leaf area index was higher in the −SH −PH than in the +SH −PH treatment (Fig. 2D), despite the two treatments having a similar plant density (Fig. 2C), and the soil moisture content was similarly low (Fig. 4A). We therefore suggest that higher water evaporation from the soil due to lower plant density in both newly established treatments is more likely responsible for the decreased soil moisture content than are differences in leaf area index. Leaf area index can still be high despite low plant densities, if there are more horizontal than vertical leaf angles. In this case, evaporation could be high, despite the shading effects of the canopy. However, we cannot exclude the possibility that physical factors operated as well. The establishment of the treatments −SH −PH and +SH −PH involved soil disturbance and homogenization similar to tillage, which is known to affect many physical soil properties (Holland, 2004), such as increased destruction of soil aggregates (Six et al., 2000), low soil structure, and thus a lower retention of soil moisture. Reduced soil moisture as well as preceding leaching might also explain the much lower concentration of plant-available N in those treatments compared to +SH +PH.

The described absence of early soil-history effects was unexpected, particularly for soil mineral N concentrations, since the arable soil of the −SH −PH treatment was derived from a conventional agricultural field with fertilizer application. In the early years of the Jena Experiment, mineral N concentrations increased shortly after land-use change from arable field to experimental grasslands and then decreased again over time due to discontinued fertilization, increasing aboveground N storage in plant biomass (Oelmann et al., 2007, 2011), and continuous removal of plant biomass by mowing (Leimer et al., 2014). However, there was a short lag phase before the initial increase in mineral N concentrations. It is possible that the current phase of the −SH −PH treatment corresponds to this lag phase where soil mineral N remained unchanged before this increase

(Oelmann et al., 2011). We found equally low concentrations of plant-available N in the +SH −PH treatment as in −SH −PH, and higher concentrations of plant-available N in the +SH +PH treatment. This may point towards a strong effect of the plant community on soil mineral N, with younger, still establishing communities having a higher demand for N during initial growth, whereas more mature, established communities may require less N and therefore deplete the pool of available N in the soil to a lesser degree. This also might be reflected in the higher biomass in the −SH −PH treatment, where pre-seeding N in the soil was presumably higher than in the +SH −PH treatment. This would be a subject to be investigated further in future studies.

Further, the mixing of the soil during the setup of the experiment might contribute to the low mineral N concentrations in the +SH −PH treatment compared to the +SH +PH treatment. Higher SOM accumulation in old soil provides additional sources for N mineralization and may, therefore, impact on mineral N concentrations in the soil (Rosenkranz et al., 2012). The increase in SOM over time in undisturbed soils was shown to be strongest in the upper 5–10 cm of the soil but much less pronounced at deeper soil layers (Gerzabek et al., 2005; Steinbeiss et al., 2008). Mixing might have caused a homogenization of SOM in the top 30 cm of the soil and therefore reduced the SOM content in upper soil layers, where we measured soil mineral N concentrations. This was further underlined by the low soil organic C content in the +SH −PH treatment, but high soil organic C content in the +SH +PH treatment, which established over time (Lange et al., 2019).

Soil biota showed significant responses to the soil-history treatment, but these were additionally affected by plant removal. We expected a much lower microbial biomass in the arable soil of −SH −PH, since microbial biomass was shown to increase only 4 years after converting an agricultural field to an experimental grassland in the Jena Experiment, and this increase was largely restricted to high-diversity plant communities (Eisenhauer et al., 2010, 2011b; Strecker et al., 2016). However, microbial biomass in the +SH −PH treatment was not higher but even lower than in the arable soil of the −SH −PH treatment. As we lack information on soil microbial community composition, we can only speculate about the reasons for this finding. One potential explanation might be that more autochthonous and less resilient microbial communities in the experimental grasslands were heavily disturbed by the topsoil mixing, while microbial communities in the agricultural soils, which may be better adapted to disturbance showed higher resilience due to the dominance of fast-growing microbial taxa (Eisenhauer et al., 2010, 2017). Future research has to explore soil microbial community changes in the different soil and plant history treatments along the plant diversity gradient, and potential feedback effects of these microbial communities on plant growth. Strong effects of soil history from the Jena Experiment on plant rhizosphere microbiomes have been found in a recent glasshouse experiment (Schmid et al., 2019).

Soil nematode abundance was reduced in both newly established treatments in comparison to the old communities of the +SH +PH treatment, but also the two new soil treatments differed marginally: the nematode abundance in the new communities (−SH −PH) was lower than in the old soil (+SH −PH). Exploring the functional composition of soil nematode communities provided further insights by showing that the low levels in the −SH −PH treatment were due to a drop in plant-feeding nematode abundances.

Our findings are in line with other studies reporting that plant removal causes a decrease in abundances of microorganisms, micro- and mesofauna (Convey and Wynn-Williams, 2002; Hirsch et al., 2009; Mikola et al., 2014; Stevenson et al., 2014; Wardle et al., 1999). Plant removal and soil homogenization can induce reductions in soil organic matter content or in releases of recently fixed carbon by plant roots and therefore diminish important food resources of soil organisms. Mikola et al. (2014) explained the decrease of microbes, nematodes, and collembolans after plant removal in arctic soils with a decrease in carbon releases by plant roots rather than decreases in soil organic matter. Even replanting could not reverse this plant-removal effect in the short-term. In our study system, soil microbial biomass increased in response to an increase in root biomass (Eisenhauer et al., 2010), while the amount of root exudates was a poor predictor of microbial biomass (Mellado-Vázquez et al., 2016, but see Eisenhauer et al., 2017). In the ΔBEF Experiment, root biomass in the new communities (−SH −PH and +SH −PH treatments) was indeed lower than in the old communities (+SH +PH treatment) in the same study year as we analysed soil microbial biomass and nematode abundance (D. Francioli, data unpublished). Thus, reduced root biomass and root-derived plant inputs to the soil might explain the lower abundances of soil biota, in addition to soil disturbance and homogenization.

However, while both, the soil treatment and the plant treatment affected the abundance of soil biota, the composition of soil nematodes responded only to the soil treatment (Fig. 5C and D). Plant removal in the above-mentioned studies also caused stronger effects on abundance than on the composition of micro- and mesofauna (Mikola et al., 2014) and microbial communities (Hirsch et al., 2009). The effect of the soil history was most pronounced when we analysed the composition of the plant-feeding nematodes only because +SH −PH and +SH +PH treatments had a more similar community composition of plant-feeding nematodes than the arable soil treatment (−SH −PH; Fig. 5D and Fig. S1 in the online version at https://doi.org/10.1016/bs.aecr.2019.06.006). Overall, the soil nematode-community analysis indicates that the soil treatments of arable versus plot-specific soil differ concerning the composition of the soil biota, while the plant treatment (plant removal, soil disturbance and reseeding) changed the abundance of soil biota. While the ΔBEF Experiment allows to study soil history effects by comparing the −SH −PH and the +SH −PH treatments, it is not possible to separate effects of plant history alone by the comparison of +SH −PH and +SH +PH, because we manipulated biotic and abiotic soil properties in addition to plant community age. Plants and their inputs drive soil communities (Leff et al., 2018), which then feedback to the plants and therefore effects of plant age or plant community age can never be separated sharply from soil feedback effects. However, such soil feedback effects, which are tightly linked to the plants that they would not operate without them, can be regarded still as plant history effects, while changes in soil properties due to the physical soil disturbance could not. The treatment of +SH +PH therefore contrasts the other two treatments of the ΔBEF Experiment by having a shared history of the soil and plant communities and hence, serves as control.

## 4.2 Treatment effects on the plant diversity–productivity relationship

In line with previous work (Guerrero-Ramírez et al., 2017; Marquard et al., 2009; Reich et al., 2012) and our expectations, we found a positive relationship between plant species richness and aboveground biomass production in all treatments. Confirming our hypothesis, we found the weakest slope of this BEF relationship in the young communities (−SH −PH and +SH −PH treatments) and the steepest slope in the old communities of the +SH +PH treatment with shared soil and plant history. Moreover, according to our hypothesis, the weakest slope of the BEF relationship in the −SH −PH treatment without plot-specific soil history and without plant history did not significantly differ from the slope in 2003, i.e., 1 year after the establishment of the Jena Experiment. Although the conditions in 2003 were similar to our study year in 2017 concerning the soil history (both, the 2003 communities and −SH −PH treatment was established on arable soil) and the age and composition of the plant communities, climatic conditions differed considerably. In summer 2003, an extreme heat wave affected Central Europe (Beniston, 2004; Zaitchik et al., 2006) and resulted in almost 2 °C higher spring and summer mean temperatures at our field site compared to the reference period of 1961–1990. The year 2017 was less extreme with only warmer spring temperatures of <1 °C higher than the average of the reference period. Experimental studies revealed that warming can change the magnitude and slope of plant diversity–productivity relationships (Cowles et al., 2016; De Boeck et al., 2008). However, this was not the case in the present study, when we compared communities of the same age but under different climate conditions. The similar magnitudes and slopes of the plant diversity–productivity relationship in the −SH −PH treatment and in 2003 suggest that community age effects play a larger role in shaping the BEF-relationship than climatic or other environmental conditions of the respective years. Therefore, we conclude that the treatments of the ΔBEF Experiment are successful in mimicking long-term trends in a cross-sectional study, at least for plant productivity-related processes. Consequently, our ΔBEF Experiment enables us to study ecological mechanisms and processes of BEF relationships independently of the environmental conditions and to transfer those insights to explain long-term trends in BEF relationships. Future studies will investigate a range of different ecosystem functions in response to the treatments of the ΔBEF Experiment, because the change in the strength of diversity effects on functioning and the underlying processes have been shown to differ among ecosystem functions (Meyer et al., 2016) and have important implications for the multifunctionality of ecosystems (Giling et al., 2019; Manning et al., 2018).

Our second aim in this experiment was to test for the contribution of soil history or plant history to the strengthening BEF relationship. As outlined above, pure soil history effects can be separated by comparing the treatments −SH −PH and +SH −PH. Plot-specific soil history decreased overall aboveground biomass production, but in contrast to our expectations, it did not strengthen the plant diversity–productivity relationship. Grassland-soil history in our experiment changed the abundance of soil organisms and the composition of herbivorous nematodes, while soil abiotic conditions did not differ significantly. We speculate that effects of soil organisms may need some time to materialize, e.g., via shifts in soil nutrient status, plant community composition, and antagonistic and mutualistic interactions (Eisenhauer, 2012; Kulmatiski et al., 2012). Notably, in the ΔBEF

Experiment, we did not fully cross soil history and plant history treatments due to logistic constraints (there is no –SH +PH treatment). This means that we can study the role of soil history effects in BEF relationships in a long-term and large-scale field experiment, while the +SH +PH treatment can serve as an undisturbed control. However, results of this long-term control and previous smaller-scale experiments (van Moorsel et al., 2018; Zuppinger-Dingley et al., 2014) indicate that plant history is an important determinant of BEF relationships, which should be studied in future work.

## 5. Conclusions

The $\Delta$BEF Experiment allows us to study mechanisms of BEF relationships in grassland plant communities by (1) isolating community age effects from climate effects and (2) exploring the role of soil history effects. Notably, the plant community properties of this experimental setup are very similar to the starting conditions of the Jena Experiment and thus enables us to compare young with old grassland communities of the same species composition. Our initial results suggest that BEF relationships are robust to climatic conditions and depend on soil and plant history. While the role of soil history may need more time to emerge, our findings indicate that future studies should explore the single and interactive effects of soil history and plant history. Moreover, plant history effects need to be studied by investigating compositional shifts (Allan et al., 2013; Marquard et al., 2009; Roscher et al., 2013) and changes in plant trait expression related to resource use (Tilman and Snell-Rood, 2014; Zuppinger-Dingley et al., 2014) and defence against plant antagonists (Latz et al., 2012; Weisser et al., 2017; Zuppinger-Dingley et al., 2016). The community composition and effects of soil biota on plant community composition and growth need to be explored to better understand soil history effects. The $\Delta$BEF Experiment provides the ideal infrastructure to perform such integrative research by considering the multifunctionality of these experimental grasslands and the linkages to plant-consumer interactions above and below the ground as well as consequences for element cycling.

## Acknowledgements

We gratefully acknowledge the help of the coordinators of the Jena Experiment as well as all gardeners, technicians and numerous student helpers, who maintained the long-term experimental plots and helped to establish the $\Delta$BEF Experiment. Furthermore, we thank Uta Gerighausen, Anja Zeuner and Odette Gonzalez for soil sampling and Ronja

Kober-Moritz, Rebekka Mandler, Janine Schmid and Erika Snjaric for soil extraction. The Jena Experiment is funded by the German Research Foundation (DFG FOR 456, FOR 1451). Moreover, A.V., A.L. and N.E. acknowledge financial support from the German Centre for Integrative Biodiversity Research Halle-Jena-Leipzig, funded by the German Research Foundation (FZT 118) and M.L. from the Zwillenberg-Tietz foundation. Comments by two anonymous reviewers helped to improve the manuscript.

## References

Allan, E., Jenkins, T., Fergus, A.J.F., Roscher, C., Fischer, M., Petermann, J., Weisser, W.W., Schmid, B., 2013. Experimental plant communities develop phylogenetically overdispersed abundance distributions during assembly. Ecology 94 (2), 465–477.

Allan, E., Weisser, W., Weigelt, A., Roscher, C., Fischer, M., Hillebrand, H., 2011. More diverse plant communities have higher functioning over time due to turnover in complementary dominant species. Proc. Natl. Acad. Sci. U. S. A. 108 (41), 17034–17039.

Anderson, M.J., Crist, T.O., Chase, J.M., Vellend, M., Inouye, B.D., Freestone, A.L., Sanders, N.J., Cornell, H.V., Comita, L.S., Davies, K.F., Harrison, S.P., Kraft, N.J.B., Stegen, J.C., Swenson, N.G., 2011. Navigating the multiple meanings of β diversity: a roadmap for the practicing ecologist. Ecol. Lett. 14 (1), 19–28.

Bai, Y.H., Xingguo, Chen, Z., Li, L., 2004. Eosystem stability and compensatory effects in the inner mongolia grassland. Nature 431, 181–184.

Barry, K.E., Mommer, L., van Ruijven, J., Wirth, C., Wright, A.J., Bai, Y., Connolly, J., De Deyn, G.B., de Kroon, H., Isbell, F., Milcu, A., Roscher, C., Scherer-Lorenzen, M., Schmid, B., Weigelt, A., 2019. The future of complementarity: disentangling causes from consequences. Trends Ecol. Evol. 34 (2), 167–180.

Beck, T., Joergensen, R.G., Kandeler, E., Makeschin, F., Nuss, E., Oberholzer, H.R., Scheu, S., 1997. An inter-laboratory comparison of ten different ways of measuring soil microbial biomass C. Soil Biol. Biochem. 29 (7), 1023–1032.

Beniston, M., 2004. The 2003 heat wave in Europe: a shape of things to come? An analysis based on Swiss climatological data and model simulations. Geophys. Res. Lett. 31 (2), 4.

Bessler, H., Temperton, V.M., Roscher, C., Buchmann, N., Schmid, B., Schulze, E.D., Weisser, W.W., Engels, C., 2009. Aboveground overyielding in grassland mixtures is associated with reduced biomass partitioning to belowground organs. Ecology 90 (6), 1520–1530.

Bongers, T., Bongers, M., 1998. Functional diversity of nematodes. Appl. Soil Ecol. 10 (3), 239–251.

Cardinale, B.J., Duffy, J.E., Gonzalez, A., Hooper, D.U., Perrings, C., Venail, P., Narwani, A., Mace, G.M., Tilman, D., Wardle, D.A., Kinzig, A.P., Daily, G.C., Loreau, M., Grace, J.B., Larigauderie, A., Srivastava, D.S., Naeem, S., 2012. Biodiversity loss and its impact on humanity. Nature 486 (7401), 59–67.

Cardinale, B.J., Matulich, K.L., Hooper, D.U., Byrnes, J.E., Duffy, E., Gamfeldt, L., Balvanera, P., O'Connor, M.I., Gonzalez, A., 2011. The functional role of producer diversity in ecosystems. Am. J. Bot. 98 (3), 572–592.

Cardinale, B.J., Wrigh, J.P., Cadotte, M.W., Carroll, I.T., Hector, A., Srivastava, D.S., Loreau, M., Weis, J.J., 2007. Impacts of plant diversity on biomass production increase through time because of species complementarity. Proc. Natl. Acad. Sci. U. S. A. 104 (46), 18123–18128.

Convey, P., Wynn-Williams, D.D., 2002. Antarctic soil nematode response to artificial climate amelioration. Eur. J. Soil Biol. 38 (3), 255–259.

Cowles, J.M., Wragg, P.D., Wright, A.J., Powers, J.S., Tilman, D., 2016. Shifting grassland plant community structure drives positive interactive effects of warming and diversity on aboveground net primary productivity. Glob. Chang. Biol. 22 (2), 741–749.

De Boeck, H.J., Lemmens, C., Zavalloni, C., Gielen, B., Malchair, S., Carnol, M., Merckx, R., Van den Berge, J., Ceulemans, R., Nijs, I., 2008. Biomass production in experimental grasslands of different species richness during three years of climate warming. Biogeosciences 5 (2), 585–594.

Diaz, S., Cabido, M., 2001. Vive la difference: plant functional diversity matters to ecosystem processes. Trends Ecol. Evol. 16 (11), 646–655.

Duffy, J.E., Godwin, C.M., Cardinale, B.J., 2017. Biodiversity effects in the wild are common and as strong as key drivers of productivity. Nature 549 (7671), 261–264.

Eisenhauer, N., 2012. Aboveground-belowground interactions as a source of complementarity effects in biodiversity experiments. Plant Soil 351 (1–2), 1–22.

Eisenhauer, N., Bessler, H., Engels, C., Gleixner, G., Habekost, M., Milcu, A., Partsch, S., Sabais, A.C.W., Scherber, C., Steinbeiss, S., Weigelt, A., Weisser, W.W., Scheu, S., 2010. Plant diversity effects on soil microorganisms support the singular hypothesis. Ecology 91 (2), 485–496.

Eisenhauer, N., Dobies, T., Cesarz, S., Hobbie, S.E., Meyer, R.J., Worm, K., Reich, P.B., 2013. Plant diversity effects on soil food webs are stronger than those of elevated $CO_2$ and N deposition in a long-term grassland experiment. Proc. Natl. Acad. Sci. U. S. A. 110 (17), 6889–6894.

Eisenhauer, N., Lanoue, A., Strecker, T., Scheu, S., Steinauer, K., Thakur, M.P., Mommer, L., 2017. Root biomass and exudates link plant diversity with soil bacterial and fungal biomass. Sci. Rep. 7, 44641.

Eisenhauer, N., Migunova, V.D., Ackermann, M., Ruess, L., Scheu, S., 2011a. Changes in plant species richness induce functional shifts in soil nematode communities in experimental grassland. PLoS One 6 (9), 9.

Eisenhauer, N., Milcu, A., Sabais, A.C.W., Bessler, H., Brenner, J., Engels, C., Klarner, B., Maraun, M., Partsch, S., Roscher, C., Schonert, F., Temperton, V.M., Thomisch, K., Weigelt, A., Weisser, W.W., Scheu, S., 2011b. Plant diversity surpasses plant functional groups and plant productivity as driver of soil biota in the long term. PLoS One 6 (1), 11.

Eisenhauer, N., Reich, P.B., Scheu, S., 2012. Increasing plant diversity effects on productivity with time due to delayed soil biota effects on plants. Basic Appl. Ecol. 13 (7), 571–578.

Eisenhauer, N., Schielzeth, H., Barnes, A.D., Barry, K.E., Bonn, A., Brose, U., Bruelheide, H., Buchmann, N., Buscot, F., Ebeling, A., Ferlian, O., Freschet, G.T., Giling, D.P., Hättenschwiler, S., Hillebrand, H., Hines, J., Isbell, F., Koller-France, E., König-Ries, B., de Kroon, H., Meyer, S.T., Milcu, A., Müller, J., Nock, C.A., Petermann, J.S., Roscher, C., Scherber, C., Scherer-Lorenzen, M., Schmid, B., Schnitzer, S.A., Schuldt, A., Tscharntke, T., Türke, M., van Dam, N.M., van der Plas, F., Vogel, A., Wagg, C., Wardle, D.A., Weigelt, A., Weisser, W.W., Wirth, C., Jochum, M., 2019. A multitrophic perspective on biodiversity–ecosystem functioning research. Adv. Ecol. Res. 61, 1–54.

Fargione, J., Tilman, D., Dybzinski, R., HilleRisLambers, J., Clark, C., Harpole, W.S., Knops, J.M.H., Reich, P.B., Loreau, M., 2007. From selection to complementarity: shifts in the causes of biodiversity-productivity relationships in a long-term biodiversity experiment. Proc. R. Soc. B Biol. Sci. 274 (1611), 871–876.

Fischer, C., Leimer, S., Roscher, C., Ravenek, J., de Kroon, H., Kreutziger, Y., Baade, J., Beßler, H., Eisenhauer, N., Weigelt, A., Mommer, L., Lange, M., Gleixner, G., Wilcke, W., Schröder, B., Hildebrandt, A., 2019. Plant species richness and functional groups have different effects on soil water content in a decade-long grassland experiment. J. Ecol. 107 (1), 127–141.

Fornara, D.A., Tilman, D., 2008. Plant functional composition influences rates of soil carbon and nitrogen accumulation. J. Ecol. 96 (2), 314–322.

Fridley, J.D., 2002. Resource availability dominates and alters the relationship between species diversity and ecosystem productivity in experimental plant communities. Oecologia 132 (2), 271–277.

Galloway, J.N., Dentener, F.J., Capone, D.G., Boyer, E.W., Howarth, R.W., Seitzinger, S.P., Asner, G.P., Cleveland, C.C., Green, P.A., Holland, E.A., Karl, D.M., Michaels, A.F., Porter, J.H., Townsend, A.R., Vöosmarty, C.J., 2004. Nitrogen cycles: past, present, and future. Biogeochemistry 70 (2), 153–226.

Gerzabek, M.H., Strebl, F., Tulipan, M., Schwarz, S., 2005. Quantification of organic carbon pools for Austria's agricultural soils using a soil information system. Can. J. Soil Sci. 85 (Special Issue), 491–498.

Giling, D.P., Beaumelle, L., Phillips, H.R.P., Cesarz, S., Eisenhauer, N., Ferlian, O., Gottschall, F., Guerra, C., Hines, J., Sendek, A., Siebert, J., Thakur, M.P., Barnes, A.D., 2019. A niche for ecosystem multifunctionality in global change research. Glob. Chang. Biol. 25 (3), 763–774.

Gubsch, M., Buchmann, N., Schmid, B., Schulze, E.D., Lipowsky, A., Roscher, C., 2011. Differential effects of plant diversity on functional trait variation of grass species. Ann. Bot. 107 (1), 157–169.

Guerrero-Ramírez, N.R., Craven, D., Reich, P.B., Ewel, J.J., Isbell, F., Koricheva, J., Parrotta, J.A., Auge, H., Erickson, H.E., Forrester, D.I., Hector, A., Joshi, J., Montagnini, F., Palmborg, C., Piotto, D., Potvin, C., Roscher, C., van Ruijven, J., Tilman, D., Wilsey, B., Eisenhauer, N., 2017. Diversity-dependent temporal divergence of ecosystem functioning in experimental ecosystems. Nat. Ecol. Evol. 1 (11), 1639–1642.

Habekost, M., Eisenhauer, N., Scheu, S., Steinbeiss, S., Weigelt, A., Gleixner, G., 2008. Seasonal changes in the soil microbial community in a grassland plant diversity gradient four years after establishment. Soil Biol. Biochem. 40 (10), 2588–2595.

He, J.S., Bazzaz, F.A., Schmid, B., 2002. Interactive effects of diversity, nutrients and elevated $CO_2$ on experimental plant communities. Oikos 97 (3), 337–348.

Hector, A., Schmid, B., Beierkuhnlein, C., Caldeira, M.C., Diemer, M., Dimitrakopoulos, P.G., Finn, J.A., Freitas, H., Giller, P.S., Good, J., Harris, R., Hogberg, P., Huss-Danell, K., Joshi, J., Jumpponen, A., Korner, C., Leadley, P.W., Loreau, M., Minns, A., Mulder, C.P.H., O'Donovan, G., Otway, S.J., Pereira, J.S., Prinz, A., Read, D.J., Scherer-Lorenzen, M., Schulze, E.D., Siamantziouras, A.S.D., Spehn, E.M., Terry, A.C., Troumbis, A.Y., Woodward, F.I., Yachi, S., Lawton, J.H., 1999. Plant diversity and productivity experiments in European grasslands. Science 286 (5442), 1123–1127.

Heisse, K., Roscher, C., Schumacher, J., Schulze, E.D., 2007. Establishment of grassland species in monocultures: different strategies lead to success. Oecologia 152 (3), 435–447.

HilleRisLambers, J., Harpole, W.S., Tilman, D., Knops, J., Reich, P.B., 2004. Mechanisms responsible for the positive diversity-productivity relationship in Minnesota grasslands. Ecology Letters 7 (8), 661–668.

Hirsch, P.R., Gilliam, L.M., Sohi, S.P., Williams, J.K., Clark, I.M., Murray, P.J., 2009. Starving the soil of plant inputs for 50 years reduces abundance but not diversity of soil bacterial communities. Soil Biol. Biochem. 41 (9), 2021–2024.

Hoffmann, K., Bivour, W., Früh, B., Koßmann, M., Voß, P.-H., 2014. Klimauntersuchungen in Jena für die Anpassung an den Klimawandel und seine erwarteten Folgen. In: Ein Ergebnisbericht, Berichte des Deutschen Wetterdienstes. Selbstverlag des Deutschen Wetterdienstes, Offenbach am Main.

Holland, J.M., 2004. The environmental consequences of adopting conservation tillage in Europe: reviewing the evidence. Agric. Ecosyst. Environ. 103 (1), 1–25.

Hooper, D.U., Adair, E.C., Cardinale, B.J., Byrnes, J.E.K., Hungate, B.A., Matulich, K.L., Gonzalez, A., Duffy, J.E., Gamfeldt, L., O'Connor, M.I., 2012. A global synthesis reveals biodiversity loss as a major driver of ecosystem change. Nature 486 (7401), 105–U129.

Huang, Y., Chen, Y., Castro-Izaguirre, N., Baruffol, M., Brezzi, M., Lang, A., Li, Y., Härdtle, W., von Oheimb, G., Yang, X., Liu, X., Pei, K., Both, S., Yang, B., Eichenberg, D., Assmann, T., Bauhus, J., Behrens, T., Buscot, F., Chen, X.-Y., Chesters, D., Ding, B.-Y., Durka, W., Erfmeier, A., Fang, J., Fischer, M., Guo, L.-D., Guo, D., Gutknecht, J.L.M., He, J.-S., He, C.-L., Hector, A., Hönig, L., Hu, R.-Y., Klein, A.-M., Kühn, P., Liang, Y., Li, S., Michalski, S., Scherer-Lorenzen, M., Schmidt, K., Scholten, T., Schuldt, A., Shi, X., Tan, M.-Z., Tang, Z., Trogisch, S., Wang, Z., Welk, E., Wirth, C., Wubet, T., Xiang, W., Yu, M., Yu, X.-D., Zhang, J., Zhang, S., Zhang, N., Zhou, H.-Z., Zhu, C.-D., Zhu, L., Bruelheide, H., Ma, K., Niklaus, P.A., Schmid, B., 2018. Impacts of species richness on productivity in a large-scale subtropical forest experiment. Science 362 (6410), 80–83.

IPCC, 2014. Climate Change 2014: Synthesis Report. Contribution of Working Groups I, II and III to the Fifth Assessment Report of the Intergovernmental Panel on Climate Change. IPCC, Geneva, Switzerland.

Isbell, F., Calcagno, V., Hector, A., Connolly, J., Harpole, W.S., Reich, P.B., Scherer-Lorenzen, M., Schmid, B., Tilman, D., van Ruijven, J., Weigelt, A., Wilsey, B.J., Zavaleta, E.S., Loreau, M., 2011. High plant diversity is needed to maintain ecosystem services. Nature 477 (7363), 199–U96.

Isbell, F., Craven, D., Connolly, J., Loreau, M., Schmid, B., Beierkuhnlein, C., Bezemer, T.M., Bonin, C., Bruelheide, H., de Luca, E., Ebeling, A., Griffin, J.N., Guo, Q.F., Hautier, Y., Hector, A., Jentsch, A., Kreyling, J., Lanta, V., Manning, P., Meyer, S.T., Mori, A.S., Naeem, S., Niklaus, P.A., Polley, H.W., Reich, P.B., Roscher, C., Seabloom, E.W., Smith, M.D., Thakur, M.P., Tilman, D., Tracy, B.F., van der Putten, W.H., van Ruijven, J., Weigelt, A., Weisser, W.W., Wilsey, B., Eisenhauer, N., 2015. Biodiversity increases the resistance of ecosystem productivity to climate extremes. Nature 526 (7574), 574–U263.

Isbell, F., Reich, P.B., Tilman, D., Hobbie, S.E., Polasky, S., Binder, S., 2013. Nutrient enrichment, biodiversity loss, and consequent declines in ecosystem productivity. Proc. Natl. Acad. Sci. U. S. A. 110 (29), 11911–11916.

IUSS Working Group WRB, 2015. World Reference Base for Soil Resources 2014, International Soil Classification System for Naming Soils and Creating Legends for Soil Maps. update 2015, FAO, Rome.

Kulmatiski, A., Beard Karen, H., Heavilin, J., 2012. Plant–soil feedbacks provide an additional explanation for diversity–productivity relationships. Proc. R. Soc. B Biol. Sci. 279 (1740), 3020–3026.

Lange, M., Eisenhauer, N., Sierra, C.A., Bessler, H., Engels, C., Griffiths, R.I., Mellado-Vázquez, P.G., Malik, A.A., Roy, J., Scheu, S., Steinbeiss, S., Thomson, B.C., Trumbore, S.E., Gleixner, G., 2015. Plant diversity increases soil microbial activity and soil carbon storage. Nat. Commun. 6, 6707.

Lange, M., Habekost, M., Eisenhauer, N., Roscher, C., Bessler, H., Engels, C., Oelmann, Y., Scheu, S., Wilcke, W., Schulze, E.-D., Gleixner, G., 2014. Biotic and abiotic properties mediating plant diversity effects on soil microbial communities in an experimental grassland. PLoS One 9 (5), e96182.

Lange, M., Koller-France, E., Hildebrandt, A., Oelmann, Y., Wilcke, W., Gleixner, G., 2019. How plant diversity impacts the coupled water, nutrient and carbon cycles. Adv. Ecol. Res. 61, 185–219.

Latz, E., Eisenhauer, N., Rall, B.C., Allan, E., Roscher, C., Scheu, S., Jousset, A., 2012. Plant diversity improves protection against soil-borne pathogens by fostering antagonistic bacterial communities. J. Ecol. 100 (3), 597–604.

Latz, E., Eisenhauer, N., Rall, B.C., Scheu, S., Jousset, A., 2016. Unravelling linkages between plant community composition and the pathogen-suppressive potential of soils. Sci. Rep. 6, 23584.

Leff, J.W., Bardgett, R.D., Wilkinson, A., Jackson, B.G., Pritchard, W.J., De Long, J.R., Oakley, S., Mason, K.E., Ostle, N.J., Johnson, D., Baggs, E.M., Fierer, N., 2018. Predicting the structure of soil communities from plant community taxonomy, phylogeny, and traits. ISME J. 12 (7), 1794–1805.

Leimer, S., Wirth, C., Oelmann, Y., Wilcke, W., 2014. Biodiversity effects on nitrate concentrations in soil solution: a Bayesian model. Biogeochemistry 118 (1), 141–157.

Lipowsky, A., Roscher, C., Schumacher, J., Michalski, S.G., Gubsch, M., Buchmann, N., Schulze, E.-D., Schmid, B., 2015. Plasticity of functional traits of forb species in response to biodiversity. Perspect. Plant Ecol. Evol. Syst. 17 (1), 66–77.

Lipowsky, A., Schmid, B., Roscher, C., 2011. Selection for monoculture and mixture genotypes in a biodiversity experiment. Basic Appl. Ecol. 12 (4), 360–371.

Londo, G., 1976. The decimal scale for releves of permanent quadrats. Vegetatio 33 (1), 61–64.

Loreau, M., Hector, A., 2001. Partitioning selection and complementarity in biodiversity experiments. Nature 412 (6842), 72–76.

Manning, P., van der Plas, F., Soliveres, S., Allan, E., Maestre, F.T., Mace, G., Whittingham, M.J., Fischer, M., 2018. Redefining ecosystem multifunctionality. Nat. Ecol. Evol. 2 (3), 427–436.

Maron, J.L., Marler, M., Klironomos, J.N., Cleveland, C.C., 2011. Soil fungal pathogens and the relationship between plant diversity and productivity. Ecol. Lett. 14 (1), 36–41.

Marquard, E., Weigelt, A., Temperton, V.M., Roscher, C., Schumacher, J., Buchmann, N., Fischer, M., Weisser, W.W., Schmid, B., 2009. Plant species richness and functional composition drive overyielding in a six-year grassland experiment. Ecology 90 (12), 3290–3302.

Mellado-Vázquez, P.G., Lange, M., Bachmann, D., Gockele, A., Karlowsky, S., Milcu, A., Piel, C., Roscher, C., Roy, J., Gleixner, G., 2016. Plant diversity generates enhanced soil microbial access to recently photosynthesized carbon in the rhizosphere. Soil Biol. Biochem. 94, 122–132.

Meyer, S.T., Ebeling, A., Eisenhauer, N., Hertzog, L., Hillebrand, H., Milcu, A., Pompe, S., Abbas, M., Bessler, H., Buchmann, N., De Luca, E., Engels, C., Fischer, M., Gleixner, G., Hudewenz, A., Klein, A.-M., de Kroon, H., Leimer, S., Loranger, H., Mommer, L., Oelmann, Y., Ravenek, J.M., Roscher, C., Rottstock, T., Scherber, C., Scherer-Lorenzen, M., Scheu, S., Schmid, B., Schulze, E.-D., Staudler, A., Strecker, T., Temperton, V., Tscharntke, T., Vogel, A., Voigt, W., Weigelt, A., Wilcke, W., Weisser, W.W., 2016. Effects of biodiversity strengthen over time as ecosystem functioning declines at low and increases at high biodiversity. Ecosphere 7 (12), e01619, n/a.

Mikola, J., Sørensen, L.I., Kytöviita, M.-M., 2014. Plant removal disturbance and replant mitigation effects on the abundance and diversity of low-arctic soil biota. Appl. Soil Ecol. 82, 82–92.

Mueller, K.E., Tilman, D., Fornara, D.A., Hobbie, S.E., 2013. Root depth distribution and the diversity–productivity relationship in a long-term grassland experiment. Ecology 94 (4), 787–793.

Naeem, S., Duffy, J.E., Zavaleta, E., 2012. The Functions of Biological Diversity in an Age of Extinction. Science 336 (6087), 1401–1406.

Oelmann, Y., Buchmann, N., Gleixner, G., Habekost, M., Roscher, C., Rosenkranz, S., Schulze, E.D., Steinbeiss, S., Temperton, V.M., Weigelt, A., Weisser, W.W., Wilcke, W., 2011. Plant diversity effects on aboveground and belowground N pools in temperate grassland ecosystems: Development in the first 5 years after establishment. Glob. Biogeochem. Cycles 25, 11.

Oelmann, Y., Wilcke, W., Temperton, V.M., Buchmann, N., Roscher, C., Schumacher, J., Schulze, E.D., Weisser, W.W., 2007. Soil and plant nitrogen pools as related to plant diversity in an experimental grassland. Soil Sci. Soc. Am. J. 71 (3), 720–729.

Oksanen, J., Blanchet, F.G., Friendly, M., Kindt, R., Legendre, P., McGlin, D., Minchin, P., O'Hara, R.B., Simpson, G., Solymos, P., et al., 2017. Vegan: Community Ecology Package. https://cran.r-project.org; https://github.com/vegandevs/vegan.

R Development Core Team, 2011. R: A Language and Environment for Statistical Computing. R Foundation for Statistical Computing, Vienna. http://www.R-project.org.

Ravenek, J.M., Bessler, H., Engels, C., Scherer-Lorenzen, M., Gessler, A., Gockele, A., De Luca, E., Temperton, V.M., Ebeling, A., Roscher, C., Schmid, B., Weisser, W.W., Wirth, C., de Kroon, H., Weigelt, A., Mommer, L., 2014. Long-term study of root biomass in a biodiversity experiment reveals shifts in diversity effects over time. Oikos 123 (12), 1528–1536.

Reich, P.B., Knops, J., Tilman, D., Craine, J., Ellsworth, D., Tjoelker, M., Lee, T., Wedin, D., Naeem, S., Bahauddin, D., Hendrey, G., Jose, S., Wrage, K., Goth, J., Bengston, W., 2001. Plant diversity enhances ecosystem responses to elevated $CO_2$ and nitrogen deposition. Nature 410 (6839), 809–812.

Reich, P.B., Tilman, D., Isbell, F., Mueller, K., Hobbie, S.E., Flynn, D.F.B., Eisenhauer, N., 2012. Impacts of biodiversity loss escalate through time as redundancy fades. Science 336 (6081), 589–592.

Roscher, C., Schmid, B., Buchmann, N., Weigelt, A., Schulze, E.D., 2011a. Legume species differ in the responses of their functional traits to plant diversity. Oecologia 165 (2), 437–452.

Roscher, C., Schumacher, J., Baade, J., Wilcke, W., Gleixner, G., Weisser, W.W., Schmid, B., Schulze, E.D., 2004. The role of biodiversity for element cycling and trophic interactions: an experimental approach in a grassland community. Basic Appl. Ecol. 5 (2), 107–121.

Roscher, C., Schumacher, J., Gubsch, M., Lipowsky, A., Weigelt, A., Buchmann, N., Schmid, B., Schulze, E.-D., 2012. Using plant functional traits to explain diversity-productivity relationships. PLoS One 7 (5), e36760.

Roscher, C., Schumacher, J., Lipowsky, A., Gubsch, M., Weigelt, A., Pompe, S., Kolle, O., Buchmann, N., Schmid, B., Schulze, E.D., 2013. A functional trait-based approach to understand community assembly and diversity-productivity relationships over 7 years in experimental grasslands. Perspect. Plant Ecol. Evol. Syst. 15 (3), 139–149.

Roscher, C., Thein, S., Schmid, B., Scherer-Lorenzen, M., 2008. Complementary nitrogen use among potentially dominant species in a biodiversity experiment varies between two years. J. Ecol. 96 (3), 477–488.

Roscher, C., Weigelt, A., Proulx, R., Marquard, E., Schumacher, J., Weisser, W.W., Schmid, B., 2011b. Identifying population- and community-level mechanisms of diversity-stability relationships in experimental grasslands. J. Ecol. 99 (6), 1460–1469.

Rosenkranz, S., Wilcke, W., Eisenhauer, N., Oelmann, Y., 2012. Net ammonification as influenced by plant diversity in experimental grasslands. Soil Biol. Biochem. 48, 78–87.

Ruess, L., 1995. Studies on the nematode fauna of an acid forest soil: spatial distribution and extraction. Nematologica 41, 229–239.

Scheu, S., 1992. Automated measurement of the respiratory response of soil microcompartments—active microbial biomass in earthworm faeces. Soil Biol. Biochem. 24 (11), 1113–1118.

Schmid, M.W., Hahl, T., van Moorsel, S.J., Wagg, C., De Deyn, G.B., Schmid, B., 2019. Feedbacks of plant identity and diversity on the diversity and community composition of rhizosphere microbiomes from a long-term biodiversity experiment. Mol. Ecol. 28 (4), 863–878.

Schnitzer, S.A., Klironomos, J.N., HilleRisLambers, J., Kinkel, L.L., Reich, P.B., Xiao, K., Rillig, M.C., Sikes, B.A., Callaway, R.M., Mangan, S.A., van Nes, E.H., Scheffer, M., 2011. Soil microbes drive the classic plant diversity-productivity pattern. Ecology 92 (2), 296–303.

Six, J., Elliott, E.T., Paustian, K., 2000. Soil macroaggregate turnover and microaggregate formation: a mechanism for C sequestration under no-tillage agriculture. Soil Biol. Biochem. 32 (14), 2099–2103.

Steinbeiss, S., Bessler, H., Engels, C., Temperton, V.M., Buchmann, N., Roscher, C., Kreutziger, Y., Baade, J., Habekost, M., Gleixner, G., 2008. Plant diversity positively affects short-term soil carbon storage in experimental grasslands. Glob. Chang. Biol. 14 (12), 2937–2949.

Stevenson, B.A., Hunter, D.W.F., Rhodes, P.L., 2014. Temporal and seasonal change in microbial community structure of an undisturbed, disturbed, and carbon-amended pasture soil. Soil Biol. Biochem. 75, 175–185.

Strecker, T., Macé, O.G., Scheu, S., Eisenhauer, N., 2016. Functional composition of plant communities determines the spatial and temporal stability of soil microbial properties in a long-term plant diversity experiment. Oikos 125 (12), 1743–1754.

ter Heerdt, G.N.J., Verweij, G.L., Bekker, R.M., Bakker, J.P., 1996. An Improved Method for Seed-Bank Analysis: Seedling Emergence After Removing the Soil by Sieving. Funct. Ecol. 10 (1), 144–151.

Thakur, M.P., Milcu, A., Manning, P., Niklaus, P.A., Roscher, C., Power, S., Reich, P.B., Scheu, S., Tilman, D., Ai, F.X., Guo, H.Y., Ji, R., Pierce, S., Ramirez, N.G., Richter, A.N., Steinauer, K., Strecker, T., Vogel, A., Eisenhauer, N., 2015. Plant diversity drives soil microbial biomass carbon in grasslands irrespective of global environmental change factors. Glob. Chang. Biol. 21 (11), 4076–4085.

Tilman, D., Lehman, C.L., Thomson, K.T., 1997. Plant diversity and ecosystem productivity: theoretical considerations. Proc. Natl. Acad. Sci. U. S. A. 94 (5), 1857–1861.

Tilman, D., Snell-Rood, E., 2014. Diversity breeds complementarity. Nature 515 (7525), 44–45.

van Moorsel, S.J., Hahl, T., Wagg, C., De Deyn, G.B., Flynn, D.F.B., Zuppinger-Dingley, D., Schmid, B., 2018. Community evolution increases plant productivity at low diversity. Ecol. Lett. 21 (1), 128–137.

Wagg, C., Barendregt, C., Jansa, J., van der Heijden, M.G.A., 2015. Complementarity in both plant and mycorrhizal fungal communities are not necessarily increased by diversity in the other. J. Ecol. 103 (5), 1233–1244.

Wardle, D.A., Bonner, K.I., Barker, G.M., Yeates, G.W., Nicholson, K.S., Bardgett, R.D., Watson, R.N., Ghani, A., 1999. Plant removals in perennial grassland: vegetation dynamics, decomposers, soil biodiversity, and ecosystem properties. Ecol. Monogr. 69 (4), 535–568.

Weisser, W.W., Roscher, C., Meyer, S.T., Ebeling, A., Luo, G., Allan, E., Beßler, H., Barnard, R.L., Buchmann, N., Buscot, F., Engels, C., Fischer, C., Fischer, M., Gessler, A., Gleixner, G., Halle, S., Hildebrandt, A., Hillebrand, H., de Kroon, H., Lange, M., Leimer, S., Le Roux, X., Milcu, A., Mommer, L., Niklaus, P.A., Oelmann, Y., Proulx, R., Roy, J., Scherber, C., Scherer-Lorenzen, M., Scheu, S., Tscharntke, T., Wachendorf, M., Wagg, C., Weigelt, A., Wilcke, W., Wirth, C., Schulze, E.-D., Schmid, B., Eisenhauer, N., 2017. Biodiversity effects on ecosystem functioning in a 15-year grassland experiment: patterns, mechanisms, and open questions. Basic and Appl. Ecol. 23 (Suppl. C), 1–73.

Wright, A.J., Ebeling, A., de Kroon, H., Roscher, C., Weigelt, A., Buchmann, N., Buchmann, T., Fischer, C., Hacker, N., Hildebrandt, A., Leimer, S., Mommer, L., Oelmann, Y., Scheu, S., Steinauer, K., Strecker, T., Weisser, W., Wilcke, W., Eisenhauer, N., 2015. Flooding disturbances increase resource availability and productivity but reduce stability in diverse plant communities. Nat. Commun. 6, 6.

Wright, A.J., Wardle, D.A., Callaway, R., Gaxiola, A., 2017. The overlooked role of facilitation in biodiversity experiments. Trends Ecol. Evol. 32 (5), 383–390.

Yachi, S., Loreau, M., 1999. Biodiversity and ecosystem productivity in a fluctuating environment: the insurance hypothesis. Proc. Natl. Acad. Sci. U. S. A. 96, 1463–1468.

Yeates, G.W., Bongers, T., De Goede, R.G.M., Freckman, D.W., Georgieva, S.S., 1993. Feeding habits in soil nematode families and genera—an outline for soil ecologists. J. Nematol. 25 (3), 315–331.

Zaitchik, B.F., Macalady, A.K., Bonneau, L.R., Smith, R.B., 2006. Europe's 2003 heat wave: A satellite view of impacts and land-atmosphere feedbacks. Int. J. Climatol. 26 (6), 743–769.

Zuppinger-Dingley, D., Flynn, D.F.B., De Deyn, G.B., Petermann, J.S., Schmid, B., 2016. Plant selection and soil legacy enhance long-term biodiversity effects. Ecology 97 (4), 918–928.

Zuppinger-Dingley, D., Schmid, B., Petermann, J.S., Yadav, V., De Deyn, G.B., Flynn, D.F.B., 2014. Selection for niche differentiation in plant communities increases biodiversity effects. Nature 515 (7525), 108–111.

# CHAPTER EIGHT

# Linking local species coexistence to ecosystem functioning: a conceptual framework from ecological first principles in grassland ecosystems

Kathryn E. Barry[a,b,*], Hans de Kroon[d], Peter Dietrich[a,c], W. Stanley Harpole[a,c,e], Anna Roeder[a,c], Bernhard Schmid[f], Adam T. Clark[a,c,g], Margaret M. Mayfield[h], Cameron Wagg[i], Christiane Roscher[a,c]

[a]German Centre for Integrative Biodiversity Research (iDiv) Halle-Jena-Leipzig, Leipzig, Germany
[b]Systematic Botany and Functional Biodiversity, Institute of Biology, Leipzig University, Leipzig, Germany
[c]Department of Physiological Diversity, Helmholtz Centre for Environmental Research (UFZ), Leipzig, Germany
[d]Department of Experimental Plant Ecology, Institute for Water and Wetland Research, Radboud University, Nijmegen, The Netherlands
[e]Institute of Biology, Martin Luther University Halle-Wittenberg, Halle (Saale), Germany
[f]Department of Geography, University of Zürich, Zürich, Switzerland
[g]Synthesis Centre for Biodiversity Sciences (sDiv), Leipzig, Germany
[h]The University of Queensland, School of Biological Sciences, Brisbane, QLD, Australia
[i]Fredericton Research and Development Centre, Agriculture and Agri-Food Canada, Fredericton, NB, Canada
*Corresponding author: e-mail address: barry.kt@gmail.com

## Contents

| | |
|---|---|
| 1. Introduction | 266 |
| 2. Jointly emerging local coexistence and ecosystem functioning from ecological first principles | 269 |
|    2.1 Abiotic conditions | 270 |
|    2.2 Biotic conditions | 272 |
| 3. Population level effects of abiotic and biotic conditions on fecundity, growth, and survival | 273 |
|    3.1 Abiotic conditions | 274 |
|    3.2 Biotic conditions | 278 |
| 4. How ecological first principles influence trade-offs between fecundity, growth, and survival and in turn influence local coexistence and ecosystem functioning | 279 |
|    4.1 Productivity | 281 |
|    4.2 Root decomposition | 284 |
| 5. Conclusion | 285 |
| Author contributions | 286 |

| | |
|---|---|
| Acknowledgements | 286 |
| References | 286 |
| Further reading | 296 |

## Abstract

One of the unifying goals of ecology is understanding the mechanisms that drive ecological patterns. For any particular observed pattern, ecologists have proposed varied mechanistic models. However, in spite of their differences, all of these mechanistic models rely on either abiotic conditions or biotic conditions, our "ecological first principles". These major components underlie all of the major mechanistic explanations for patterns of diversity like the latitudinal gradient in diversity, the maintenance of diversity, and the (often positive) biodiversity-ecosystem functioning relationship. These components and their interactions alter the dynamics of plant populations, which ultimately determine local coexistence at the community level, and functioning at the ecosystem level. We present a review, starting from ecological first principles of the ways in which ecosystem functioning may be linked to local coexistence in plant communities via mutual effects on and reactions to the abiotic and biotic conditions in which they are imbedded.

## 1. Introduction

Humans are driving drastic environmental changes including unprecedented global biodiversity loss (Carneiro da Cunha et al., 2019; Millenium Ecosystem Assessment Ecosystem and human well-being: synthesis, 2005; Newbold et al., 2015; Tittensor et al., 2014). Many studies predict that the rate of biodiversity loss will accelerate in the coming decades (e.g., Pereira et al., 2010; Pimm et al., 2014). Small-scale biodiversity experiments have shown that certain ecosystem functions typically decline after random species loss in experimental communities (reviewed by Tilman et al., 2014). This decline of ecosystem functioning at the local community level is often interpreted to mean that regional or global biodiversity loss will also result in declines in ecosystem functioning at global scales (e.g., Hooper et al., 2012). The effect and extent of diversity loss and compositional change in nature, however, are also thought to be determined by the interplay between local and regional processes that affect local abundance and thereby local ecosystem functioning. Understanding how these processes interact to affect ecosystem functioning is crucial for predicting the consequences of global biodiversity loss and local and regional compositional shifts (Barry et al., 2019).

At the local scale, coexistence between species arises from population-level interactions with abiotic conditions (including resources and other abiotic factors) and the biotic community (reviewed by Holt, 2013, see also Chase and Leibold, 2003). These interactions are characterized by feedbacks that stabilize local coexistence between populations by preventing dominance and allowing rare species to persist (local stable coexistence, Chesson, 2000, 2018; Ellner et al., 2019; Holt, 2013). These conditions may also allow for long term persistence rather than long term stable coexistence. Two populations of species may coexist because they partition resources, mediate stressful abiotic conditions for each other, or are controlled by species-specific enemies (Bertness and Callaway, 1994; Brooker et al., 2008; Bruno et al., 2003; Holt, 2013; Palmer, 1994; Wright, 2002). Here, we do not differentiate between long term stable coexistence and unstable persistence, rather we refer to both as "local coexistence" throughout. We define this local coexistence as "the state of two or more species being found in the same place at the same time" (Holt, 2013). We exclude, however, coexistence supported by dispersal from outside of the local community; we consider these "meta-community dynamics" to be beyond the scope of this paper.

While local coexistence is a necessary condition for locally diverse communities to exist, it is not a sufficient condition for said biodiversity to enhance ecosystem functioning. Rather, circumstances may result in local coexistence that variably increases or decreases ecosystem functioning with increasing species richness (Becker et al., 2012; Huston, 1997; Loreau, 2004; Tilman et al., 1997). Importantly, the likelihood of a positive or negative relationship between biodiversity and ecosystem functioning depends on the function that is being measured and the coexistence mechanisms at play. Some evidence suggests that negative relationships between ecosystem functioning and species richness may more commonly result from meta-community dynamics which are not examined here (Leibold and Chase, 2018; Mouquet and Loreau, 2003; Vandermeer, 1981). Further, Turnbull et al. (2013) demonstrated that diversity can enhance productivity even when long-term stable local coexistence is not possible because stabilizing resource niche (sensu Chesson, 2000) differences between species were not sufficient to overcome fitness differences between species. That is, biodiversity may enhance productivity when species persist together in the short-term but do not stably coexist.

We propose that local coexistence between species and the effects of species on ecosystem functioning are inherently linked by mutual dependence and effects on abiotic and biotic conditions operating at the population level. Here, we describe a framework for the link between local coexistence and ecosystem functioning where both jointly emerge from three major population-level life-history variables (fecundity, growth, and survival) that are simultaneously influenced by abiotic and biotic conditions (ecological first principles, Fig. 1). To do so, we first review how abiotic and biotic conditions have traditionally been used to describe local coexistence and enhanced ecosystem functioning in diverse natural communities. Second, we review how abiotic and biotic conditions determine the fecundity, growth, and survival of populations. Finally, we discuss how trade-offs within and between populations may lead to local coexistence between species and how ecosystem functioning may be altered by these population-level dynamic trade-offs.

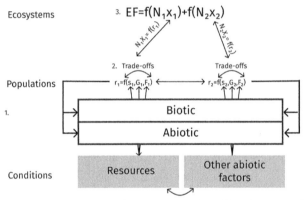

**Fig. 1** Plant population growth and fitness are regulated by abiotic and biotic conditions. Resources are needed to build and maintain tissues but these processes are influenced by competitors, consumers, mutualists and abiotic factors such as temperature and pH. Different vital rates (r, e.g., growth—denoted with a "G" in the above figure, fecundity—denoted with an "F" in the above figure, survival—denoted with an "S" in the above figure, or age or stage classes) may be sensitive to different factors, and when population growth is not negative on average, the population will persist (population size = "N" above). Diversity arises if multiple populations differ in their traits and their tradeoffs for various combinations of resources, biotic and abiotic factors. The process of growth also sets up feedbacks between populations and the environment. Local coexistence is characterized by feedbacks that tend to stabilize population growth: preventing dominance and buffering rarity. That is, all populations have positive or at least zero average fitness across the range of resources and conditions they experience. The functioning of individual populations, modified by interactions and environmental factors ("X" above) emerges from the collective population-level processes and how abiotic and biotic conditions affect basic vital demographic rates. The consequence of interacting and coexisting species is the aggregate ecosystem function ("EF" above).

Reviewing the links to the abiotic and biotic conditions as a mutual cause for local coexistence and ecosystem functioning is the major contribution of our review. Formerly, local coexistence mechanisms and mechanisms for biodiversity-ecosystem functioning relationships have been separately discussed in spite of their high overlap. When these links have been made explicit (e.g., Carroll et al., 2011; Turnbull et al., 2013, 2016), the focus was largely on a single local coexistence mechanism and ecosystem function. Our review provides a general overview for how local coexistence and ecosystem functioning can inform each other and be informed by abiotic and biotic conditions. Thinking of these sets of mechanisms in a more synthetic way allows us to begin to explore how environmental conditions may mutually drive both local coexistence and ecosystem functioning simultaneously, rather than independently. Further, these conditions are one of the key differences between experiments, from which the majority of our understanding of biodiversity-ecosystem functioning relationships come and the natural systems in which nonrandom biodiversity loss is problematic. Finally, understanding how abiotic and biotic conditions influence population level trade-offs that may favour coexistence while simultaneously decreasing ecosystem functioning and vice versa is crucial for predicting the consequences of biodiversity change on ecosystem functioning.

## 2. Jointly emerging local coexistence and ecosystem functioning from ecological first principles

Various mechanisms have been proposed to explain species coexistence (Aarssen, 1983; Chesson, 2000; Gause, 1934; Holt, 2013; MacArthur, 1969; Macarthur and Levins, 1967; MacArthur and Wilson, 1963, 2001; Palmer, 1994; Shmida and Wilson, 1985; Tilman, 1982; Wright, 2002). Many of these local coexistence mechanisms are also cited as potential drivers of BEF relationships (reviewed in Barry et al., 2019; Tilman et al., 2014; Turnbull et al., 2016). While these mechanisms may influence local coexistence and ecosystem functioning independently, these mechanisms are inherently linked by their dependence on their abiotic and biotic conditions. We consider primary effects of the abiotic environment on plant species that occur within a trophic level to fall under the category of "abiotic conditions" while interactions with organisms at other trophic levels to fall under the category of "biotic conditions". These categories are similar to those proposed by Chase and Leibold (2003) but we believe that our more general categories better capture the diversity of dynamics within the categories. Further, using these categories we avoid the term "stress" which is difficult to define and does not capture the full spectrum of ways in which nonresource abiotic conditions can affect population dynamics.

## 2.1 Abiotic conditions
### 2.1.1 Resources
Here, we define resources as anything that is needed by a plant and where use by the plant precludes other individuals from using the same unit of resources (Chase and Leibold, 2003). Heterogeneity in resources may stabilize interactions between plant species allowing them to coexist via resource partitioning in space, time, or on different resource types (Chesson, 2000; Tilman, 1982). There is some empirical evidence that this type of resource partitioning may allow for local coexistence between populations. For example, studies have found that the addition of nutrients can reduce limitation of multiple resources and thus reduce the number of coexisting species, indicating that limitation of multiple resources allowed these species to coexist (Harpole and Tilman, 2006; Harpole et al., 2016). Theoretical work allowing populations of interacting species to develop according to Lotka–Volterra models demonstrated that local coexistence resulting from lower inter- than intraspecific competition among species can be related to differentiation between species in their resource use (Loreau, 2004; Vandermeer, 1992).

Partitioning resources in space, time, or by type (resource partitioning sensu Schoener, 1970, 1974, also called resource complementarity) is also hypothesized to drive enhanced ecosystem functioning in more diverse plant mixtures. Using a mechanistic resource-competition model, Tilman et al. (1997) showed that allocation trade-offs that favour local coexistence cause more efficient resource exploitation, leading to higher productivity in more diverse communities. Similarly, Carroll et al. (2011)) used a modified version of MacArthur's consumer-resource model (MacArthur, 1970, 1972) and showed that promoting local coexistence by increasing differences in terms of resource use also increases overyielding in more diverse communities. However, empirical evidence for resource partitioning in biodiversity-ecosystem functioning experiments is limited (reviewed by Barry et al., 2019). Mueller et al. (2013) and Ravenek et al. (2014) found evidence that plants allocate belowground biomass in a way that is consistent with resource partitioning. However, using nitrogen tracers, von Felten et al. (2009) found evidence that species may partition nitrogen across a diversity gradient but that this resource partitioning was not associated with increases in certain ecosystem functions (notably nitrogen uptake). Finally, Jesch et al. (2018) found that community resource uptake of nitrogen and potassium likely increases with biodiversity but this was not associated with evidence of resource partitioning across a species-richness gradient.

### *2.1.2 Other abiotic factors*
Other abiotic conditions may allow for local coexistence between species, especially if these conditions vary over time and space. If species differ in their performance across environmental conditions like temperature, and if these conditions vary across space or time, then these species can coexist (Chesson, 2000, 2018; Holt, 2013). In addition to their effect on resources, disturbances over time may also allow for local coexistence between populations via alterations to abiotic conditions that are not traditional resources (but see Fox, 2013a for a discussion of how such local coexistence might be attributed to changes in average mortality rates and the subsequent responses Fox, 2013b; Sheil and Burslem, 2013). For instance, in tropical forests, gap formation causes heterogeneity in abiotic conditions, such as light availability, allowing species that may be outcompeted under high shade to recruit and persist (Connell, 1978; Schnitzer and Carson, 2001). Denslow (1995) provided support for this idea, showing that plant diversity in tropical forests is higher in areas of high stand turnover.

Interactions between populations of plants and their abiotic conditions are also hypothesized to drive enhanced ecosystem functioning in more diverse mixtures. Yachi and Loreau (2007) and Isbell et al. (2018) suggest that species richness can provide a type of temporal or spatial insurance effect against environmental fluctuations and disturbances and that this can result in higher stability over time and an increase in mean functioning. Further, Loreau et al. (2003) suggest that if locally adapted species are able to migrate across an abiotically heterogeneous landscape then local ecosystem functioning may be enhanced. There is some empirical evidence that these types of insurance effects may occur in grasslands (Isbell et al., 2011) where species redundancy was shown to decrease over time as the production of biomass increased in the face of environmental variability.

In addition to the heterogeneity of abiotic conditions, their spatial and temporal extent can also result in increases in the number of species that can coexist and the ecosystem functioning derived from the increased species richness. For example, longer duration of conditions beneficial for plant growth can lead to more asynchronous phenologies. Additionally, deeper soils can accommodate more nonoverlapping rooting depths of different plant species by providing vertical space for roots to occupy (Dimitrakopoulos and Schmid, 2004). On a temporal scale, Oehri et al. (2017) found that increases in growing-season length in Switzerland due to global warming was correlated with plant species richness (Oehri et al., 2017).

Within the same trophic level, positive interactions between species (facilitation) promote local coexistence by promoting population growth when rare (Bruno et al., 2003). Empirical examples of this dynamic are common in stressful environments (Bertness and Callaway, 1994). For example, in arid landscapes nurse plant effects whereby one plant ameliorates the microclimate for a whole community are relatively common (reviewed by Brooker et al., 2008). This microclimate amelioration in diverse mixtures also allows mixtures to perform better than lower diversity communities (Barry et al., 2019; Wright et al., 2017b).

## 2.2 Biotic conditions

Biotic interactions between members of different trophic levels, which we refer to here as a plant's "biotic conditions", also allow for coexistence between plant species in a community. For example, mycorrhizae or other beneficial interaction partners of plants may also ameliorate the local microclimate or increase resource availability (Ferlian et al., 2018; Khan and Kim, 2007; Latz et al., 2012; de la Peña et al., 2006; Wagg et al., 2011). Between trophic levels, specialist pests and pathogens may reduce the abundance of common species, while indirectly supporting others. Rare species are less likely to encounter species-specific pests and pathogens than common species (the Janzen-Connell effect, reviewed by Carson et al., 2008; Comita et al., 2014; see also Connell, 1971; Janzen, 1970; Mangan et al., 2010; Mitchell et al., 2002), indirectly promoting population growth for rare species, a necessary condition for local coexistence. Empirical evidence suggests such cross-trophic density-dependent effects drive local coexistence in lakes, deserts, grasslands, marine ecosystems, and temperate and tropical forests (Anderson, 2001; Comita et al., 2010; Goldberg et al., 2001; Johnson et al., 2012, 2014; Ledo and Schnitzer, 2014; Lorenzen and Enberg, 2002; Mangan et al., 2010; Petermann et al., 2008; Schnitzer et al., 2011).

The influence of biotic conditions can also determine enhanced ecosystem functioning. In experimental systems, biotic feedbacks from other trophic levels have been shown to drive enhanced ecosystem functioning for general belowground biota (Eisenhauer, 2012; Eisenhauer et al., 2012; Maron et al., 2011; Schnitzer et al., 2011; Seabloom et al., 2017), mycorrhizae (Klironomos et al., 2000; Van der Heijden et al., 1998; Wagg et al., 2011), earthworms (Eisenhauer et al., 2009), nematodes (Eisenhauer et al., 2010, 2011), and general aboveground biota (Ebeling et al., 2008; Seabloom et al., 2017).

## 3. Population level effects of abiotic and biotic conditions on fecundity, growth, and survival (Fig. 2)

For many of the above examples of how local coexistence and ecosystem functioning can simultaneously be driven by abiotic and biotic conditions, local coexistence and ecosystem functioning are determined by trade-offs between and within populations in terms of their growth, survival, and fecundity. Because fecundity, growth, and survival are considered to be three fundamental components of plant life-history, we refer to them as plant "vital rates" (Franco and Silvertown, 2004). Their means and variances determine the fitness of individuals and populations. Based on the assumption that plant species must allocate a certain portion of their limited resources to each of them, life-history theory predicts trade-offs among the different components of species life cycles (Obeso, 2002; Roff, 2000; Stearns, 2000). These trade-offs determine local coexistence and ecosystem functioning at the community level. Importantly, all coexistence mechanisms can affect all of these vital rates. The sum total of these effects determines coexistence within and between populations.

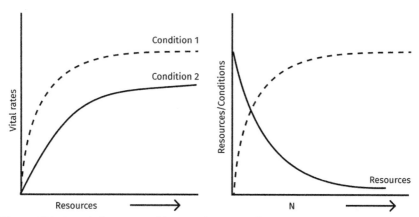

**Fig. 2** Abiotic conditions modify population vital rates. As the population size (N) increases this will create feedbacks on resources and other abiotic conditions.

## 3.1 Abiotic conditions
### 3.1.1 Resources
Resource availability may largely influence fecundity through allocation to reproductive tissues (Jongejans et al., 2006; Sugiyama and Bazzaz, 1998; Waite and Hutchings, 1982). Levels and ratios of mineral nutrients, light, and water may have direct physiological effects on plant growth and development (Weiner, 1988). Higher nutrient availability may lead to coupled growth responses involving increases in plant size and therefore higher seed production (Jongejans et al., 2006). Independent of these size-related responses, nutrient availability may also directly affect reproductive allocation or the mode of reproduction (sexual reproduction or clonal growth). Examples in the literature relating resource availability to plant reproduction show mixed results indicating that the effects are species-specific and modified by competition (e.g., Nicholls, 2011; Suter, 2009). The effects of elevated $CO_2$ on reproductive output have been documented in various studies at the stand level and at the level of individual plants. Further, reproductive allocation (He et al., 2005; Jablonski et al., 2002) and seed production decrease at elevated levels of $CO_2$, although these effects vary significantly among species (HilleRisLambers et al., 2009; van Kleunen et al., 2006). Resource availability may also affect the timing of maturity. Biere (1995) showed in a study along a soil fertility gradient, for instance, that plants mature earlier in in less fertile sites. Other studies have shown that individuals that postpone reproduction gained an increase in fecundity at first reproduction at fertile sites.

Resource availability is also a clear determinant of plant growth (Clarkson and Hanson, 1980). Plants preferentially allocate biomass to access limiting resources and thus grow more slowly if they have to invest more into roots (low nutrients) or stems (low light) than into assimilating leaves (Müller et al., 2000). Furthermore, the response of different species to resource availability may vary according to their life-history strategy. Species adapted to low resource availability often express a "slow" life-history strategy with low rates of resource uptake, high resource-use efficiency, and slow growth. In contrast, species adapted to high resource availability often exhibit a "fast" life-history strategy with high rates of resource acquisition and fast growth (Reich et al., 2003). Consequently, these "fast" life-history species are more successful in high-resource environments than in low-resource environments. One possible reason promoting the success of fast-growing exploitative species in favourable environments is their greater ability to respond plastically to variation in nutrient availability

(Crick and Grime, 1987). Further, plants may respond to elevated $CO_2$ by adjusting their photosynthetic processes and therefore leaf production and growth. However, these responses are species-specific due to different growth forms, patterns of resource allocation, different photosynthesis pathways ($C_3$ vs. $C_4$ plants) or organ and meristem development (Morison and Lawlor, 1999), all of which may change with time (Reich et al., 2018).

Changes in the pattern of supply and the amount of available resources can accelerate local species extinctions (e.g., Berendse and Elberse, 1990; Harpole et al., 2016; Tilman, 1987). For instance, persistent fertilization has been shown to decrease competition for nutrients belowground, while increasing competition for light by driving the development of denser and taller canopies of more productive plants (Harpole et al., 2017; Hautier et al., 2009). Such changes to plant communities are associated with more asymmetric competition for light and a higher probability that competitive exclusion will drive plant species with smaller statures or slower growth to local extinction (Tilman, 1987). Resource availability may also affect survival under different abiotic stresses, for example, a drought may increase mortality in well-fertilized plants due to low root-shoot ratios whereas increased $CO_2$ may allow plants to save water and thus reduce mortality risk under dry conditions (Reich et al., 2004; Wei et al., 2017).

Competition between plant species for resources may also indirectly impact fecundity, for example, interference competition for resources from neighbouring individuals. Such competition, if resources are limited, can result in delays in the onset of reproductive activity, alter the mode of reproduction between vegetative and sexual reproduction, or change the proportion of individuals in a population that produce offspring (van Kleunen et al., 2001; Wayne et al., 1999; Weiner, 1988). The effects of competition for resources on reproductive allocation are likely indirect. For example, Aarssen (2005) suggests that competition between plants can reduce growth and therefore reproductive output because plants take longer to reach size thresholds for reproduction.

### 3.1.2 Other abiotic factors
Apart from resources, other abiotic conditions (e.g., light and temperature) determine together with plant traits (e.g., size, age) the probability of reproduction and thereby fecundity (de Jong et al., 1998). For example, Hansen et al. (2013) showed that dry conditions increased investment in reproductive biomass and that trade-offs between vegetative growth and investment

into reproduction are more pronounced under more stressful dry conditions. A study of the effects of light availability (Jacquemyn et al., 2010) found that the threshold size for initial flowering was three times larger in shaded than sunny environments. Further, some evidence suggests that plants flower more frequently over time and produce more fruits in brighter environments. Similarly, higher temperature has been documented to advance flowering phenology and reproductive allocation (He et al., 2005; Whittington et al., 2015).

Solar radiation, soil and air temperature also may have direct effects on plant growth while also interacting with each other and influencing resource availability. For example, under drought and heat stress, biological processes related to growth, such as rates of photosynthesis and respiration, are reduced. Drought may also reduce plant growth by decreasing soil-nitrogen and -phosphorus uptake (He and Dijkstra, 2014). In spite of the likely effect of these abiotic conditions, a meta-analysis across studies in arid environments showed that the effect of neighbouring plant species on growth of target plants was not dependent on abiotic stress level (in most cases water availability, Maestre et al., 2005).

In experimental plant communities, increased productivity of more diverse plant communities correlates with variable growth responses of individual species in terms of biomass production and plant size (HilleRisLambers et al., 2004; Marquard et al., 2009; Roscher and Schumacher, 2016; Roscher et al., 2007). Several studies in the Jena Experiment demonstrated that grassland species with clearly distinguishable individuals produce fewer shoots per genet in more diverse plant communities (Roscher et al., 2008b, 2011a,b; Thein et al., 2008) resulting in smaller genet sizes. Nevertheless, more shoots are produced per unit area in these diverse communities (Marquard et al., 2009; Thein et al., 2008) suggesting a trade-off between reduced growth and increased survival at the genet level induced by high community diversity. Analyses of foliar nitrate and carbohydrate concentrations as indicators of plant nutritional status indicated that increased light competition at increasing species richness correlated with decreased growth of individual plants in spite of increased productivity at the community level (Roscher et al., 2011a).

Abiotic factors are especially important for plant survival in early stages of the life cycle. For example, it has been shown that water availability is crucial for the survival of seedlings, with additional watering increased seedling survival most in open vegetation (de Jong and Klinkhamer, 1988; Eckstein,

2005). Multiple mechanisms have been suggested to explain the mortality of plants suffering from drought (McDowell et al., 2008) from hydraulic failure due to stomatal closure and subsequent carbon starvation to reduced resistance against biotic "mortality agents" such as pathogens and herbivores. Differential mortality of species in response to drought might have large demographic impacts on populations and shift species composition in response to climate change (Mueller et al., 2005).

For most plant species starting their life cycle from seeds, the risk of mortality is particularly high at the seedling or juvenile stages due to strong competition or abiotic stress (Harper, 1977). In most plant communities, positive (facilitation) and negative (competition) interactions occur simultaneously and their relative impacts on survival probability change with life stage (Eckstein, 2005; Kelemen et al., 2015; Wright et al., 2014). For example, survival of young plants has been shown to be positively influenced by the surrounding vegetation. This positive effect suggests that shading by neighbouring plants may be beneficial for some young plants because the protection against desiccation outweighs the reduction in photosynthesis (Semchenko et al., 2012). This effect may also occur because neighbouring plants reduce soil irradiance and therefore maintain more stable temperatures under severe temperature conditions (Wright et al., 2015). Alternatively, older plants in the same systems experience more negative effects (e.g., in wet meadow vegetation, Kelemen et al., 2015, in a grassland Wright et al., 2014). In experimental grasslands, for instance, it was found that at higher plant diversity where interactions between species of the same trophic level are more intense, grass tussocks were smaller. Furthermore, over a period of 4 years, changes in population sizes (i.e., numbers of individuals per area) indicated a negative population growth rate (i.e., decreasing number of plants) at high plant diversity suggesting that poor performing species were more likely to die and/or new individuals did not establish (Roscher et al., 2011a) It should be noted that in this case the experimental design was not substitutive with regard to the test species and included the test species in all communities at all diversity levels. This experimental design led to higher neighbour diversity being associated with more neighbours with similar resource requirements (Roscher et al., 2008a). In contrast, resident species in substitutive randomized biodiversity experiments on average will have more neighbours with similar resource requirements at lower than at higher diversity where they are planted at reduced density and may increase in size (e.g., Dimitrakopoulos and Schmid, 2004).

## 3.2 Biotic conditions

Interactions between trophic levels, such as herbivory, also influence plant fecundity (Obeso, 1993, 2002). Herbivory pressure can redirect resources to chemical defences from reproductive organs, resulting in decreases in the numbers of flowers and seeds produced (e.g., Louda and Potvin, 1995; Maron, 1998). Further, even when herbivores do not directly feed on reproductive tissue they may reduce seed production. Plant interactions with mycorrhizal fungi may also impact resource availability and allocation. For instance, these associations can reduce limitation for some resources while increasing limitation of others (like carbon). These associations depend on the biotic and abiotic context of the plant and therefore increase fecundity by allowing plants to reach a reproductive size threshold more quickly. However, the effects of mycorrhizal associations on seed production are not consistent, but rather vary with environmental conditions and plant density (Koide and Dickie, 2002).

Different biotic conditions such as plant competition, associations with microorganisms, herbivory, or fungal infestations, do not act in isolation, but exhibit complex interactions affecting plant growth. For example, Nitschke et al. (2010) found that reducing insect herbivory by spraying insecticides increased growth of transplants of *Centaurea jacea* in monocultures but not in plant communities of higher diversity. Plant antagonists such as herbivores or fungal pathogens are thought to promote plant community diversity through negative density-dependence reducing the abundance of dominant species. Faster-growing plant species are expected to be more susceptible to pathogens because of their lower investment in defence (Coley et al., 1985). Parker and Gilbert (2018) demonstrated in an experimental study that faster-growing species experienced greater fungal infestation. However, the impact of fungal infestation on growth was less severe, possibly because these faster-growing species were better able to compensate for fungal damage. Various microbial interactions with plants increase the ability of plants to acquire nutrients in different ways. First, mycorrhizae increase the surface area of roots by extending existing root length. Second, rhizobacteria enhance root growth and branching. Third, phosphate-mobilizing microorganisms (among others) stimulate metabolic processes that mobilize nutrients or nutrient supply. Finally, symbiotic rhizobacteria supply nitrogen via $N_2$ fixation (Richardson et al., 2009). Further, a number of studies have demonstrated increased mortality under competition due to indirect effects of increased herbivory or diseases (Bell et al., 2006).

## 4. How ecological first principles influence trade-offs between fecundity, growth, and survival and in turn influence local coexistence and ecosystem functioning

Long-term biodiversity–ecosystem function experiments, like the Jena Experiment, provide excellent opportunities to explore the mechanistic links between ecosystem functioning and local coexistence dynamics. Understanding how changes to fecundity, growth, and survival change across diversity gradients at the population level enables us to understand how these factors influence local coexistence and ecosystem functioning simultaneously (see above). While there are many publications that confirm that increasing diversity alters community level biomass production which may be a reasonable proxy for growth, few studies examine population-level growth, survival, and fecundity across diversity gradients. Trade-offs between investment in fecundity, growth, and survival likely alter local coexistence between species in these experiments as well as ecosystem functioning simultaneously.

Community-level growth of resident plants generally increases with diversity via changes to community shoot density and biomass production (e.g., Marquard et al., 2009; Roscher et al., 2007 for grassland biodiversity experiments, Barrufol et al., 2013 for forests, Roscher and Schumacher, 2016 for communities with arable weeds). However, fecundity may decrease as plants grow more slowly, or flower and fruit development are delayed, and plants form fewer inflorescences per shoot in more diverse neighbourhoods. This may result in delayed reproduction as has been shown for transplants, i.e., colonizing plants (Mwangi et al., 2007; Nitschke et al., 2010; Scherber et al., 2006). In a biodiversity experiment with arable weeds (mostly annual species), the proportion of reproductive individuals decreased with increasing species richness and the associated increase in community density, i.e., a higher proportion of individuals failed to reproduce completely (Roscher and Schumacher, 2016). Finally, plant diversity and the presence of particular functional groups (i.e., legumes) also decrease the growth and survival of colonizers (but not residents) such as transplants of *Festuca pratensis, Plantago lanceolata, Knautia arvensis, Trifolium pratense* (Mwangi et al., 2007) or *Centaurea jacea* (Nitschke et al., 2010 in the Jena Experiment). These negative plant diversity effects on colonizers were partly attributed to higher community biomass indicating a negative impact of competition by a diverse neighbourhood. As an exception, Scherber et al. (2006) found positive effects of plant diversity on the survival of *Rumex acetosa* transplants in the same biodiversity experiment suggesting that the effect of plant diversity on survival is species-specific.

Trade-offs in investment toward fecundity, growth, and survival may determine species' ability to persist within a community and their potential to enhance ecosystem functioning (Fig. 3). For example, at the Jena Experiment, *Taraxacum officinale,* sown in 16 plant communities of different plant diversity, does not change shoot density, biomass production and the number of inflorescences per area along the diversity gradient (Fig. 4, Table 1). Investing heavily in aboveground growth (in terms of shoot density and biomass production) may come at the cost of reduced survival (in terms of lifespan, e.g. because of reduced shoot-root ratios) or fecundity (in terms of the % of seeds that germinate). Although such trade-offs are not evident in this field example, they have previously been found in common garden experiments with offspring of seed families of *T. officinale* collected in plant communities of different diversity in the Jena Experiment. Cuttings of plants originating from resident populations growing at higher diversity produced more leaf and root biomass and had fewer inflorescences, but heavier seeds (Lipowsky et al., 2012). Consequently, a higher investment in growth rather than in survival and fecundity increased the contribution of *T. officinale* to enhanced biomass production in mixture. These trade-offs likely occur at the population level and, in nature, may cause variation in population contributions to ecosystem functioning (Wohlgemuth et al., 2017). That is, at the population level, if trade-offs favour investment in reproduction, which may or may not be related to biomass production, then biomass production may not increase with increasing species richness.

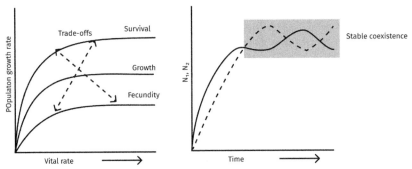

**Fig. 3** Trade-offs within populations in investment toward survival growth and fecundity determine whether populations of N size can coexist.

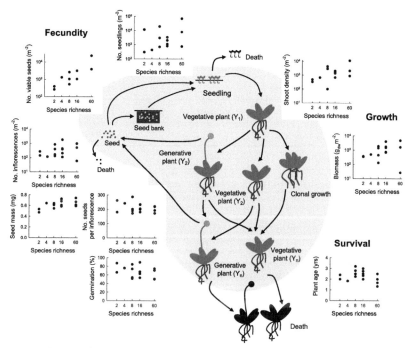

**Fig. 4** Vital rates of *Taraxacum officinale* change across diversity at the Jena Experiment. Statistics presented in Table 1.

Whether or not these trade-offs between populations result in enhanced ecosystem functioning at the community level may depend on several factors (Fig. 5). First, biomass is a common proxy for ecosystem functioning in biodiversity-ecosystem functioning experiments. Trade-offs between populations that favour investment in growth over investment in fecundity and survival may result in enhanced biomass production in more diverse grasslands. However, the extent to which trade-offs result in enhanced ecosystem functioning at the community level depends on the function(s) of focus. Below, we give two examples of how our framework may inform local coexistence and predict ecosystem functioning between two species positively, for productivity, and negatively for root decomposition.

## 4.1 Productivity

The positive relationship between species richness and productivity is well documented (Balvanera et al., 2006; Liang et al., 2016; Reich et al., 2012; Tilman et al., 1996). Further, this relationship may be derived from how

**Table 1** Results of analysis of variance on resident *Taraxacum officinale* vital rates across the diversity gradient at the Jena Experiment.

| Vital rate | Measure | DF | F | P |
| --- | --- | --- | --- | --- |
| Fecundity | # Seedlings | 1,11 | 0.05 | 0.833 |
| | # Viable seeds | 1,11 | 0.01 | 0.909 |
| | # Of seeds per inflorescence | 1,11 | 0.24 | 0.631 |
| | # Of inflorescences | 1,11 | 0.30 | 0.595 |
| | Seed mass | 1,11 | 3.71 | 0.080 |
| | % Germination | 1,11 | 0.60 | 0.455 |
| Growth | Shoot density | 1,11 | <0.01 | 0.954 |
| | Population biomass | 1,11 | 0.24 | 0.636 |
| Survival | Mean age | 1,11 | 1.33 | 0.274 |

All models are vital rate ~ block + species richness. Data were collected in a long-term grassland biodiversity experiment (Jena Experiment; Roscher et al., 2004). *Taraxacum officinale* belongs to the sown species combinations in 16 plots of the Jena Experiment covering different species-richness levels (2, 4, 8, 16, and 60 species). The biodiversity experiment was established in 2002 by sowing. Sowing density was 1000 germinable seeds per $m^2$ distributed equally among species in mixture, i.e., for single species 500 seeds per $m^2$ were sown in a 2-species mixture, and 17 seeds per $m^2$ were sown in a 60-species mixture.

The age of adult plant individuals was determined by analysing growth rings in the root crown of five individuals of *T. officinale* collected in each of the 16 populations in the 12-year old experimental grasslands (for details see Roeder et al., 2017). Other data shown in Fig. 4 were collected in 2014. The viable seed bank in the topsoil (to 5 cm depth) was determined by taking soil samples, which were sieved and spread on heat-sterilized substrate and cultivated for one growing season to determine the number of germinating *T. officinale* seeds. The number of seedlings emerging on the plots was counted on a permanently marked subplot and summed up from three censi (spring, summer, autumn). Growth-related variables were determined by harvesting two strips (100 × 10 cm) on each plot at estimated peak biomass before mowing (late May, late August). The number of shoots was counted and averaged between both harvests to get shoot density. Biomass samples of *T. officinale* were dried and weighed; annual biomass production was derived as the sum of both harvests. The number of inflorescences of *T. officinale* was counted on two permanently marked subplots of 1 $m^2$ size and averaged between both subplots. Three seed heads of *T. officinale* were collected on each plot, when they had ripe seeds. The numbers of seeds per seed head were counted, and 50 seeds were taken to determine seed mass and germination rates under standardized conditions.

To account for the lower number seeds sown for individual species at higher plant diversity, we multiplied all variables, which are related to the sown density (i.e., number of viable seeds in the seedbank, number of seedlings, shoot density, biomass, and number of inflorescences) by species richness for statistical analyses and data presented in Fig. 4).

plant species interact with abiotic and biotic conditions. There is evidence that each of these influences how species richness can enhance productivity. At the Jena Experiment, plants may partition resources (Ravenek et al., 2014 but see Jesch et al., 2018; Oram et al., 2018) that are likely limiting in this context (Oelmann et al., 2011). Similarly, abiotic conditions such as stress from flooding alter biodiversity-productivity relationships

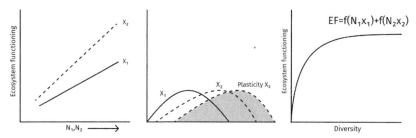

**Fig. 5** Trade-offs within and between populations determine how increase species diversity alters ecosystem functioning. Ecosystem functioning is a function of the population sizes (N) and the environmental conditions (x). Plasticity at the population level determines the effect of x.

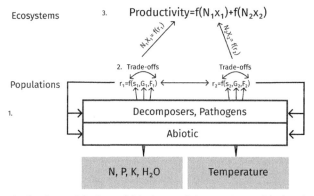

**Fig. 6** Example for how abiotic and biotic conditions may enable local coexistence between two populations of two species and enhance productivity simultaneously. Plant productivity is influenced by the available soil nutrients, temperature and light availability, and the presences of decomposers, pathogens, and other biotic interaction partners. These factors simultaneously allow populations to coexist and result in a positive relationship between species richness and productivity.

(Fischer et al., 2016; Wright et al., 2017a,b) via their influence on biotic interactions between species at the same trophic level and between species at different trophic levels (Eisenhauer et al., 2009). These interacting factors may enhance productivity by allowing for local coexistence between species (Fig. 6).

In natural systems, this biodiversity-productivity relationship may be reinforced by a positive feedback of productivity on biodiversity (Grace et al., 2016). Further, many other abiotic and biotic conditions may contribute to increased productivity and these factors may not equally influence local coexistence between species (Grace et al., 2016). For example, higher nutrient availability may decouple local coexistence from biomass production. If nutrient addition reduces the need for plants to partition resources to avoid competition then species richness will decline while biomass production simultaneously increases (as demonstrated by Harpole et al., 2016).

## 4.2 Root decomposition

Biodiversity can both increase (Handa et al., 2014) and decrease litter decomposition (Hättenschwiler and Gasser, 2005). However, for root litter which makes up more than half of plant litter input in grassland ecosystems (Poorter et al., 2012) the majority of evidence suggests increasing species richness results in slower decomposition (Chen et al., 2017a,b; Hättenschwiler and Gasser, 2005). Evidence from the Jena Experiment suggests that biodiversity may impact root litter decomposition via three pathways (Chen et al., 2017a,b): (1) increasing diversity of the litter components may alter the quality of the litter and therefore resource availability (e.g., increasing the carbon to nitrogen ratio, Chen et al., 2017b), (2) the increased diversity of the environment may alter the abiotic conditions of the community, and (3) the increased diversity of the environment may alter the biotic conditions of the community. Not coincidentally – these categories also enable species local coexistence and alterations to decomposition may arise as an outcome of local coexistence between species (Fig. 7).

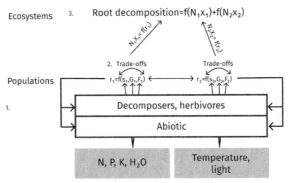

**Fig. 7** Example for how abiotic and biotic conditions may enable local coexistence between two populations of two species and decrease root decomposition simultaneously. Root decomposition depends on the available soil resources, temperature and light availability, as well as the presence of decomposers and herbivores. These factors simultaneously influence local coexistence between populations and root decomposition which decreases with increasing species richness.

Importantly, productivity and root decomposition in a plant community also provide feedbacks on the populations that comprise them and the biotic and abiotic conditions and resources that enable their local coexistence. These feedbacks may reinforce local coexistence by stabilizing population dynamics. For example, higher diversity root litter has a higher K concentration than lower diversity plant litter. This higher K concentration enhances root decomposition, increasing nutrient availability for plants. This enhanced nutrient availability may allow plants to produce more biomass and this increased productivity decreases surface light penetration cooling the soil surface and potentially ameliorating the local microclimate reinforcing local coexistence between species via abiotic facilitation (Milcu et al., 2016). However, while this microclimate amelioration may act to buffer seedlings from harmful abiotic conditions increasing population-level fecundity it may simultaneously decrease the decomposition rate of root litter (Hättenschwiler and Gasser, 2005) and therefore decrease root growth rate and carbon sequestration.

## 5. Conclusion

Abiotic and biotic conditions—ecological first principles—underlie all of the varied mechanisms that may maintain biodiversity and/or enhanced ecosystem functioning in more diverse systems. Starting from these first principles allows us to unify theory and empirical results obtained for originally unconnected reasons. Theory from local coexistence, ecosystem functioning, and plant population biology enables us to understand how abiotic and biotic conditions affect population dynamics and vice versa. Many empirical studies from various contexts in natural environments show how differently the major components of our ecological first principles influence vital rates of populations. Further, although the majority of biodiversity-ecosystem functioning research is conducted at the community level, these effects are species specific. In the context of biodiversity-ecosystem functioning research, such population-level data are often restricted to above-ground biomass. Our review of the literature suggests that deeper mechanistic understanding of the linkage between species coexistence and ecosystem functioning requires more information about population-level trade-offs between growth, survival, and fecundity. This information is especially sparse in terms of their response to varying biotic conditions. Understanding how abiotic and biotic conditions affect population level trade-offs between species across diversity gradients may allow us to combine complementary theories from different field to better explain how different local coexistence processes can interact to influence ecosystem functioning in grassland ecosystems and how these ecosystem functions provide feedbacks onto local coexistence.

## Author contributions

The idea for this paper was conceptualized during the "BEF-Coexist" workshop organized by C.R. and W.S·H. during discussions among C.R., W.S·H., K.E.B., H.dK., and B.S. These ideas were discussed with many others during the "BEF-Coexist" workshop and this input informed discussion and all drafts of the manuscript. The first draft of the paper was written by K.E.B. and C.R. with contributions from W.S.H, P.D., and B.S. Data on *Taraxacum officinale* presented in Table 1 and Fig. 4 was collected by A.R. and C.R. All authors contributed significantly to the revision of the manuscript prior to submission.

## Acknowledgements

This work arose as part of the "BEF-Coexist" workshop, organized by Yanhao Feng, C.R. and W.S.H. with funding provided by the German Research Foundation (RO2397/8) in the framework of Jena Experiment (FOR1451). We thank all workshop participants for discussion. KEB was funded by the Flexible Pool of the German Centre for Integrative Biodiversity Research Halle-Jena-Leipzig (grant # 34600900).

## References

Aarssen, L.W., 1983. Ecological combining ability and competitive combining ability in plants: toward a general evolutionary theory of coexistence in systems of competition. Am. Nat. 122, 707–731.

Aarssen, L.W., 2005. Why don't bigger plants have proportionately bigger seeds? Oikos 111, 199–207.

Anderson, T.W., 2001. Predator responses, prey refuges, and density-dependent mortality of a marine fish. Ecology 82, 245–257.

Balvanera, P., Pfisterer, A.B., Buchmann, N., He, J.-S., Nakashizuka, T., Raffaelli, D., Schmid, B., 2006. Quantifying the evidence for biodiversity effects on ecosystem functioning and services. Ecol. Lett. 9, 1146–1156.

Barrufol, M., Schmid, B., Bruelheide, H., Chi, X., Hector, A., Ma, K., Michalski, S., Tang, Z., Niklaus, P.A., 2013. Biodiversity promotes tree growth during succession in subtropical forest. PLoS One 8 e81246. https://doi.org/10.1371/journal.pone.0081246.

Barry, K.E., Mommer, L., van Ruijven, J., Wirth, C., Wright, A.J., Bai, Y., Connolly, J., Deyn, G.B.D., de Kroon, H., Isbell, F., Milcu, A., Roscher, C., Scherer-Lorenzen, M., Schmid, B., Weigelt, A., 2019. The future of complementarity: disentangling causes from consequences. Trends Ecol. Evol. 34, 167–180.

Becker, J., Eisenhauer, N., Scheu, S., Jousset, A., 2012. Increasing antagonistic interactions cause bacterial communities to collapse at high diversity. Ecol. Lett. 15, 468–474.

Bell, T., Freckleton, R.P., Lewis, O.T., 2006. Plant pathogens drive density-dependent seedling mortality in a tropical tree. Ecol. Lett. 9, 569–574.

Berendse, F., Elberse, W.T., 1990. Competition and nutrient availability in heathland and grassland ecosystems. In: Grace, J., Tilman, D.T. (Eds.), Perspectives on Plant Competition. Academic Press, San Diego, CA, USA, pp. 93–115.

Bertness, M.D., Callaway, R., 1994. Positive interactions in communities. Trends Ecol. Evol. 9, 191–193.

Biere, A., 1995. Genotypic and plastic variation in plant size: effects on fecundity and allocation patterns in lychnis flos-cuculi along a gradient of natural soil fertility. J. Ecol. 83, 629–642.

Brooker, R.W., Maestre, F.T., Callaway, R.M., Lortie, C.L., Cavieres, L.A., Kunstler, G., Liancourt, P., Tielbörger, K., Travis, J.M.J., Anthelme, F., Armas, C., Coll, L., Corcket, E., Delzon, S., Forey, E., Kikvidze, Z., Olofsson, J., Pugnaire, F., Quiroz, C.L., Saccone, P., Schiffers, K., Seifan, M., Touzard, B., Michalet, R., 2008. Facilitation in plant communities: the past, the present, and the future. J. Ecol. 96, 18–34.

Bruno, J.F., Stachowicz, J.J., Bertness, M.D., 2003. Inclusion of facilitation into ecological theory. Trends Ecol. Evol. 18, 119–125.

Carneiro da Cunha, M., Mace, G., Mooney, H. (Eds.), 2019. Summary for Policymakers of the Global Assessment Report on Biodiversity and Ecosystem Services of the Intergovernmental Science-Policy Platform on Biodiversity and Ecosystem Services (IPBES). Advanced Working Copy, United Nations.

Carroll, I.T., Cardinale, B.J., Nisbet, R.M., 2011. Niche and fitness differences relate the maintenance of diversity to ecosystem function. Ecology 92, 1157–1165.

Carson, W.P., Anderson, J.T., Leigh, E.G., Schnitzer, S.A., 2008. Challenges associated with testing and falsifying the Janzen-Connell hypothesis: a review and critique. In: Tropical Forest Community Ecology. Wiley-Blackwell Publishing, Chichester; Malden, MA, pp. 210–241.

Chase, J.M., Leibold, M.A., 2003. Ecological Niches: Linking Classical and Contemporary Approaches. University of Chicago Press, Chicago and London.

Chen, H., Mommer, L., van Ruijven, J., de Kroon, H., Fischer, C., Gessler, A., Hildebrandt, A., Scherer-Lorenzen, M., Wirth, C., Weigelt, A., 2017a. Plant species richness negatively affects root decomposition in grasslands. J. Ecol. 105, 209–218.

Chen, H., Oram, N.J., Barry, K.E., Mommer, L., van Ruijven, J., de Kroon, H., Ebeling, A., Eisenhauer, N., Fischer, C., Gleixner, G., Gessler, A., Macé, O.G., Hacker, N., Hildebrandt, A., Lange, M., Scherer-Lorenzen, M., Scheu, S., Oelmann, Y., Wagg, C., Wilcke, W., Wirth, C., Weigelt, A., 2017b. Root chemistry and soil fauna, but not soil abiotic conditions explain the effects of plant diversity on root decomposition. Oecologia 185, 499–511.

Chesson, P., 2000. Mechanisms of maintenance of species diversity. Annu. Rev. Ecol. Syst. 31, 343–366.

Chesson, P., 2018. Updates on mechanisms of maintenance of species diversity. J. Ecol. 106, 1773–1794.

Clarkson, D.T., Hanson, J.B., 1980. The mineral nutrition of higher plants. Annu. Rev. Plant Physiol. 31, 239–298.

Coley, P.D., Bryant, J.P., Chapin III, F.S., 1985. Resource availability and plant antiherbivore defense. Science (Washington) 230, 895–899.

Comita, L.S., Muller-Landau, H.C., Aguilar, S., Hubbell, S.P., 2010. Asymmetric density dependence shapes species abundances in a tropical tree community. Science 329, 330–332.

Comita, L.S., Queenborough, S.A., Murphy, S.J., Eck, J.L., Xu, K., Krishnadas, M., Beckman, N., Zhu, Y., 2014. Testing predictions of the Janzen-Connell hypothesis: a meta-analysis of experimental evidence for distance- and density-dependent seed and seedling survival. J. Ecol. 102, 845–856.

Connell, J.H., 1971. On the role of natural enemies in preventing competitive exclusion in some marine animals and in rain forest trees. Dyn. Popul. 298, 312.

Connell, J.H., 1978. Diversity in tropical rain forests and coral reefs. Science 199, 1302–1310.

Crick, J.C., Grime, J.P., 1987. Morphological plasticity and mineral nutrient capture in two herbaceous species of contrasted ecology. New Phytol. 107, 403–414.

De Jong, T.J., Klinkhamer, P.G.L., 1988. Seedling establishment of the biennials cirsium vulgare and cynoglossum officinale in a Sand-Dune area: the importance of water for differential survival and growth. J. Ecol. 76, 393–402.

Denslow, J.S., 1995. Disturbance and diversity in tropical rain forests: the density effect. Ecol. Appl. 5, 962–968.

Dimitrakopoulos, P.G., Schmid, B., 2004. Biodiversity effects increase linearly with biotope space. Ecol. Lett. 7, 574–583.

Ebeling, A., Klein, A.-M., Schumacher, J., Weisser, W.W., Tscharntke, T., 2008. How does plant richness affect pollinator richness and temporal stability of flower visits? Oikos 117, 1808–1815.

Eckstein, R.L., 2005. Differential effects of interspecific interactions and water availability on survival, growth and fecundity of three congeneric grassland herbs. New Phytol. 166, 525–535.

Eisenhauer, N., 2012. Aboveground–belowground interactions as a source of complementarity effects in biodiversity experiments. Plant Soil 351, 1–22.

Eisenhauer, N., Milcu, A., Nitschke, N., Sabais, A.C.W., Scherber, C., Scheu, S., 2009. Earthworm and belowground competition effects on plant productivity in a plant diversity gradient. Oecologia 161, 291–301.

Eisenhauer, N., Beßler, H., Engels, C., Gleixner, G., Habekost, M., Milcu, A., Partsch, S., Sabais, A.C.W., Scherber, C., Steinbeiss, S., Weigelt, A., Weisser, W.W., Scheu, S., 2010. Plant diversity effects on soil microorganisms support the singular hypothesis. Ecology 91, 485–496.

Eisenhauer, N., Migunova, V.D., Ackermann, M., Ruess, L., Scheu, S., 2011. Changes in plant species richness induce functional shifts in soil nematode communities in experimental grassland. PLoS One 6 e24087.

Eisenhauer, N., Reich, P.B., Scheu, S., 2012. Increasing plant diversity effects on productivity with time due to delayed soil biota effects on plants. Basic Appl. Ecol. 13, 571–578.

Ellner, S.P., Snyder, R.E., Adler, P.B., Hooker, G., 2019. An expanded modern coexistence theory for empirical applications. Ecol. Lett. 22, 3–18.

Ferlian, O., Cesarz, S., Craven, D., Hines, J., Barry, K.E., Bruelheide, H., Buscot, F., Haider, S., Heklau, H., Herrmann, S., Kühn, P., Pruschitzki, U., Schädler, M., Wagg, C., Weigelt, A., Wubet, T., Eisenhauer, N., 2018. Mycorrhiza in tree diversity–ecosystem function relationships: conceptual framework and experimental implementation. Ecosphere 9 e02226.

Fischer, F.M., Wright, A.J., Eisenhauer, N., Ebeling, A., Roscher, C., Wagg, C., Weigelt, A., Weisser, W.W., Pillar, V.D., 2016. Plant species richness and functional traits affect community stability after a flood event. Philos. Trans. R. Soc. Lond. B Biol. Sci. 371, 20150276.

Fox, J.W., 2013a. The intermediate disturbance hypothesis should be abandoned. Trends Ecol. Evol. 28, 86–92.

Fox, J.W., 2013b. The intermediate disturbance hypothesis is broadly defined, substantive issues are key: a reply to Sheil and Burslem. Trends Ecol. Evol. 28, 572–573.

Franco, M., Silvertown, J., 2004. A comparative demography of plants based upon elasticities of vital rates. Ecology 85, 531–538.

Gause, G.F., 1934. The Struggle for Coexistence. Williams and Wilkins.

Goldberg, D.E., Turkington, R., Olsvig-Whittaker, L., Dyer, A.R., 2001. Density dependence in an annual plant community: variation among life history stages. Ecol. Monogr. 71, 423–446.

Grace, J.B., Anderson, T.M., Seabloom, E.W., Borer, E.T., Adler, P.B., Harpole, W.S., Hautier, Y., Hillebrand, H., Lind, E.M., Pärtel, M., Bakker, J.D., Buckley, Y.M., Crawley, M.J., Damschen, E.I., Davies, K.F., Fay, P.A., Firn, J., Gruner, D.S.,

Hector, A., Knops, J.M.H., MacDougall, A.S., Melbourne, B.A., Morgan, J.W., Orrock, J.L., Prober, S.M., Smith, M.D., 2016. Integrative modelling reveals mechanisms linking productivity and plant species richness. Nature 529, 390–393.

Handa, I.T., Aerts, R., Berendse, F., Berg, M.P., Bruder, A., Butenschoen, O., Chauvet, E., Gessner, M.O., Jabiol, J., Makkonen, M., McKie, B.G., Malmqvist, B., Peeters, E.T.H.M., Scheu, S., Schmid, B., van Ruijven, J., Vos, V.C.A., Hättenschwiler, S., 2014. Consequences of biodiversity loss for litter decomposition across biomes. Nature 509, 218–221.

Hansen, C.F., García, M.B., Ehlers, B.K., 2013. Water availability and population origin affect the expression of the tradeoff between reproduction and growth in Plantago coronopus. J. Evol. Biol. 26, 993–1002.

Harper, J.L., 1977. Population Biology of Plants. Academic Press, London.

Harpole, W.S., Tilman, D., 2006. Non-neutral patterns of species abundance in grassland communities. Ecol. Lett. 9, 15–23.

Harpole, W.S., Sullivan, L.L., Lind, E.M., Firn, J., Adler, P.B., Borer, E.T., Chase, J., Fay, P.A., Hautier, Y., Hillebrand, H., MacDougall, A.S., Seabloom, E.W., Williams, R., Bakker, J.D., Cadotte, M.W., Chaneton, E.J., Chu, C., Cleland, E.E., D'Antonio, C., Davies, K.F., Gruner, D.S., Hagenah, N., Kirkman, K., Knops, J.M.H., La Pierre, K.J., McCulley, R.L., Moore, J.L., Morgan, J.W., Prober, S.M., Risch, A.C., Schuetz, M., Stevens, C.J., Wragg, P.D., 2016. Addition of multiple limiting resources reduces grassland diversity. Nature 537, 93–96.

Harpole, W.S., Sullivan, L.L., Lind, E.M., Firn, J., Adler, P.B., Borer, E.T., Chase, J., Fay, P.A., Hautier, Y., Hillebrand, H., MacDougall, A.S., Seabloom, E.W., Bakker, J.D., Cadotte, M.W., Chaneton, E.J., Chu, C., Hagenah, N., Kirkman, K., Pierre, K.J.L., Moore, J.L., Morgan, J.W., Prober, S.M., Risch, A.C., Schuetz, M., Stevens, C.J., 2017. Out of the shadows: multiple nutrient limitations drive relationships among biomass, light and plant diversity. Funct. Ecol. 31, 1839–1846.

Hättenschwiler, S., Gasser, P., 2005. Soil animals alter plant litter diversity effects on decomposition. Proc. Natl. Acad. Sci. U. S. A. 102, 1519–1524.

Hautier, Y., Niklaus, P.A., Hector, A., 2009. Competition for light causes plant biodiversity loss after eutrophication. Science 324, 636–638.

He, M., Dijkstra, F.A., 2014. Drought effect on plant nitrogen and phosphorus: a meta-analysis. New Phytol. 204, 924–931.

He, J., Wolfe-Bellin, K.S., Bazzaz, F.A., 2005. Leaf-level physiology, biomass, and reproduction of phytolacca americana under conditions of elevated $CO_2$ and altered temperature regimes. Int. J. Plant Sci. 166, 615–622.

HilleRisLambers, J., Harpole, W.S., Tilman, D., Knops, J., Reich, P.B., 2004. Mechanisms responsible for the positive diversity–productivity relationship in Minnesota grasslands. Ecol. Lett. 7, 661–668.

HilleRisLambers, J., Harpole, W.S., Schnitzer, S., Tilman, D., Reich, P.B., 2009. $CO_2$, nitrogen, and diversity differentially affect seed production of prairie plants. Ecology 90, 1810–1820.

Holt, R.D., 2013. Species coexistence. In: Levin, S.A. (Ed.), Encyclopedia of Biodiversity, second ed. Academic Press, Waltham, pp. 667–678.

Hooper, D.U., Adair, E.C., Cardinale, B.J., Byrnes, J.E., Hungate, B.A., Matulich, K.L., Gonzalez, A., Duffy, J.E., Gamfeldt, L., O'Connor, M.I., 2012. A global synthesis reveals biodiversity loss as a major driver of ecosystem change. Nature 486, 105–108.

Huston, M.A., 1997. Hidden treatments in ecological experiments: re-evaluating the ecosystem function of biodiversity. Oecologia 110, 449–460.

Isbell, F., Calcagno, V., Hector, A., Connolly, J., Harpole, W.S., Reich, P.B., Scherer-Lorenzen, M., Schmid, B., Tilman, D., van Ruijven, J., Weigelt, A., Wilsey, B.J., Zavaleta, E.S., Loreau, M., 2011. High plant diversity is needed to maintain ecosystem services. Nature 477, 199–202.

Isbell, F., Cowles, J., Dee, L.E., Loreau, M., Reich, P.B., Gonzalez, A., Hector, A., Schmid, B., 2018. Quantifying effects of biodiversity on ecosystem functioning across times and places. Ecol. Lett. 21, 763–778.

Jablonski, L.M., Wang, X., Curtis, P.S., 2002. Plant reproduction under elevated CO2 conditions: a meta-analysis of reports on 79 crop and wild species. New Phytol. 156, 9–26.

Jacquemyn, H., Brys, R., Jongejans, E., 2010. Size-dependent flowering and costs of reproduction affect population dynamics in a tuberous perennial woodland orchid. J. Ecol. 98, 1204–1215.

Janzen, D.H., 1970. Herbivores and the number of tree species in tropical forests. Am. Nat. 104, 501–528.

Jesch, A., Barry, K.E., Ravenek, J.M., Bachmann, D., Strecker, T., Weigelt, A., Buchmann, N., de Kroon, H., Gessler, A., Mommer, L., Roscher, C., Scherer-Lorenzen, M., 2018. Below-ground resource partitioning alone cannot explain the biodiversity–ecosystem function relationship: a field test using multiple tracers. J. Ecol. 106, 2002–2018.

Johnson, D.J., Beaulieu, W.T., Bever, J.D., Clay, K., 2012. Conspecific negative density dependence and forest diversity. Science 336, 904–907.

Johnson, D.J., Bourg, N.A., Howe, R., McShea, W.J., Wolf, A., Clay, K., 2014. Conspecific negative density-dependent mortality and the structure of temperate forests. Ecology 95, 2493–2503.

Jong, T.J.D., Roo, L.G.-D., Klinkhamer, P.G.L., 1998. Is the threshold size for flowering in Cynoglossum officinale fixed or dependent on environment? New Phytol. 138, 489–496.

Jongejans, E., De Kroon, H., Berendse, F., 2006. The interplay between shifts in biomass allocation and costs of reproduction in four grassland perennials under simulated successional change. Oecologia 147, 369–378.

Kelemen, A., Lazzaro, L., Besnyöi, V., Albert, Á., Konečná, M., Dobay, G., Memelink, I., Adamec, V., Goetzenberger, L., de Bello, F., 2015. Net outcome of competition and facilitation in a wet meadow changes with plant's life stage and community productivity. Preslia 87, 347–361.

Khan, Z., Kim, Y.H., 2007. A review on the role of predatory soil nematodes in the biological control of plant parasitic nematodes. Appl. Soil Ecol. 35, 370–379.

Kleunen, M.V., Stephan, M.A., Schmid, B., 2006. [CO$_2$]$^-$ and density-dependent competition between grassland species. Glob. Change Biol. 12, 2175–2186.

Klironomos, J.N., McCune, J., Hart, M., Neville, J., 2000. The influence of arbuscular mycorrhizae on the relationship between plant diversity and productivity. Ecol. Lett. 3, 137–141.

Koide, R.T., Dickie, I.A., 2002. Effects of mycorrhizal fungi on plant populations. In: Smith, S.E., Smith, F.A. (Eds.), Diversity and Integration in Mycorrhizas: Proceedings of the 3rd International Conference on Mycorrhizas (ICOM3) Adelaide, Australia, 8–13 July 2001, Developments in Plant and Soil Sciences. Springer, Netherlands, Dordrecht, pp. 307–317.

Latz, E., Eisenhauer, N., Rall, B.C., Allan, E., Roscher, C., Scheu, S., Jousset, A., 2012. Plant diversity improves protection against soil-borne pathogens by fostering antagonistic bacterial communities. J. Ecol. 100, 597–604.

Ledo, A., Schnitzer, S.A., 2014. Disturbance and clonal reproduction determine liana distribution and maintain liana diversity in a tropical forest. Ecology 95, 2169–2178.

Leibold, M.A., Chase, J.M., 2018. In: Levin, S.A., Horn, H.S. (Eds.), Metacommunity Ecology. Princeton University Press.

Liang, J., Crowther, T.W., Picard, N., Wiser, S., Zhou, M., Alberti, G., Schulze, E.-D., McGuire, A.D., Bozzato, F., Pretzsch, H., de- Miguel, S., Paquette, A., Hérault, B., Scherer-Lorenzen, M., Barrett, C.B., Glick, H.B., Hengeveld, G.M., Nabuurs, G.-J.,

Pfautsch, S., Viana, H., Vibrans, A.C., Ammer, C., Schall, P., Verbyla, D., Tchebakova, N., Fischer, M., Watson, J.V., Chen, H.Y.H., Lei, X., Schelhaas, M.-J., Lu, H., Gianelle, D., Parfenova, E.I., Salas, C., Lee, E., Lee, B., Kim, H.S., Bruelheide, H., Coomes, D.A., Piotto, D., Sunderland, T., Schmid, B., Gourlet-Fleury, S., Sonké, B., Tavani, R., Zhu, J., Brandl, S., Vayreda, J., Kitahara, F., Searle, E.B., Neldner, V.J., Ngugi, M.R., Baraloto, C., Frizzera, L., Bałazy, R., Oleksyn, J., Zawiła-Niedźwiecki, T., Bouriaud, O., Bussotti, F., Finér, L., Jaroszewicz, B., Jucker, T., Valladares, F., Jagodzinski, A.M., Peri, P.L., Gonmadje, C., Marthy, W., O'Brien, T., Martin, E.H., Marshall, A.R., Rovero, F., Bitariho, R., Niklaus, P.A., Alvarez-Loayza, P., Chamuya, N., Valencia, R., Mortier, F., Wortel, V., Engone-Obiang, N.L., Ferreira, L.V., Odeke, D.E., Vasquez, R.M., Lewis, S.L., Reich, P.B., 2016. Positive biodiversity-productivity relationship predominant in global forests. Science 354 aaf8957.

Lipowsky, A., Roscher, C., Schumacher, J., Schmid, B., 2012. Density-independent mortality and increasing plant diversity are associated with differentiation of Taraxacum officinale into r- and K-strategists. PLoS One 7 e28121.

Loreau, M., 2004. Does functional redundancy exist? Oikos 104, 606–611.

Loreau, M., Mouquet, N., Gonzalez, A., 2003. Biodiversity as spatial insurance in heterogeneous landscapes. Proc. Natl. Acad. Sci. U. S. A. 100, 12765–12770.

Lorenzen, K., Enberg, K., 2002. Density-dependent growth as a key mechanism in the regulation of fish populations: evidence from among-population comparisons. Proc. R. Soc. Lond. B Biol. Sci. 269, 49–54.

Louda, S.M., Potvin, M.A., 1995. Effect of inflorescence-feeding insects on the demography and lifetime of a native plant. Ecology 76, 229–245.

MacArthur, R.H., 1969. Patterns of communities in the tropics. Biol. J. Linn. Soc. 1, 19–30.

MacArthur, R., 1970. Species packing and competitive equilibrium for many species. Theor. Popul. Biol. 1, 1–11.

MacArthur, R.H., 1972. Geographical Ecology. Harper & Row, New York.

Macarthur, R., Levins, R., 1967. The limiting similarity, convergence, and divergence of coexisting species. Am. Nat. 101, 377–385.

MacArthur, R.H., Wilson, E.O., 1963. An equilibrium theory of insular zoogeography. Evolution 17, 373–387.

MacArthur, R.H., Wilson, E.O., 2001. The Theory of Island Biogeography. Princeton University Press.

Maestre, F.T., Valladares, F., Reynolds, J.F., 2005. Is the change of plant–plant interactions with abiotic stress predictable? A meta-analysis of field results in arid environments. J. Ecol. 93, 748–757.

Mangan, S.A., Schnitzer, S.A., Herre, E.A., Mack, K.M.L., Valencia, M.C., Sanchez, E.I., Bever, J.D., 2010. Negative plant–soil feedback predicts tree-species relative abundance in a tropical forest. Nature 466, 752–755.

Maron, J.L., 1998. Insect herbivory above- and belowground: individual and joint effects on plant fitness. Ecology 79, 1281–1293.

Maron, J.L., Marler, M., Klironomos, J.N., Cleveland, C.C., 2011. Soil fungal pathogens and the relationship between plant diversity and productivity. Ecol. Lett. 14, 36–41.

Marquard, E., Weigelt, A., Temperton, V.M., Roscher, C., Schumacher, J., Buchmann, N., Fischer, M., Weisser, W.W., Schmid, B., 2009. Plant species richness and functional composition drive overyielding in a six-year grassland experiment. Ecology 90, 3290–3302.

McDowell, N., Pockman, W.T., Allen, C.D., Breshears, D.D., Cobb, N., Kolb, T., Plaut, J., Sperry, J., West, A., Williams, D.G., Yepez, E.A., 2008. Mechanisms of plant survival and mortality during drought: why do some plants survive while others succumb to drought? New Phytol. 178, 719–739.

Milcu, A., Eugster, W., Bachmann, D., Guderle, M., Roscher, C., Gockele, A., Landais, D., Ravel, O., Gessler, A., Lange, M., Ebeling, A., Weisser, W.W., Roy, J., Hildebrandt, A., Buchmann, N., 2016. Plant functional diversity increases grassland productivity-related water vapor fluxes: an Ecotron and modeling approach. Ecology 97, 2044–2054.

Millenium Ecosystem Assessment, 2005. Ecosystem and Human Well-Being: Synthesis. Island Press, Washington, DC.

Mitchell, C.E., Tilman, D., Groth, J.V., 2002. Effects of grassland plant species diversity, abundance, and composition on foliar fungal disease. Ecology 83, 1713–1726.

Morison, J.I.L., Lawlor, D.W., 1999. Interactions between increasing $CO_2$ concentration and temperature on plant growth. Plant Cell Environ. 22, 659–682.

Mouquet, N., Loreau, M., 2003. Community patterns in source-sink metacommunities. Am. Nat. 162 (5), 544–557.

Mueller, R.C., Scudder, C.M., Porter, M.E., Trotter, R.T., Gehring, C.A., Whitham, T.G., 2005. Differential tree mortality in response to severe drought: evidence for long-term vegetation shifts. J. Ecol. 93, 1085–1093.

Mueller, K.E., Tilman, D., Fornara, D.A., Hobbie, S.E., 2013. Root depth distribution and the diversity–productivity relationship in a long-term grassland experiment. Ecology 94, 787–793.

Müller, I., Schmid, B., Weiner, J., 2000. The effect of nutrient availability on biomass allocation patterns in 27 species of herbaceous plants. Perspect. Plant Ecol. Evol. Syst. 3, 115–127.

Mwangi, P.N., Schmitz, M., Scherber, C., Roscher, C., Schumacher, J., Scherer-Lorenzen, M., Weisser, W.W., Schmid, B., 2007. Niche pre-emption increases with species richness in experimental plant communities. J. Ecol. 95, 65–78.

Newbold, T., Hudson, L.N., Hill, S.L.L., Contu, S., Lysenko, I., Senior, R.A., Börger, L., Bennett, D.J., Choimes, A., Collen, B., Day, J., Palma, A.D., Díaz, S., Echeverria-Londoño, S., Edgar, M.J., Feldman, A., Garon, M., Harrison, M.L.K., Alhusseini, T., Ingram, D.J., Itescu, Y., Kattge, J., Kemp, V., Kirkpatrick, L., Kleyer, M., Correia, D.L.P., Martin, C.D., Meiri, S., Novosolov, M., Pan, Y., Phillips, H.R.P., Purves, D.W., Robinson, A., Simpson, J., Tuck, S.L., Weiher, E., White, H.J., Ewers, R.M., Mace, G.M., Scharlemann, J.P.W., Purvis, A., 2015. Global effects of land use on local terrestrial biodiversity. Nature 520, 45–50.

Nicholls, A.M., 2011. Size-dependent analysis of allocation to sexual and clonal reproduction in penthorum sedoides under contrasting nutrient levels. Int. J. Plant Sci. 172, 1077–1086.

Nitschke, N., Ebeling, A., Rottstock, T., Scherber, C., Middelhoff, C., Creutzburg, S., Weigelt, A., Tscharntke, T., Fischer, M., Weisser, W.W., 2010. Time course of plant diversity effects on Centaurea jacea establishment and the role of competition and herbivory. J. Plant Ecol. 3, 109–121.

Obeso, J.R., 1993. Does defoliation affect reproductive output in herbaceous perennials and woody plants in different ways? Funct. Ecol. 7, 150–155.

Obeso, J.R., 2002. The costs of reproduction in plants. New Phytol. 155, 321–348.

Oehri, J., Schmid, B., Schaepman-Strub, G., Niklaus, P.A., 2017. Biodiversity promotes primary productivity and growing season lengthening at the landscape scale. Proc. Natl. Acad. Sci. U. S. A. 114, 10160–10165.

Oelmann, Y., Buchmann, N., Gleixner, G., Habekost, M., Roscher, C., Rosenkranz, S., Schulze, E.-D., Steinbeiss, S., Temperton, V.M., Weigelt, A., Weisser, W.W., Wilcke, W., 2011. Plant diversity effects on aboveground and belowground N pools in temperate grassland ecosystems: development in the first 5 years after establishment. Glob. Biogeochem. Cycles 25 GB2014.

Oram, N.J., Ravenek, J.M., Barry, K.E., Weigelt, A., Chen, H., Gessler, A., Gockele, A., de Kroon, H., van der Paauw, J.W., Scherer-Lorenzen, M., Smit-Tiekstra, A., van Ruijven, J., Mommer, L., 2018. Below-ground complementarity effects in a grassland biodiversity experiment are related to deep-rooting species. J. Ecol. 106, 265–277.

Palmer, M.W., 1994. Variation in species richness: towards a unification of hypotheses. Folia Geobot. Phytotaxon. 29, 511–530.

Parker, I.M., Gilbert, G.S., 2018. Density-dependent disease, life-history trade-offs, and the effect of leaf pathogens on a suite of co-occurring close relatives. J. Ecol. 106, 1829–1838.

Peña, E.D.L., Echeverría, S.R., Putten, W.H.V.D., Freitas, H., Moens, M., 2006. Mechanism of control of root-feeding nematodes by mycorrhizal fungi in the dune grass Ammophila arenaria. New Phytol. 169, 829–840.

Pereira, H.M., Leadley, P.W., Proença, V., Alkemade, R., Scharlemann, J.P.W., Fernandez-Manjarrés, J.F., Araújo, M.B., Balvanera, P., Biggs, R., Cheung, W.W.L., Chini, L., Cooper, H.D., Gilman, E.L., Guénette, S., Hurtt, G.C., Huntington, H.P., Mace, G.M., Oberdorff, T., Revenga, C., Rodrigues, P., Scholes, R.J., Sumaila, U.R., Walpole, M., 2010. Scenarios for global biodiversity in the 21st century. Science 330, 1496–1501.

Petermann, J.S., Fergus, A.J.F., Turnbull, L.A., Schmid, B., 2008. Janzen-Connell effects are widespread and strong enough to maintain diversity in grasslands. Ecology 89, 2399–2406.

Pimm, S.L., Jenkins, C.N., Abell, R., Brooks, T.M., Gittleman, J.L., Joppa, L.N., Raven, P.H., Roberts, C.M., Sexton, J.O., 2014. The biodiversity of species and their rates of extinction, distribution, and protection. Science 344, 1246752.

Poorter, H., Niklas, K.J., Reich, P.B., Oleksyn, J., Poot, P., Mommer, L., 2012. Biomass allocation to leaves, stems and roots: meta-analyses of interspecific variation and environmental control. New Phytol. 193, 30–50.

Ravenek, J.M., Bessler, H., Engels, C., Scherer-Lorenzen, M., Gessler, A., Gockele, A., De Luca, E., Temperton, V.M., Ebeling, A., Roscher, C., Schmid, B., Weisser, W.W., Wirth, C., de Kroon, H., Weigelt, A., Mommer, L., 2014. Long-term study of root biomass in a biodiversity experiment reveals shifts in diversity effects over time. Oikos 123, 1528–1536.

Reich, P.B., Wright, I.J., Cavender-Bares, J., Craine, J.M., Oleksyn, J., Westoby, M., Walters, M.B., 2003. The evolution of plant functional variation: traits, spectra, and strategies. Int. J. Plant Sci. 164, S143–S164.

Reich, P.B., Tilman, D., Naeem, S., Ellsworth, D.S., Knops, J., Craine, J., Wedin, D., Trost, J., 2004. Species and functional group diversity independently influence biomass accumulation and its response to $CO_2$ and N. Proc. Natl. Acad. Sci. U. S. A. 101, 10101–10106.

Reich, P.B., Tilman, D., Isbell, F., Mueller, K., Hobbie, S.E., Flynn, D.F., Eisenhauer, N., 2012. Impacts of biodiversity loss escalate through time as redundancy fades. Science 336, 589–592.

Reich, P.B., Hobbie, S.E., Lee, T.D., Pastore, M.A., 2018. Unexpected reversal of C3 versus C4 grass response to elevated $CO_2$ during a 20-year field experiment. Science 360, 317–320.

Richardson, A.E., Barea, J.-M., McNeill, A.M., Prigent-Combaret, C., 2009. Acquisition of phosphorus and nitrogen in the rhizosphere and plant growth promotion by microorganisms. Plant Soil 321, 305–339.

Roeder, A., Schweingruber, F.H., Fischer, M., Roscher, C., 2017. Growth ring analysis of multiple dicotyledonous herb species—a novel community-wide approach. Basic Appl. Ecol. 21, 23–33.

Roff, D.A., 2000. Trade-offs between growth and reproduction: an analysis of the quantitative genetic evidence. J. Evol. Biol. 13, 434–445.

Roscher, C., Schumacher, J., 2016. Positive diversity effects on productivity in mixtures of arable weed species as related to density–size relationships. J. Plant Ecol. 9, 792–804.

Roscher, C., Schumacher, J., Baade, J., Wilcke, W., Gleixner, G., Weisser, W.W., Schmid, B., Schulze, E.-D., 2004. The role of biodiversity for element cycling and trophic interactions: an experimental approach in a grassland community. Basic Appl. Ecol. 5, 107–121.

Roscher, C., Schumacher, J., Weisser, W.W., Schmid, B., Schulze, E.-D., 2007. Detecting the role of individual species for overyielding in experimental grassland communities composed of potentially dominant species. Oecologia 154, 535–549.

Roscher, C., Schumacher, J., Weisser, W.W., Schulze, E.-D., 2008a. Genetic identity affects performance of species in grasslands of different plant diversity: an experiment with Lolium perenne cultivars. Ann. Bot. 102, 113–125.

Roscher, C., Thein, S., Schmid, B., Scherer-Lorenzen, M., 2008b. Complementary nitrogen use among potentially dominant species in a biodiversity experiment varies between two years. J. Ecol. 96, 477–488.

Roscher, C., Kutsch, W.L., Schulze, E.-D., 2011a. Light and nitrogen competition limit Lolium perenne in experimental grasslands of increasing plant diversity. Plant Biol. 13, 134–144.

Roscher, C., Schmid, B., Buchmann, N., Weigelt, A., Schulze, E.-D., 2011b. Legume species differ in the responses of their functional traits to plant diversity. Oecologia 165, 437–452.

Scherber, C., Milcu, A., Partsch, S., Scheu, S., Weisser, W.W., 2006. The effects of plant diversity and insect herbivory on performance of individual plant species in experimental grassland. J. Ecol. 94, 922–931.

Schnitzer, S.A., Carson, W.P., 2001. Treefall gaps and the maintenance of species diversity in a tropical forest. Ecology 82, 913–919.

Schnitzer, S.A., Klironomos, J.N., HilleRisLambers, J., Kinkel, L.L., Reich, P.B., Xiao, K., Rillig, M.C., Sikes, B.A., Callaway, R.M., Mangan, S.A., Van Nes, E.H., Scheffer, M., 2011. Soil microbes drive the classic plant diversity-productivity pattern. Ecology 92, 296–303.

Schoener, T.W., 1970. Nonsynchronous spatial overlap of lizards in patchy habitats. Ecology 51, 408–418.

Schoener, T.W., 1974. Resource partitioning in ecological communities. Science 185, 27–39. https://doi.org/10.1126/science.185.4145.27.

Seabloom, E.W., Kinkel, L., Borer, E.T., Hautier, Y., Montgomery, R.A., Tilman, D., 2017. Food webs obscure the strength of plant diversity effects on primary productivity. Ecol. Lett. 20, 505–512.

Semchenko, M., Lepik, M., Götzenberger, L., Zobel, K., 2012. Positive effect of shade on plant growth: amelioration of stress or active regulation of growth rate? J. Ecol. 100, 459–466.

Sheil, D., Burslem, D.F.R.P., 2013. Defining and defending Connell's intermediate disturbance hypothesis: a response to fox. Trends Ecol. Evol. 28, 571–572.

Shmida, A., Wilson, M.V., 1985. Biological determinants of species diversity. J. Biogeogr. 1–20.

Stearns, S.C., 2000. Life history evolution: successes, limitations, and prospects. Naturwissenschaften 87, 476–486.

Sugiyama, S., Bazzaz, F.A., 1998. Size dependence of reproductive allocation: the influence of resource availability, competition and genetic identity. Funct. Ecol. 12, 280–288.

Suter, M., 2009. Reproductive allocation of Carex flava reacts differently to competition and resources in a designed plant mixture of five species. In: Van der Valk, A.G. (Ed.), Herbaceous Plant Ecology: Recent Advances in Plant Ecology. Springer Netherlands, Dordrecht, pp. 117–125.

Thein, S., Roscher, C., Schulze, E.-D., 2008. Effects of trait plasticity on aboveground biomass production depend on species identity in experimental grasslands. Basic Appl. Ecol. 9, 475–484.

Tilman, D., 1982. Resource Competition and Community Structure. (Mpb-17). Princeton University Press.

Tilman, D., 1987. Secondary succession and the pattern of plant dominance along experimental nitrogen gradients. Ecol. Monogr. 57, 189–214.

Tilman, D., Wedin, D., Knops, J., 1996. Productivity and sustainability influenced by biodiversity in grassland ecosystems. Nature 379, 718–720.

Tilman, D., Lehman, C.L., Thomson, K.T., 1997. Plant diversity and ecosystem productivity: theoretical considerations. Proc. Natl. Acad. Sci. U. S. A. 94, 1857–1861.

Tilman, D., Isbell, F., Cowles, J.M., 2014. Biodiversity and ecosystem functioning. Annu. Rev. Ecol. Evol. Syst. 45, 471–493.

Tittensor, D.P., Walpole, M., Hill, S.L.L., Boyce, D.G., Britten, G.L., Burgess, N.D., Butchart, S.H.M., Leadley, P.W., Regan, E.C., Alkemade, R., Baumung, R., Bellard, C., Bouwman, L., Bowles-Newark, N.J., Chenery, A.M., Cheung, W.W.L., Christensen, V., Cooper, H.D., Crowther, A.R., Dixon, M.J.R., Galli, A., Gaveau, V., Gregory, R.D., Gutierrez, N.L., Hirsch, T.L., Höft, R., Januchowski-Hartley, S.R., Karmann, M., Krug, C.B., Leverington, F.J., Loh, J., Lojenga, R.K., Malsch, K., Marques, A., Morgan, D.H.W., Mumby, P.J., Newbold, T., Noonan-Mooney, K., Pagad, S.N., Parks, B.C., Pereira, H.M., Robertson, T., Rondinini, C., Santini, L., Scharlemann, J.P.W., Schindler, S., Sumaila, U.R., Teh, L.S.L., van Kolck, J., Visconti, P., Ye, Y., 2014. A mid-term analysis of progress toward international biodiversity targets. Science 346, 241–244.

Turnbull, L.A., Levine, J.M., Loreau, M., Hector, A., 2013. Coexistence, niches and biodiversity effects on ecosystem functioning. Ecol. Lett. 16, 116–127.

Turnbull, L.A., Isbell, F., Purves, D.W., Loreau, M., Hector, A., 2016. Understanding the value of plant diversity for ecosystem functioning through niche theory. Proc. R. Soc. B Biol. Sci. 283, 20160536.

Van der Heijden, M.G., Klironomos, J.N., Ursic, M., Moutoglis, P., Streitwolf-Engel, R., Boller, T., Wiemken, A., Sanders, I.R., 1998. Mycorrhizal fungal diversity determines plant biodiversity, ecosystem variability and productivity. Nature 396, 69–72.

van Kleunen, M., Fischer, M., Schmid, B., 2001. Effects of intraspecific competition on size variation and reproductive allocation in a clonal plant. Oikos 94, 515–524.

Vandermeer, J.H., 1981. The interference production principle: an ecological theory for agriculture. BioScience 31 (5), 361–364.

Vandermeer, J.H., 1992. The Ecology of Intercropping. Cambridge University Press.

von Felten, S., Hector, A., Buchmann, N., Niklaus, P.A., Schmid, B., Scherer-Lorenzen, M., 2009. Belowground nitrogen partitioning in experimental grassland plant communities of varying species richness. Ecology 90, 1389–1399.

Wagg, C., Jansa, J., Schmid, B., van der Heijden, M.G.A., 2011. Belowground biodiversity effects of plant symbionts support aboveground productivity. Ecol. Lett. 14, 1001–1009.

Waite, S., Hutchings, M.J., 1982. Plastic energy allocation patterns in plantago coronopus. Oikos 38, 333–342.

Wayne, P.M., Carnelli, A.L., Connolly, J., Bazzaz, F.A., 1999. The density dependence of plant responses to elevated $CO_2$. J. Ecol. 87, 183–192.

Wei, X., Reich, P.B., Hobbie, S.E., Kazanski, C.E., 2017. Disentangling species and functional group richness effects on soil N cycling in a grassland ecosystem. Glob. Change Biol. 23, 4717–4727.

Weiner, J., 1988. The influence of competition on plant reproduction. In: Lovett-Doust, J., Lovett-Doust, L. (Eds.), Plant Reproductive Ecology: Patterns and Strategies. Oxford University Press, Oxford, UK, pp. 228–245.

Whittington, H.R., Tilman, D., Wragg, P.D., Powers, J.S., 2015. Phenological responses of prairie plants vary among species and year in a three-year experimental warming study. Ecosphere 6 art208.

Wohlgemuth, D., Solan, M., Godbold, J.A., 2017. Species contributions to ecosystem process and function can be population dependent and modified by biotic and abiotic setting. Proc. Biol. Sci. 284, 20162805.

Wright, J.S., 2002. Plant diversity in tropical forests: a review of mechanisms of species coexistence. Oecologia 130, 1–14.

Wright, A., Schnitzer, S.A., Reich, P.B., 2014. Living close to your neighbors: the importance of both competition and facilitation in plant communities. Ecology 95, 2213–2223.

Wright, A., Schnitzer, S.A., Reich, P.B., 2015. Daily environmental conditions determine the competition–facilitation balance for plant water status. J. Ecol. 103, 648–656.

Wright, A.J., de Kroon, H., Visser, E.J.W., Buchmann, T., Ebeling, A., Eisenhauer, N., Fischer, C., Hildebrandt, A., Ravenek, J., Roscher, C., Weigelt, A., Weisser, W., Voesenek, L.A.C.J., Mommer, L., 2017a. Plants are less negatively affected by flooding when growing in species-rich plant communities. New Phytol. 213, 645–656.

Wright, A.J., Wardle, D.A., Callaway, R., Gaxiola, A., 2017b. The overlooked role of facilitation in biodiversity experiments. Trends Ecol. Evol. 32, 383–390.

Yachi, S., Loreau, M., 2007. Does complementary resource use enhance ecosystem functioning? A model of light competition in plant communities. Ecol. Lett. 10, 54–62.

## Further reading

Chesson, P., Kuang, J.J., 2008. The interaction between predation and competition. Nature 456, 235–238.

Poisot, T., Mouquet, N., Gravel, D., 2013. Trophic complementarity drives the biodiversity–ecosystem functioning relationship in food webs. Ecol. Lett. 16, 853–861.

CHAPTER NINE

# Mapping change in biodiversity and ecosystem function research: food webs foster integration of experiments and science policy

Jes Hines[a,b,]*, Anne Ebeling[c], Andrew D. Barnes[a,c,d], Ulrich Brose[a,c], Christoph Scherber[e], Stefan Scheu[f], Teja Tscharntke[g], Wolfgang W. Weisser[h], Darren P. Giling[a,b,c], Alexandra M. Klein[i], Nico Eisenhauer[a,b]

[a]German Centre for Integrative Biodiversity Research (iDiv) Halle-Jena-Leipzig, Leipzig, Germany
[b]Institute of Biology, Leipzig University, Leipzig, Germany
[c]Institute of Ecology and Evolution, Friedrich Schiller University Jena, Jena, Germany
[d]School of Science, University of Waikato, Hamilton, New Zealand
[e]Institute of Landscape Ecology, University of Münster, Münster, Germany
[f]J.F. Blumenbach Institute of Zoology and Anthropology, University of Göttingen, Göttingen, Germany
[g]Agroecology, Department of Crop Sciences, University of Göettingen, Göettingen, Germany
[h]Terrestrial Ecology Research Group, Department of Ecology and Ecosystem Management, School of Life Sciences Weihenstephan, Technical University of Munich, Freising, Germany
[i]Faculty of Environment and Natural Resources, University of Freiburg, Freiburg, Germany
*Corresponding author: e-mail address: jessica.hines@idiv.de

## Contents

| | |
|---|---:|
| 1. Topic networks as a way to visualize global conversation about biodiversity and ecosystem functioning | 298 |
|    1.1 Core research domains persist through time: 'BEF experiments' and 'Science policy' | 299 |
|    1.2 Integrative research domains connect the scientific landscape: Aquatic food webs and agricultural landscapes | 305 |
| 2. Divisions among research domains: influences on food webs | 307 |
|    2.1 Baseline comparisons across research domains: Random, null, and gradient based hypotheses | 307 |
|    2.2 Scaling multi-trophic diversity | 310 |
|    2.3 Currency across domains: Biomass, energy, valuation | 311 |
| 3. Summary and outlook: towards integrative food-web ecology | 312 |
| Acknowledgements | 314 |
| References | 315 |

## Abstract

Human activities are causing major changes in biological communities worldwide. Due to concern about the consequences of these changes, an academic conversation about biodiversity and ecosystem functioning (BEF) has emerged over the last few decades. Here we use a keyword co-occurrence analysis to characterize and review 28 years of research focused on these terms. We find that the rapidly growing literature has developed in four research domains. The first two domains "BEF Experiments" and "Science Policy" emerge early, and persist through time, as core research areas with emphases on experiments and management, respectively. The second two domains, "Agricultural Landscapes" and "Aquatic Food Webs", arise as integrative domains that connect divisions in scientific discussion surrounding BEF Experiments and Science Policy. Terms related to species interactions (i.e. pollinator, predator, food web) appear more commonly in the two integrative domains reflecting shared interests of many scientists focusing on biodiversity and ecosystem functioning. Despite shared interests in food webs, research in the four domains differ with respect to their spatial scale, baseline comparisons, and currency of measurements. Food-web research that bridges these divides should be pushed to the forefront of biodiversity and ecosystem functioning research priorities.

## 1. Topic networks as a way to visualize global conversation about biodiversity and ecosystem functioning

One of the most remarkable features of our planet is the incredible diversity of life (Wilson, 1999). Biological diversity appeals to many people, yet it is perceived and valued in strikingly different ways (Chan et al., 2016; Díaz et al., 2015; Pascual et al., 2017). This is, in part, because biodiversity and species interactions are fundamental to so many aspects of human well-being (Díaz et al., 2006; IPBES, 2019; MEA, 2005; Naeem et al., 2009). For example, we rely on biological communities for provisioning of food, water, energy, and medicines as well as regulating climate and diseases and contributing to our culture and traditional beliefs (Carpenter et al., 2009; Chan et al., 2016). Growing concern about human influences on the functioning of natural systems has promoted debate about how to best understand, conserve, and promote biodiversity (Eisenhauer et al., 2019; IPBES, 2019; Manning et al., 2019).

Scientists debate concerns about biodiversity by publishing papers in a global conversation. The ideas that they share shape an idea space, which is formed by groups of thematically related words (Börner et al., 2005; Cobo et al., 2011). By carefully choosing to include or exclude words from documents, authors link concepts and give a structure to idea space, which influences our thinking and motivates our research and conservation priorities (Dretske, 1981). Scientists have repeatedly joined the words 'Biodiversity' and 'Ecosystem Functioning', and the frequency that these words co-occur in research papers implies there is a special semantic relationship between them. What does the map of our conversation about Biodiversity and Ecosystem Functioning look like? What are the main lines of inquiry, and how has research on these topics changed through time? Do common words reflect common thoughts, or are scientists approaching research from fundamentally different perspectives? Mapping knowledge domains can help us visualize the scientific landscape, identify core and integrative research areas, and consider where potential innovations lie.

In the sections that follow, we describe temporal trends in the structure of the scientific literature on biodiversity and ecosystem functioning (Section 1; Box 1). Our results show that food webs and consumer-driven processes in terrestrial and aquatic habitats have emerged as important topics that bridge classic divides between basic research in biodiversity and ecosystem-functioning experiments and science policy. Therefore, in the next section, we discuss role of food webs and consumer-driven process as linking concepts with particular emphasis on how three factors (spatial scale, baselines for comparisons, and currency of assessments) differ among research areas (Section 2). These differences simultaneously reflect unique expertise of researchers in each domain as well as knowledge gaps that may limit knowledge transfer. Therefore, we conclude by discussing an integrative food-web research agenda as a focal priority for biodiversity and ecosystem functioning research across the globe (Section 3).

## 1.1 Core research domains persist through time: 'BEF experiments' and 'Science policy'

Our analysis reveals some striking trends in the literature discussing biodiversity and ecosystem functioning. As in most other research areas, there is rapid growth of the literature through time (Fig. 1A). Our search identified 12,300 papers published between 1990 and 2018. There has been a greater than 18-fold increase in the number of publications during the time span covered in our search, and the contribution of these papers grew from less than 1% to greater than 6% to the total papers published in the 'Lifescience and Biomedicine' research area. This proportion is likely to be higher at the end of the decade, because despite the slightly shorter time span nearly as many papers were published in 2016–2018 (4375 papers) as in 2011–2015 (4455 papers). Associated with increases in number of papers,

## BOX 1 Mapping trends in biodiversity-ecosystem function research: Keyword co-occurrence analysis reveals emergence of food-web interaction terms as integrative topics

To characterize the development of the scientific landscape surrounding the terms 'Biodiversity' AND 'Ecosystem Function*', we conducted a topic search using the ISI Web of Science search core collection on 26 February 2019. To examine temporal trends in topics, we divided the literature corpus into 5-year segments (1990–1995, 1996–2000, 2001–2005, 2006–2010, 2011–2015, 2016–2018). We used Visualization of Similarities (VOS) algorithms that rely on Apache Open natural language processing to identify parts of speech, apply stemming, and exclude stop words (Van Eck and Waltman, 2014). We established a list of potential keywords from binary counting of terms present in the titles and abstracts of papers (Van Eck and Waltman, 2010). We refined the data set to keywords that occured more than 10 times and achieved a greater than 60% relevance score based on rankings of nouns that non-randomly occur together (Van Eck and Waltman, 2014). A thesaurus was applied to ensure that abbreviations referred to the terms they describe (i.e. sla refers to specific leaf area) and plural words were synonyms with their singular counterparts (i.e. communities is considered equivalent to community). To allow detection of subtle semantic differences among research areas, potential synonyms were not manipulated otherwise (i.e. ecosystem function and ecosystem process are considered as separate terms even when they are used to describe the same concept).

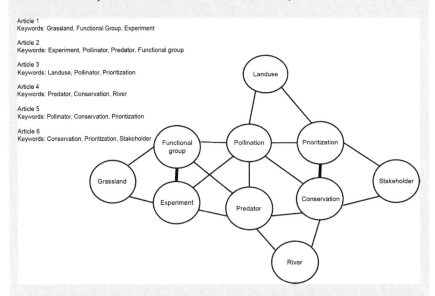

To examine connections between words, we conducted a keyword co-occurrence analysis. In this analysis, keywords are nodes, and keywords that appear together in a document are joined by a link. The figure above depicts

## BOX 1 Mapping trends in biodiversity-ecosystem function research: Keyword co-occurrence analysis reveals emergence of food-web interaction terms as integrative topics—cont'd

connections among keywords in six hypothetical articles. The number of co-occurrences are indicated by the thickness of the link, and words that more commonly occur together (i.e. 'Functional group and Experiment', or 'Prioritization and Conservation') appear closer together in space (Boyack et al., 2005). Subsequently, we identified communities of words that form research domains using the LinLog/modularity normalization (Newman, 2004; Noack, 2007). 'Pollinator' and 'Predator' appear as integrative terms that connect articles that focus on 'Experiments' and 'Conservation'. The four distinct word communities in our analysis (see Fig. 2) reflect an emergent property of the network structure, and we did not personally assign words to research domains. However, the names of the four domains (BEF Experiments, Science Policy, Agricultural Landscapes, Aquatic Food Webs) are a qualitative description based on authors' intuition.

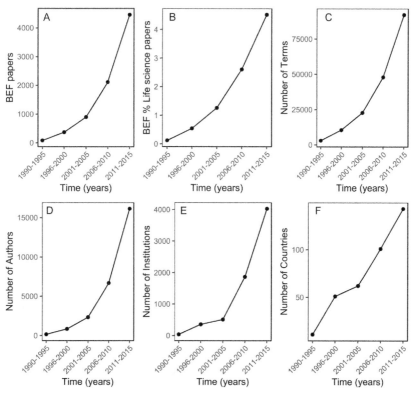

**Fig. 1** The number of papers, terms, authors, and affiliated countries obtained in a Web of Science core collection search for the terms 'Biodiversity' AND 'Ecosystem function*'.

there has been concomitant growth in the number of terms, authors, institutions, and countries associated with biodiversity and ecosystem functioning research (Fig. 1B–F). Considered together, these metrics suggest that research on biodiversity and ecosystem functioning is one of the most prominent topics investigated by ecologists in the last three decades.

Our keyword co-occurrence analysis reveals that a strong division between communities of words emerged in the 1990s, and it persists over the course of three decades resulting in the development of two core research domains (Fig. 2). On one side, we see a conversation about experimental research (BEF experiments: green domain in Fig. 2). This research domain emerged in response to a heated debate regarding whether biodiversity per se influenced ecosystem functioning, and which experimental designs were appropriate to test for mechanisms driving potential influences (Hooper et al., 2005; Naeem et al., 2012; Tilman et al., 1996). Predominantly associated with grasslands and plant communities, research in this domain consistently emphasizes the terms 'species richness', 'experiment', 'abundance', and 'biomass' (Fig. 3). Although there is comparatively less emphasis on litter decomposition and communities composed of non-plant species, soil is among the top five terms in four of the six periods, and it appears as the most frequent term in 'BEF Experiments' 2016–2018. Over time, the number of terms in the 'BEF Experiments' domain has expanded and included more terms that suggest more integrated measurements from different areas of biology (primary production, enzyme activity, soil fauna). The central focus of this domain continues to be rigorous field experiments designed to test mechanisms underlying the relationship between biodiversity and ecosystem functioning (Eisenhauer et al., 2019).

On the other side of scientific idea space, we see terms related to science policy (red domain in Fig. 2). This research domain emerged based on interest in developing strategies to conserve, manage, and restore landscapes confronted by increasing human demands (Hobbs et al., 2009; Suding et al., 2015; Vitousek et al., 1997). There is a stronger emphasis on 'landscape', 'landuse', and 'landcover' as influences on biodiversity, ecosystem functioning, and human-wellbeing in this research domain (Tscharntke et al., 2012a). In the Science Policy domain, monitoring of biodiversity is typically considered as an essential indicator of ecosystem integrity (Mace and Baillie, 2007; Pereira et al., 2013) and ecosystem functions are valued in terms of how they make contributions to human society (Costanza et al., 2014; Pascual et al., 2017). Over time, research in this domain has

# Mapping change in biodiversity and ecosystem function research 303

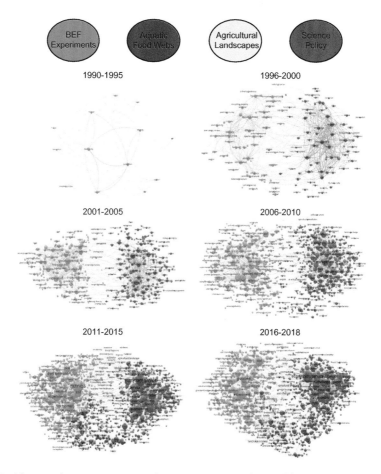

**Fig. 2** A keyword co-occurrence analysis using papers obtained from an ISI Web of Science Search for the terms 'Biodiversity' AND 'Ecosystem function*'. In each plot, nodes are keywords, and keywords that co-occur more frequently appear closer together in space (Boyack et al., 2005; Van Eck and Waltman, 2010). Links are formed when two keywords co-occur in the same document. For visual clarity, however, only the top 50% strongest links are depicted in the plots shown here. Communities of words which represent groupings of thematically related ideas are identified by different colours (Newman, 2004): BEF Experiments (green), Aquatic Food Webs (blue), Agricultural Landscapes (yellow), and Science Policy (red).

expanded to include terms such as 'remote sensing', 'valuation', and 'environmental degradation' suggesting that Science Policy research is integrating perspectives of multiple disciplines (i.e. geographers, economists, and biologists). A central challenge of this domain continues to be developing

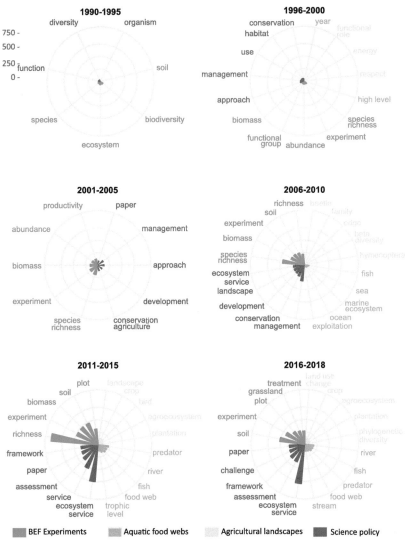

**Fig. 3** Temporal trends in the occurrence and frequency of keywords from 1990 to 2018. Bars report the frequency of the five most common words in each research domain: BEF Experiments, Aquatic Food Webs, Agricultural Landscapes, and Science Policy. Standard axes across plots allow for visual comparison of the absolute frequencies in term use in domains through time as well as changes in the relative frequency of the top five words within and across domains.

assessment tools for decision support to evaluate potential conflicts among multiple stakeholders (Sharp et al., 2015), with the goal of developing win-win scenarios for biodiversity and landscape productivity as well as economic profit (Griscom et al., 2017).

The core focus of research in BEF Experiments and Science Policy domains has persisted through time. Particularly for Science Policy; however, there has been some turn-over in the dominant terms used, which provides insight into how research within the domains is being refined and developed. For example, 'conservation', 'management', and 'sustainable' appear more prominently in early years of our time series when the literature was smaller overall. During these most recent years (2016–2018), there were increases in the terms 'prioritization' and 'multifunctionality', possibly reflecting the progressively multivariate challenges imposed by a larger literature reaching a wider audience with diverse demands. Some turnover in terms can be attributed to the Millennium Ecosystem Assessment (MEA, 2005), which had lasting impacts on the development and visibility of whole research domains (Carpenter et al., 2009; Mulder et al., 2015). Indeed, starting in 2006 the year after MEA was published, there are dramatic increases in the term 'Ecosystem service'. Ecosystem service has now become the most prominent science policy term used in the last 8 years (2011–2018) (Fig. 3), and it is a key aspect of intergovernmental science-policy processes (TEEB Foundations, 2010; IPBES, 2019). The location of 'Ecosystem Services' in our keyword co-occurrence analysis reflects that they are more commonly discussed in association with other terms in the Science Policy domain, not that they are completely excluded from research in other domains. There also exemplary papers that demonstrate interest in multivariate approaches, ecosystem services and the implications of biodiversity for human well-being expressed by scientists evaluating results of BEF experiments (Balvanera et al., 2006; Byrnes et al., 2014; Cardinale et al., 2012; Meyer et al., 2018; Naeem et al., 2012). Turnover of terms reflect growth and development of research domains, and emerging frontiers in the idea space surrounding BEF Experiments and Science Policy.

## 1.2 Integrative research domains connect the scientific landscape: Aquatic food webs and agricultural landscapes

Two new domains (Aquatic Food Webs and Agricultural Landscapes) arose in the last decade (2011–2018). These domains represent integrative research that bridges the gap between BEF experiments and Science Policy (Fig. 2). There is clear division among vocabulary describing terrestrial and aquatic realms, particularly with respect to human influences (Fig. 2). In the aquatic realm domain-specific descriptors like 'sea', 'river', 'ocean', and 'marine ecosystems' consistently appear. Only in 2006–2010 did a human influence (exploitation) enter the top five terms in the Aquatic Food Web domain

(Fig. 3). Attribution of a singular driver to the frequency of the term 'exploitation' cannot be justified, but there was a contentious prediction of collapse of marine fisheries published by Worm et al. (2006). One reason this paper garnered attention was that it linked concepts from BEF experiments with global fisheries policy, and the implications of model predictions were still being discussed 10 years later (Worm, 2016).

On the terrestrial side, the realm-specific terminology has comparatively stronger emphasis on human influences than in the Aquatic Food Web domain. Land-uses such as 'crop', 'agro-ecosystem', and 'plantation' appear among the most frequently used terms (Fig. 3). Agricultural intensification is a leading driver of biodiversity loss and reflects humanity's oldest and most fundamental use of ecological science to manage nature (Lichtenberg et al., 2017; Tilman et al., 2011). Maintenance of biodiversity and ecosystem functioning in agricultural landscapes requires cooperation between farmers and foresters, and we see that the words 'forest fragment', 'agroforest', and 'plantation' frequently co-occur with 'crop yield' and 'agricultural practice' (Fig. 2). Accordingly there have been recent increases in the number of BEF experiments focused on trees (Huang et al., 2018; Paquette et al., 2018; Verheyen et al., 2016), although forests have more typically been associated with science policy research (Nelson et al., 2009).

Notably, terms related to food webs and consumer-driven processes surfaced more prominently in both integrative research domains, where our analysis shows words reflecting interest in animals and consumer species interactions (Figs 2 and 3). In agricultural landscapes there is particular emphasis on invertebrates ('beetle', 'hymenoptera', 'butterfly', 'spider') and 'bird', as well as some of the processes they drive (pollination) (Figs 2 and 3). Long-standing interest in diversity of invertebrate species in agro-ecosystems reflects the importance of invertebrate food web interactions for regulating services like natural pest control, pollination, and decomposition (Altieri, 1999; Hatt et al., 2018; Simons and Weisser, 2017; Tscharntke et al., 2005). In the Aquatic Food-Web domain, animals, namely, fish, are a consistent focus, and there is more emphasis on terms related to species interactions (i.e. 'predator', 'food web', 'trophic'; Fig. 3). The central location of these two research domains reflects common word use and interests of many scientists. In particular, focus on diversity of animal taxa (i.e. 'macro-invertebrate community', 'ant diversity') and their trophic interactions (i.e. 'interacting species', 'parasite') provide a common theme among both of the integrative domains (Aquatic Food Webs and Agricultural Landscapes).

## 2. Divisions among research domains: influences on food webs

The influence of integrative research domains can be seen in both of the core research areas where there have been trends towards including food web theory in BEF experiments (Hines et al., 2015b) and management decisions in science policy (McDonald-Madden et al., 2016). Yet, the intermediate location of these terms in our analysis also suggests that both domains have not fully integrated key food web principles (Dee et al., 2017; Hines et al., 2015b). Such lack of integration does not necessarily reflect weaknesses. Instead, the approaches, activities and interpretations of research in each domain contribute diverse knowledge to a broad audience. In the following section, we describe qualitative interpretations of how the research domains differ from one another with respect to baselines that serve the basis for comparison, scale used to assess the influence of interactions, and currency used to quantify the magnitude of responses (Table 1). These differences reflect archetypic characteristics of research approaches in each domain that are well ingrained in our understanding of food webs and consumer-driven ecosystem processes.

### 2.1 Baseline comparisons across research domains: Random, null, and gradient based hypotheses

The overarching goal of research in all four domains is to understand, conserve, and manage biodiversity. To reach that goal scientists identify patterns

**Table 1** Archetypal descriptors of differences among biodiversity and ecosystem functioning research domains.

|  | BEF experiments | Aquatic food webs | Agricultural landscapes | Science policy |
|---|---|---|---|---|
| Baseline | Monocultures/polycultures Replicated gradients | Null model random matrix | Non-agricultural habitat | Pristine environment Pre-disturbance Disturbance gradient |
| Scale | Field plot | Location | Landscape | Region |
| Currency | Plant biomass | Fish biomass | Crop production | Value to stakeholder |

in species assemblages and evaluate whether changes in those patterns will influence ecosystem functioning. Across domains, however, there are strong differences regarding what constitutes a pattern, and which baseline comparisons appropriately illustrate the consequences of change (Mihoub et al., 2017) (Table 1). Without explicit appreciation of how and why baselines differ across domains, protracted debates about the relevance of each research domain will persist (Cardinale et al., 2018; Eisenhauer et al., 2016; Huston, 1997) and we will forsake meaningful integration of food-web concepts into our understanding of relationships between biodiversity and ecosystem functioning (Hines et al., 2015b; Thompson et al., 2012).

In BEF experiments, scientists create experimental gradients in species richness to characterize ecosystem function responses to changes in species richness per se. To correct for any artefacts resulting from particular species assemblages, random draws from a designed species pool are used to assemble communities with particular species richness (Ebeling et al., 2014; Roscher et al., 2004). The implications of BEF experimental designs have been thoroughly debated with respect to manipulations of one trophic level (Eisenhauer et al., 2016; Wardle, 2016). However, plant communities typically are manipulated, and rigorous experimental designs thwart simultaneous efforts to manipulate higher trophic levels (but see exceptions described in Hines et al., 2015b). Therefore, the nature of random community assemblages changes across trophic levels when changes in food webs are assessed in BEF experiments. That is, *non-random consumer communities* are allowed to naturally assemble on *random plant community assemblages*. The extremes of the diversity gradient (i.e. either monocultures or the highest diversity level) then serve as baselines for comparisons of diversity effects on ecosystem functioning (Hector et al., 1999) and food web structure (Ebeling et al., 2018; Rzanny and Voigt, 2012; Scherber et al., 2010).

In contrast, complex but un-replicated food webs have been constructed, predominately for aquatic food web studies, with the goal of evaluating relationships between the diversity (network size) and stability of natural ecosystems (Martinez, 1991). In these cases, null models often serve as a baseline (Martinez, 1991; Williams and Martinez, 2000), and cross-system comparisons are used to speculate about generality of findings (Jacob et al., 2011). Two main classes of criticisms have emerged regarding these null model approaches. First, the nature of the null models have been debated (Gotelli and Graves, 1996). Completely random matrices are thought to be 'too null' to reflect any aspect of reality. Yet, the alternative models that contain some realism (i.e. cascade, niche, nested hierarchy) may overgeneralize food web structure rendering these null models insensitive to potential changes in trophic interactions (Fox, 2006). Second, lack of standardized methods, for species sampling and food-web construction, has resulted in questions about the suitability of original data for cross system comparisons (Pascual and Dunne, 2006). Considered together, these two critiques suggest that one-off comparisons of un-replicated food webs may be more useful for generating hypotheses than for testing them.

Using a more selective subset of cross-system comparisons, research in the science policy domain evaluates which non-random sets of species and their interactions will be vulnerable to anthropogenic drivers (Valiente-Banuet et al., 2014). Perspectives papers have suggested that undisturbed habitats serve as an ideal, but rarely realized, baseline for identifying human impacts on communities, food webs, and ecosystem functions (Dobson, 2005; Perino et al., 2019). Because we often lack information about pristine conditions, less disturbed neighbouring habitats are often used as baseline for comparative monitoring (Knowlton and Jackson, 2008). Given that many ecosystems have high levels of natural temporal and spatial variation in diversity and functioning that cannot be attributed to human influences, establishing appropriate baselines that allow us to recognize non-random changes in food webs is a persistent challenge (Dobson, 2005; Eisenhauer et al., 2016).

Residual expectations of scientific training will dictate whether scientists prefer to consider if food web structure deviates from pristine conditions, null model matrices, or if they favour empirical assessments demonstrating how networks assemble on random plant communities. Integrative food-web research will build upon strengths of analyses developed in each domain. In particular we expect there to be increased efforts to test for the causes and consequences of interaction complexity in replicated gradients (i.e. BEF experiments, salinity, urbanization, restoration) (Hines et al., 2015b; Pellissier et al., 2018; Tylianakis and Morris, 2017). Such gradient based approaches allow for explicit tests of the focal driver using multiple baseline comparisons (i.e. experimental gradients, random matrices, and null models) (Giling et al., 2019; Hines et al., 2019). Without more systematic compilations of food webs, and cross-system comparisons that evaluate random and non-random changes in food web structure, debate will persist about the importance of food web structure for the functioning of ecosystems (Pascual and Dunne, 2006; Pellissier et al., 2018).

## 2.2 Scaling multi-trophic diversity

A second key aspect dividing research domains is the scale at which research is carried out (Table 1). One of the first multi-trophic BEF experiments was carried out in a small-scale, highly controlled Ecotron (Naeem et al., 1994). Subsequently, several multi-collaborator BEF experiments were established using field plots with sizes that were supposed to be large enough to host plot-specific consumer communities above- and below-ground (Borer et al., 2012; Ebeling et al., 2018; Koricheva et al., 2000). In these studies, plots size were intended to be large enough to reflect cycling of matter and energy in a closed system. However, a persistent challenge is that higher trophic levels typically have at least one life stage that is more mobile and uses resources at larger spatial scale than belowground species and lower trophic levels (De Deyn and van der Putten, 2005; Holt, 2002). Therefore, movement of organisms between plots may explain differences in responses between in silico simulations where a closed system is assumed (Stouffer et al., 2007), and open field plots where immigration and emigration are possible (Giling et al., 2019). To evaluate these differences requires explicit consideration of animal movement between plots as a mechanism influencing food web structure and ecosystem functioning (Jeltsch et al., 2013; Schmitz et al., 2018).

Research in Agricultural Landscapes and Aquatic Food Webs cannot be accurately pinned as being conducted at any given scale. This research is conducted in field plots and mesocosms that are more similar to spatial scale of BEF experiments, as well as at the landscape and catchment scale that is more typical of science policy endeavours (Tscharntke et al., 2005, 2012b). In these two integrative domains, heterogeneous habitat patches often are considered as important sources of biodiversity in landscapes (Polis et al., 1997). Mobile species can connect habitat patches, and locations of sedentary species may indicate more isolated food web compartments (Fuentes-Montemayor et al., 2017; Martinson et al., 2012).

Importantly, socio-political boundaries typically do not operate at the same spatial scales as food web compartments. For example, reduced trophic connections between species predominately dwelling in either benthic or pelagic zones can lead to compartmentalization in the Chesapeake Bay food web (Krause et al., 2003), but ecosystem management decisions are applied at watershed-scale rather than by water depth (Lefcheck et al., 2018; Leslie, 2018). Similarly, spatial compartmentalization of invertebrates in agricultural food web interactions can result from variation in fine-scale vegetation structure that influences predator movement in the centres and edges of agricultural fields (Macfadyen et al., 2011). Concerted effort to pair spatial structure of food webs with land-use recommendations can be used to manage ecosystem services, such as biological control of pests, but will require coordinated efforts by multiple land-owners (Landis, 2017). Accordingly, diverse stake-holders will need to recognize how small scale factors, which influence local food web interactions, will translate to enhanced ecosystem services in designed landscapes.

## 2.3 Currency across domains: Biomass, energy, valuation

The predominant metric or currency used to quantify the ecosystem implications of diversity change also vary across domains (Table 1). Early research in BEF focused on biomass, particularly of plants, as a proxy for productivity, a key aspect of ecosystem functioning (Tilman and Downing, 1994). More recently, there has been interest in assessing multivariate ecosystem responses to diversity change (Fig. 2). Consequently, quantification of energy flux through food webs has been proposed as a way to standardize the ecosystem consequences of species interactions for multiple ecosystem functions related to trophic interactions (Barnes et al., 2018; Gauzens et al., 2019). Energy flux between any two populations of organisms describes the feeding activity of the consumer on its resource. These fluxes of energy describe the weighted structure of food webs, and also the main function that each population in a food web carries out (de Ruiter et al., 1995; Neutel and Thorne, 2014). Calculations of energy flux have some cross domain appeal, and have been used to assess the effects of diversity change in aquatic, aboveground, and soil food webs (Barnes et al., 2014; de Vries et al., 2012; Hildrew et al., 2007).

Despite the appeal of standardized quantification, however, energy flux methods rarely are adopted by scientists focused on valuation of multiple ecosystem services in the science policy domain (Costanza et al., 2014). Instead, some ecosystem services like pollination and natural biological control have clear economic value to crop production. In such cases, monetary value can be translated directly from biomass estimates (IPBES, 2016; Naranjo et al., 2014). Nonetheless, non-monetary valuation is also important, and such assessments often focus on cultural values of humans rather than multi-trophic interactions (Hill et al., 2019; Hungate et al., 2017; IPBES, 2016). Multi-level networks that quantify energy flux as well as monetary and non-monetary value of ecosystem services will facilitate cross domain communication about ecosystem consequences of diversity changes in food webs (Dee et al., 2017; Potts et al., 2016).

## 3. Summary and outlook: towards integrative food-web ecology

Over the last three decades, thousands of scientists have written tens of thousands of words and papers about biodiversity and ecosystem functioning. The authors come from over 140 countries and present research from study systems that span the globe. One of the central problems that unites this diverse group of people is the desire to understand how we can live and grow in a way that will not have a negative influence on future generations. However, the juxtaposition of basic research conducted in BEF Experiments and applied research focused on ecosystem management has resulted in conceptual divisions among research in this field. Our keyword co-occurrence analysis suggests that food web research is poised to bridge conceptual divides. We reason that food web studies will be most effective when they specifically integrate perspectives from multiple domains (i.e. by including multiple baselines, spatial scales, or valuations). Failure to acknowledge differences across research domain will promote classic conceptual divides and limit progress towards solving central problems.

Here we briefly summarize changes in reasoning across domains in the biodiversity ecosystem functioning research landscape (Fig. 2). At each stage, we highlight a research outlook that will lead to more integrated food web ecology (Fig. 4). Calls for integrative food web ecology are not new (Hines et al., 2015b; Polis et al., 1997), but valuable insights can be gained by considering them together in light of conceptual divides among research domains.

Specifically, many classic BEF experiments are grounded by the idea that losses in biodiversity are happening faster now than ever before in the fossil record (Barnosky et al., 2011), but previous efforts have focused predominately on plants and plant productivity. Integrative approaches will assess *scenarios for multi-trophic diversity change* including, but not limited to, systematic evaluation of random and non-random species loss, species gains via invasion and range expansion, and turnover in species identity during community assembly (Fig. 4, column A). Such changes in species diversity also change the topology of trophic interaction, which influence species coexistence and stability (Koh et al., 2004). Integrative food web ecology will use standardized methods to construct food web topologies and evaluate *food web structural changes using multiple baselines* such as gradient based diversity change

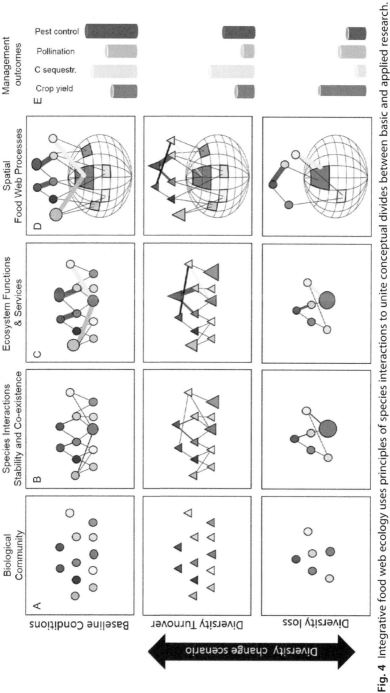

**Fig. 4** Integrative food web ecology uses principles of species interactions to unite conceptual divides between basic and applied research. (A) Changes in the composition of biological communities composed of primary producers (green), herbivores (blue), pollinators (yellow), predators (red) involve multiple types of deviations from baseline conditions (species turnover, species losses, and species gains). (B) Changes in community composition influence species interactions as well as (C) the flux of nutrients and energy through food webs. (D) Food web interactions occur in space, (E) which can be managed based on valuation of diversity change scenarios.

scenarios (Fig. 4 column B). Changes in species abundance and interaction topology will have implications for ecosystem functions and services that can be easily associated with trophic interactions (pest suppression, pollination, productivity) (Roslin and et al., 2017). Integrative food web approaches will quantify how diversity change scenarios alter the flux of nutrients and energy through food webs to influence multiple ecosystem functions and services (Fig. 4 column C). Notably, species interactions occur in space, and community assembly processes that influence ecosystem functioning include species movement among local habitat patches (immigration, emigration, foraging range) (Jeltsch et al., 2013) as well broader-scale environmental filtering (Baiser et al., 2019). Integrative food web research will test spatial mechanisms underlying empirical relationships between biodiversity and ecosystem functioning (Fig. 4 column D). Food webs assembled in space can then be used to evaluate the outcome of management decisions for diversity change and ecosystem service valuation by multiple-stake holders (Fig. 4 column E)(Chan et al., 2016; Díaz et al., 2015; Pascual et al., 2017). Each of these steps towards integrative food web ecology (Fig. 4 column A–E) bridges a conceptual divide between research conducted in the four research domains (BEF Experiments, Agricultural Landscapes, Aquatic Food Webs, and Science Policy). Additionally, the necessarily broad agenda reflects interests of scientists spanning the scientific landscape of biodiversity ecosystem functioning (Fig. 2).

In conclusion, changes in animal diversity are expected to influence ecosystem functioning and human well-being (Dirzo et al., 2014). Consumers dominate global biodiversity and drive several ecosystem processes (Wilson, 1987) and services (e.g. Potts et al., 2016; Soliveres et al., 2016). Moreover, higher trophic levels may be particularly sensitive to environmental variation and change (Hines et al., 2015a; Voigt et al., 2003). Therefore, we expect that food web approaches will continue to shape future dialogue about basic and applied biodiversity-ecosystem functioning research in the coming decades.

## Acknowledgements

This chapter was initiated at the 'Past, Present, and Future of Biodiversity and Ecosystem functioning workshop' in the frame of the Jena Experiment which is funding by the German Research Foundation (DFG, FOR 1451) with additional support from the Max Planck Society and the University of Jena. Further support to J.E., A.B., U.B., D.G., and N.E. came from the German Centre for Integrative Biodiversity Research (iDiv) Halle-Jena-Leipzig, funded by the German Research Foundation (FZT 118).

# References

Altieri, M.A., 1999. The ecological role of biodiversity in agroecosystems. Agric. Ecosyst. Environ. 74, 19–31.
Baiser, B., Gravel, D., Cirtwill, A.R., Dunne, J.A., Fahimipour, A.K., Gilarranz, L.J., Grochow, J.A., Li, D., Martinez, N.D., McGrew, A., Poisot, T., Romanuk, T.N., Stouffer, D.B., Trotta, L.B., Valdovinos, F.S., Williams, R.J., Wood, S.A., Yeakel, J.D., 2019. Ecogeographical rules and the macroecology of food webs. Glob. Ecol. Biogeogr. 1–15. https://doi.org/10.1111/geb.12925.
Balvanera, P., Pfisterer, A.B., Buchmann, N., He, J.-S., Nakashizuka, T., Raffaelli, D.G., Schmid, B., 2006. Quantifying the evidence for biodiversity effects on ecosystem functioning and services. Ecol. Lett. 9, 1146–1156.
Barnes, A.D., Jochum, M., Mumme, S., Haneda, N.F., Farajallah, A., Widarto, T.H., Brose, U., 2014. Consequences of tropical land use for multitrophic biodiversity and ecosystem functioning. Nat. Commun. 5, 5351.
Barnes, A.D., Jochum, M., Lefcheck, J.S., Eisenhauer, N., Scherber, C., O'Connor, M.I., de Ruiter, P.C., Brose, U., 2018. Energy flux: the link between multitrophic biodiversity and ecosystem functioning. Trends Ecol. Evol. 33, 186–197.
Barnosky, A.D., Matzke, N., Tomiya, S., Wogan, G.O.U., Swartz, B., Quental, T.B., Marshall, C., McGuire, J.L., Lindsey, E.L., Maguire, K.C., Mersey, B., Ferrer, E.A., 2011. Has the Earth's sixth mass extinction already arrived? Nature 471, 51–57.
Borer, E.T., Seabloom, E.W., Tilman, D., Novotny, V., 2012. Plant diversity controls arthropod biomass and temporal stability. Ecol. Lett. 15, 1457–1464.
Börner, K., Chen, C., Boyack, K.W., 2005. Visualizing knowledge domains. Annu. Rev. Inf. Sci. Technol. 37, 179–255.
Boyack, K.W., Klavans, R., Börner, K., 2005. Mapping the backbone of science. Scientometrics 64, 351–374.
Byrnes, J.E.K., Gamfeldt, L., Isbell, F., Lefcheck, J.S., Griffin, J.N., Hector, A., Cardinale, B.J., Hooper, D.U., Dee, L.E., Duffy, J.E., 2014. Investigating the relationship between biodiversity and ecosystem multifunctionality: challenges and solutions. Methods Ecol. Evol. 5, 111–124.
Cardinale, B.J., Duffy, J.E., Gonzalez, A., Hooper, D.U., Perrings, C., Venail, P., Narwani, A., Mace, G.M., Tilman, D., Wardle, D., Kinzig, A.P., Daily, G.C., Loreau, M., Grace, J.B., Larigauderie, A., Srivastava, D.S., Naeem, S., 2012. Biodiversity loss and its impact on humanity. Nature 486, 59–67.
Cardinale, B.J., Gonzalez, A., Allington, G.R.H., Loreau, M., 2018. Is local biodiversity declining or not? A summary of the debate over analysis of species richness time trends. Biol. Conserv. 219, 175–183.
Carpenter, S.R., Mooney, H.A., Agard, J., Capistrano, D., DeFries, R.S., Díaz, S., Dietz, T., Duraiappah, A.K., Oteng-Yeboah, A., Pereira, H.M., Perrings, C., Reid, W.V., Sarukhan, J., Scholes, R.J., Whyte, A., 2009. Science for managing ecosystem services: beyond the millenium ecosystem assessment. Proc. Natl. Acad. Sci. U. S. A. 106, 1305–1312.
Chan, K.M.A., Balvanera, P., Benessaiah, K., Chapman, M., Díaz, S., Gómez-Baggethun, E., Gould, R., Hannahs, N., Jax, K., Klain, S., Luck, G.W., Martín-López, B., Muraca, B., Norton, B., Ott, K., Pascual, U., Satterfield, T., Tadaki, M., Taggart, J., Turner, N., 2016. Opinion: why protect nature? Rethinking values and the environment. Proc. Natl. Acad. Sci. U. S. A. 113, 1462–1465.
Cobo, M.J., López-Herrera, A.G., Herrera-Viedma, E., Herrera, F., 2011. An approach for detecting, quantifying, and visualizing the evolution of a research field: a practical application to the fuzzy sets theory field. J. Informet. 5, 146–166.

Costanza, R., De Groot, R., Sutton, P., van der Ploeg, S., Anderson, S.J., Kubiszewski, I., Farber, S., Turner, R.K., 2014. Changes in the global value of ecosystem services. Glob. Environ. Chang. 26, 152–158.

De Deyn, G.B., van der Putten, W.H., 2005. Linking aboveground and belowground diversity. Trends Ecol. Evol. 20, 625–633.

de Ruiter, P.C., Neutel, A.M., Moore, J.C., 1995. Energetics, patterns of interaction strengths, and stability in real ecosystems. Science 269, 1257–1260.

de Vries, F.T., Liiri, M.E., Bjørnlund, L., Bowker, M.A., Christensen, S., Setälä, H., Bardgett, R.D., 2012. Land use alters the resistance and resilience of soil food webs to drought. Nat. Clim. Chang. 2, 276–280.

Dee, L., Allesina, S., Bonn, A., Eklöf, A., Gaines, S.D., Hines, J., Jacob, U., McDonald-Madden, E., Possingham, H., Schröter, M., Thompson, R.M., 2017. Operationalizing network theory for ecosystem service assessments. Trends Ecol. Evol. 32, 118–130.

Díaz, S., Fargione, J., Chapin III, F.S., Tilman, D., 2006. Biodiversity loss threatens human well-being. PLoS Biol. 4, e277.

Díaz, S., Demissew, S., Carabias, J., Joly, C., Lonsdale, M., Ash, N., Larigauderie, A., Adhikari, J.R., Arico, S., Báldi, A., Bartuska, A., Baste, I., Bilgin, A., Brondizio, E., Chan, K., Figueroa, V., Duraiappah, A., Fischer, M., Hill, R., Koetz, T., Leadley, P., Lyver, P., Mace, G.M., Martin-Lopez, B., Okumura, M., Pacheco, D., Pascual, U., Pérez, E., Reyers, B., Roth, E., Saito, O., Scholes, R., Sharma, N., Tallis, H., Thaman, R., Watson, R., Yahara, T., Hamid, Z.A., Akosi, C., Al-Hafedh, Y., Allahverdiyev, R., Amankwah, E., Asah, S.T., Asfaw, Z., Bartus, G., Brooks, L.A., Caillaux, J., Dalle, G., Darnaedi, D., Driver, A., Erpul, G., Escobar-Eyzaguirre, P., Failler, P., MokhtarFouda, A.M.M., Fu, B., Gundimeda, H., Hashimoto, S., Homer, F., Lavorel, S., Lichtenstein, G., Mala, W.A., Mandivenyi, W., Matczak, P., Mbizvo, C., Mehrdadi, M., Metzger, J.P., Mikissa, J.B., Moller, H., Mooney, H.A., Mumby, P., Nagendra, H., Nesshover, C., Oteng-Yeboah, A.A., Pataki, G., Roué, M., Rubis, J., Schultz, M., Smith, P., Sumaila, R., Takeuchi, K., Thomas, S., Verma, M., Yeo-Chang, Y., Zlatanova, D., 2015. The IPBES conceptual framework—connecting nature and people. Curr. Opin. Environ. Sustain. 14, 1–16.

Dirzo, R., Young, H.S., Galetti, M., Ceballos, G., Isaac, N.J., Collen, B., 2014. Defaunation in the anthropocene. Science 345, 401–406.

Dobson, A., 2005. Monitoring global rates of biodiversity change: challenges that arise in meeting the convention on biological diversity (CBD) 2010 goals. Philos. Trans. R. Soc. Lond. B Biol. Sci. 360, 229–241.

Dretske, F., 1981. Knowledge and the Flow of Information. MIT Press.

Ebeling, A., Pompe, S., Baade, S., Eisenhauer, N., Hillebrand, H., Proulx, R., Roscher, C., Schmid, B., Wirth, C., Weisser, W.W., 2014. A trait-based experimental approach to understand the mechanisms underlying biodiversity–ecosystem functioning relationships. Basic Appl. Ecol. 15, 229–240.

Ebeling, A., Hines, J., Hertzog, L., Lange, M., Meyer, S.T., Simons, N.K., Weisser, W.W., 2018. Plant diversity effects on arthropods and arthropod-dependent ecosystem functions in a biodiversity experiment. Basic Appl. Ecol. 26, 50–63.

Eisenhauer, N., Barnes, A., Cesarz, C., Craven, D., Ferlian, O., Gottschall, F., Hines, J., Sendek, A., Siebert, J., Thakur, M., T. M., 2016. Biodiversity-ecosystem function experiments reveal the mechanisms underlying the consequences of biodiversity change in real world ecosystems. J. Veg. Sci. 27, 1061–1070.

Eisenhauer, N., Schielzeth, H., Barnes, A.D., Barry, K.E., Bonn, A., Brose, U., Bruelheide, H., Buchmann, N., Buscot, F., Ebeling, A., Ferlian, O., Freschet, G.T., Giling, D.P., Hättenschwiler, S., Hillebrand, H., Hines, J., Isbell, F., Koller-France, E., König-Ries, B., de Kroon, H., Meyer, S.T., Milcu, A., Müller, J., Nock, C.A., Petermann, J.S., Roscher, C., Scherber, C., Scherer-Lorenzen, M., Schmid, B.,

Schnitzer, S.A., Schuldt, A., Tscharntke, T., Türke, M., van Dam, N.M., van der Plas, F., Vogel, A., Wagg, C., Wardle, D.A., Weigelt, A., Weisser, W.W., Wirth, C., Jochum, M., 2019. A multitrophic perspective on biodiversity–ecosystem functioning research. Adv. Ecol. Res. 61, 1–54.

Fox, J.W., 2006. Current food web models cannot explain the overall topological structure of observed food webs. Oikos 115, 97–109.

Fuentes-Montemayor, E., Watts, K., Macgregor, N.A., Lopez-Gellego, Z., Park, K.J., 2017. Species mobility and landscape context determines the importance of local and landscape-level attributes. Ecol. Appl. 27, 1541–1554.

Gauzens, B., Barnes, A., Jochum, M., Hines, J., Wang, S., Giling, D.P., Rosebaum, B., Brose, U., 2019. Fluxweb: an R package to easily estimate energy fluxes in food webs. Methods Ecol. Evol. 10, 270–279.

Giling, D., Ebeling, A., Eisenhauer, N., Meyer, S.T., Roscher, C., Rzanny, M., Voigt, W., Weisser, W.W., Hines, J., 2019. Plant diversity alters the representation of motifs in food webs. Nat. Commun. 10, 1226.

Gotelli, N.J., Graves, G.R., 1996. Null Models in Ecology. Smithsonian Institution Press, Washington and London, pp. 273–301.

Griscom, B.W., Adams, J., Ellis, P.W., Houghton, R.A., Lomax, G., Miteva, D.A., Schlesinger, W.H., Shoch, D., Siikamäki, J.V., Smith, P., Woodbury, P., Zganjar, C., Blackman, A., Campari, J., Conant, R.T., Delgado, C., Elias, P., Gopalakrishna, T., Hamsik, M., Herrero, M., Kiesecker, J., Landis, E., Laestadius, L., Leavitt, S.M., Minnemeyer, S., Polasky, S., Potapov, P., Putz, F.E., Sanderman, J., Silvius, M., Wollenberg, E., Fargione, J., 2017. Natural climate solutions. Proc. Natl. Acad. Sci. U. S. A. 114, 11645–11650.

Hatt, S., Boeraeve, F., Artru, S., Dufrêne, M., Francis, F., 2018. Spatial diversification of agroecosystems to enhance biological control and other regulating services: an agroecological perspective. Sci. Total Environ. 621, 600–611.

Hector, A., Schmid, B., Beierkuhnlein, C., Caldeira, M.C., Diemer, M., Dimitrakopoulos, P.G., Finn, J.A., Freitas, H., Giller, P.S., Good, J., Harris, R., Högberg, P., Huss-Danell, K., Joshi, J., Jumpponen, A., Körner, C., Leadley, P.W., Loreau, M., Minns, A., Mulder, C.P.H., O'Donovan, G., Otway, S.J., Pereira, J.S., Prinz, A., Read, D.J., Scherer-Lorenzen, M., Schulze, E.-D., Siamantziouras, A.-S.D., Spehn, E.M., Terry, A.C., Troumbis, A.Y., Woodward, F.I., Yachi, S., Lawton, J.H., 1999. Plant diversity and productivity experiments in European grasslands. Science 286, 1123–1127.

Hildrew, A.G., Raffaelli, D.G., Edmonds-Brown, R. (Eds.), 2007. Body Size: The Structure and Function of Aquatic Ecosystems. Cambridge University Press.

Hill, R., Nates-Parra, G., Quezada-Euán, J.J.G., Buchori, D., LeBuhn, G., Maués, M.M., et al., 2019. Biocultural approaches to pollinator conservation. Nat. Sustain. 2, 214–222.

Hines, J., Eisenhauer, N., Drake, B.G., 2015a. Inter-annual changes in detritus based food chains can enhance plant growth response to elevated atmospheric $CO_2$. Glob. Chang. Biol. 21, 4642–4650.

Hines, J., van der Putten, W.H., De Deyn, G.B., Wagg, C., Voigt, W., Mulder, C., Weisser, W.W., Engel, J., Melian, C., Scheu, S., Birkhofer, K., Ebeling, A., Scherber, C., Eisenhauer, N., 2015b. Towards an integration of biodiversity-ecosystem functioning and food web theory to evaluate relationships between multiple ecosystem services. Adv. Ecol. Res. 53, 161–199.

Hines, J., Giling, D., Rzanny, M., Voigt, W., Meyer, S.T., Weisser, W.W., Eisenhauer, N., Ebeling, A., 2019. A meta food web for invertebrate species collected in an experimental grassland. Ecology 100, e02679.

Hobbs, R.J., Higgs, E., Harris, J.A., 2009. Novel ecosystems: implications for conservation and restoration. Trends Ecol. Evol. 24, 599–605.

Holt, R.D., 2002. Food webs in space: on the interplay of dynamic instability and spatial processes. Ecol. Res. 17, 261–273.
Hooper, D.U., Chapin III, F.S., Ewel, J.J., Hector, A., Inchausti, P., Lavorel, S., Lawton, J.H., Lodge, D.M., Loreau, M., Naeem, S., Schmid, B., Setälä, H., Symstad, A.J., Vandermeer, J., Wardle, D.A., 2005. Effects of biodiversity on ecosystem functioning: a consensus of current knowledge. Ecol. Monogr. 75, 3–35.
Huang, Y., Chen, Y., Castro-Izaguirre, N., Baruffol, M., Brezzi, M., Lang, A., Li, Y., Härdtle, W., von Oheimb, G., Yang, X., Liu, X., Pei, K., Both, S., Yang, B., Eichenberg, D., Assmann, T., Bauhus, J., Behrens, T., Buscot, F., Chen, X.-Y., Chesters, D., Ding, B.-Y., Durka, W., Erfmeier, A., Fang, J., Fischer, M., Guo, L.-D., Guo, D., Gutknecht, J.L.M., He, J.-S., He, C.-L., Hector, A., Hönig, L., Hu, R.-Y., Klein, A.-M., Kühn, P., Liang, Y., Li, S., Michalski, S., Scherer-Lorenzen, M., Schmidt, K., Scholten, T., Schuldt, A., Shi, X., Tan, M.-Z., Tang, Z., Trogisch, S., Wang, Z., Welk, E., Wirth, C., Wubet, T., Xiang, W., Yu, M., Yu, X.-D., Zhang, J., Zhang, S., Zhang, N., Zhou, H.-Z., Zhu, C.-D., Zhu, L., Bruelheide, H., Ma, K., Niklaus, P.A., Schmid, B., 2018. Impacts of species richness on productivity in a large-scale subtropical forest experiment. Science 362, 80–83.
Hungate, B.A., Barbier, E.B., Ando, A.W., Marks, S.P., Reich, P.B., van Gestel, N., Tilman, D., Knops, J.M.H., Hooper, D.U., Butterfield, B.J., Cardinale, B.J., 2017. The economic value of grassland species for carbon storage. Sci. Adv. 3, e1601880.
Huston, M.A., 1997. Hidden treatments in ecological experiments: re-evaluating the ecosystem function of biodiversity. Oecologia 110, 449–460.
IPBES, 2016. The Assessment Report of the Intergovernmental Science-Policy Platform on Biodiversity and Ecosystem Services on Pollinators, Pollination and Food Production. Secretariat of the Intergovernmental Science-Policy Platform on Biodiversity and Ecosystem ServicesBonn, Germany.
IPBES, 2019. Global assessment report on biodiversity and ecosystem services. Intergovernmental Science-Policy Platform on Biodiversity and Ecosystem Services (IPBES).
Jacob, U., Thierry, A., Brose, U., Arntz, W.E., Berg, S., Brey, T., Fetzer, I., Jonsson, T., Mintenbeck, K., Möllmann, C., Petchey, O.L., Riede, J.O., Dunne, J.A., 2011. The role of body size in complex food webs: a cold case. Adv. Ecol. Res. 45, 181–223.
Jeltsch, F., Bonte, D., Pe'er, G., Reineking, B., Leimgruber, P., Balkenhol, N., Schröder, B., Buchmann, C.M., Mueller, T., Blaum, N., Zurell, D., Böhning-Gaese, K., Wiegand, T., Eccard, J.A., Hofer, H., Reeg, J., Eggers, U., Bauer, S., 2013. Integrating movement ecology with biodiversity research—exploring new avenues to address spatiotemporal biodiversity dynamics. Mov. Ecol. 1, 6.
Knowlton, N., Jackson, J.B.C., 2008. Shifting baselines, local impacts, and global change on coral reefs. PLoS Biol. 6, e54.
Koh, L.P., Dunn, R.R., Sodhi, N.S., Colwell, R.K., Proctor, H.C., Smith, V.S., 2004. Species co-existence and the biodiversity crisis. Science 305, 1632–1634.
Koricheva, J., Mulder, C.P.H., Schmid, B., Joshi, J., Huss-Danell, K., 2000. Numerical responses of different trophic groups of invertebrates to manipulations of plant diversity in grasslands. Oecologia 125, 271–282.
Krause, A.E., Frank, K.A., Mason, D.M., Ulanowicz, R.E., Taylor, W.W., 2003. Compartments revealed in food-web structure. Nature 426, 282–285.
Landis, D.A., 2017. Designing agricultural landscapes for biodiversity-based ecosystem services. Basic Appl. Ecol. 18, 1–12.
Lefcheck, J.S., Orth, R.J., Dennison, W.C., Wilcox, D.J., Murphy, R.R., Keisman, J., Gurbisz, C., Hannam, M., Landry, J.B., Moore, K.A., Patrick, C.J., Testa, J., Weller, D.E., Batiuk, R.A., 2018. Long-term nutrient reductions lead to the unprecedented recovery of a temperate coastal region. Proc. Natl. Acad. Sci. U. S. A. 115, 3658–3662.

Leslie, H., 2018. Value of ecosystem-based management. Proc. Natl. Acad. Sci. U. S. A. 115, 3518–3520.
Lichtenberg, E.M., Kennedy, C.M., Kremen, C., Batáry, P., Berendse, F., Bommarco, R., Bosque-Pérez, N.A., Carvalheiro, L.G., Snyder, W.E., Williams, N.M., Winfree, R., Klatt, B.K., Åström, S., Benjamin, F., Brittain, C., Chaplin-Kramer, R., Clough, Y., Danforth, B., Diekötter, T., Eigenbrode, S.D., Ekroos, J., Elle, E., Freitas, B.M., Fukuda, Y., Gaines-Day, H.R., Grab, H., Gratton, C., Holzschuh, A., Isaacs, R., Isaia, M., Jha, S., Jonason, D., Jones, V.P., Klein, A.-M., Krauss, J., Letourneau, D.K., Macfadyen, S., Mallinger, R.E., Martin, E.A., Martinez, E., Memmott, J., Morandin, L., Neame, L., Otieno, M., Park, M.G., Pfiffner, L., Pocock, M.J.O., Ponce, C., Potts, S.G., Poveda, K., Ramos, M., Rosenheim, J.A., Rundlöf, M., Sardiñas, H., Saunders, M.E., Schon, N.L., Sciligo, A.R., Sidhu, C.S., Steffan-Dewenter, I., Tscharntke, T., Veselý, M., Weisser, W.W., Wilson, J.K., Crowder, D.W., 2017. A global synthesis of the effects of diversified farming systems on arthropod diversity within fields and across agricultural landscapes. Glob. Chang. Biol. 23, 4946–4957.
Mace, G.M., Baillie, J.E.M., 2007. The 2010 biodiversity indicators: challenges for science and policy. Conserv. Biol. 21, 1406–1413.
Macfadyen, S., Gibson, R.H., Symondson, W.O.C., Memmott, J., 2011. Landscape structure influences modularity patterns in farm food webs: consequences for pest control. Ecol. Appl. 21, 516–524.
Manning, P., Loos, J., Barnes, A.D., Batáry, P., Bianchi, F.J.J.A., Buchmann, N., De Deyn, G.B., Ebeling, A., Eisenhauer, N., Fischer, M., Fründ, J., Grass, I., Isselstein, J., Jochum, M., Klein, A.M., Klingenberg, E.O.F., Landis, D.A., Lepš, J., Lindborg, R., Meyer, S.T., Temperton, V.M., Westphal, C., Tscharntke, T., 2019. Transferring biodiversity-ecosystem function research to the management of 'real-world' ecosystems. Adv. Ecol. Res. 61, 323–356.
Martinez, N.D., 1991. Artifacts or attributes—effects of resolution on the little-rock lake food web. Ecol. Monogr. 61, 367–392.
Martinson, H.M., Fagan, W.F., Denno, R.F., 2012. Critical patch sizes for food-web modules. Ecology 93, 1779–1786.
McDonald-Madden, E., Sabbadin, R., Game, E.T., Baxter, P.W.J., Chadès, I., Possingham, H.P., 2016. Using food-web theory to conserve ecosystems. Nat. Commun. 7, 10245.
MEA, 2005. Millennium Ecosystem Assessment: Ecosystems and Human Well-being: Synthesis. Island Press, Washington, DC.
Meyer, S.T., Ptacnik, R., Hillebrand, H., Bessler, H., Buchmann, N., Ebeling, A., Eisenhauer, N., Engels, C., Fischer, M., Halle, S., Klein, A.M., Oelmann, Y., Roscher, C., Rottstock, T., Scherber, C., Scheu, S., Schmid, B., Schulze, E.D., Temperton, V.M., Tscharntke, T., Voigt, W., Weigelt, A., Wilcke, W., Weisser, W.W., 2018. Biodiversity-multifunctionality relationships depend on identity and number of measured functions. Nat. Ecol. Evol. 2, 44–49.
Mihoub, J.-B., Henle, K., Titeux, N., Brotons, L., Brummitt, N.A., Schmeller, D.S., 2017. Setting temporal baselines for biodiversity: the limits of available monitoring data for capturing the full impact of anthropogenic pressures. Sci. Rep. 7, 41591.
Mulder, C., Bennett, E.M., Bohan, D.A., Bonkowski, M., Carpenter, S.R., Chalmers, R., Cramer, W., Durance, I., Eisenhauer, N., Haughton, A.J., Hettelingh, J.-P., Hines, J., Huston, M.A., Jeppesen, E., Krumins, J.A., Ma, A., Mace, G., Mancinelli, M.G., McLaughlin, Ó., Naeem, S., Pascual, U., Peñuelas, J., Pettorelli, N., Pocock, M.J.O., Raffaelli, D., Rasmussen, J.J., Rusch, G.M., Scherber, C., Setälä, H.H., Vacher, C., Voigt, W., Vonk, J.A., Wood, S.A., Woodward, G., 2015. 10 years later: networking 35 priorities for science and society after the millennium assessment. Adv. Ecol. Res. 53, 1–53.

Naeem, S., Thompson, L.J., Lawler, S.P., Lawton, J.H., Woodfin, R.M., 1994. Declining biodiversity can alter the performance of ecosystems. Nature 368, 734–737.
Naeem, S., Bunker, D.E., Hector, A., Loreau, M., Perrings, C., 2009. Biodiversity, Ecosystem Functioning, and Human Well-being. Oxford University Press.
Naeem, S., Duffy, J.E., Zavaleta, E., 2012. The functions of biological diversity in an age of extinction. Science 336, 1401–1406.
Naranjo, S.E., Ellsworth, P.C., Frisvold, G.B., 2014. Economic value of biological control in integrated pest management of managed plant systems. Annu. Rev. Entomol. 60, 621–645.
Nelson, E., Mendoza, G., Regetz, J., Polasky, S., Tallis, H., Richard, C.D., Chan, K.M.A., Daily, G.C., Goldstein, J., Kareiva, P.M., Lonsdorf, E., Naidoo, R., Ricketts, T.H., Shaw, M.R., 2009. Modeling multiple ecosystem services, biodiversity conservation, commodity production, and tradeoffs at landscape scales. Front. Ecol. Environ. 7, 4–11.
Neutel, A.-M., Thorne, M.A.S., 2014. Interaction strengths in balanced carbon cycles and the absence of a relation between ecosystem complexity and stability. Ecol. Lett. 17, 651–661.
Newman, M.E.J., 2004. Fast algorithm for detecting community structure in networks. Phys. Rev. E 69, 066133.
Noack, A., 2007. Energy models for graph clustering. J. Graph Algorithms Appl. 11, 453–480.
Paquette, A., Hector, A., Castagneyrol, B., Vanhellemont, M., Koricheva, J., Scherer-Lorenzen, M., Verheyen, K., TreeDivNet, 2018. A million and more trees for science. Nat. Ecol. Evol. 2, 763–766.
Pascual, M., Dunne, J.A., 2006. Ecological Networks: Linking Structure to Dynamics in Food Webs. Oxford University Press, New York, New York.
Pascual, U., Balvanera, P., Díaz, S., Pataki, G., Roth, E., Stenseke, M., Watson, R.T., Dessane, E.B., Islar, M., Kelemen, E., Maris, V., Quaas, M., Subramanian, S.M., Wittmer, H., Adlan, A., Ahn, S.E., Al-Hafedh, Y., Amankwah, E., Asah, S.T., Berry, P., Bilgin, A., Breslow, S.J., Bullock, C., Cáceres, D., Daly-Hassen, H., Figueroa, E., Golden, C.D., Gómez-Baggethun, E., González-Jiménez, D., Houdet, J., Keune, H., Kumar, R., Ma, K., May, P.H., Mead, A., O'Farrell, P., Pandit, R., Pengue, W., Pichis-Madruga, R., Popa, F., Preston, S., Pacheco-Balanza, D., Saarikoski, H., Strassburg, B.B., Marjanvan den Belt, M., Verma, M., Wickson, F., Yagi, N., 2017. Valuing natures contributions to people: the IPBES approach. Curr. Opin. Environ. Sustain. 26, 7–16.
Pellissier, L., Albouy, C., Bascompte, J., Farwig, N., Graham, C., Loreau, M., Maglianesi, M.A., Melián, C., Pitteloud, C., Roslin, T., Rohr, R., Saavedra, S., Thuiller, W., Woodward, G., Zimmermann, N.E., Gravel, D., 2018. Comparing species interaction networks along environmental gradients. Biol. Rev. 93, 785–800.
Pereira, H.M., Ferrier, S., Walters, M., Geller, G.N., Jongman, R.H.G., Scholes, R.J., Bruford, M.W., Brummitt, N., Butchart, S.H.M., Cardoso, A.C., Coops, N.C., Dulloo, E., Faith, D.P., Freyhof, J., Gregory, R.D., Heip, C., Höft, R., Hurtt, G., Jetz, W., Karp, D.S., McGeoch, M.A., Obura, D., Onoda, Y., Pettorelli, N., Reyers, B., Sayre, R., Scharlemann, J.P.W., Stuart, S.N., Turak, E., Walpole, M., Wegmann, M., 2013. Essential biodiversity variables, a global system of harmonized observation is needed to inform scientist and policy-makers. Science 339, 277–278.
Perino, A., Pereira, H.M., Navarro, L., Néstor Fernández, N., James, J.M., Bullock, M., Silvia Ceaușu, S., Ainara Cortés-Avizanda, A., van Klink, R.R., Kuemmerle, T., Lomba, A., Pe'er, G., Plieninger, T., Benayas, J.M.R., Sandom, C., Svenning, J.-C.C., Wheeler, H., 2019. Rewilding complex ecosystems. Science 364 eaav5570.

Polis, G.A., Anderson, W.B., Holt, R.D., 1997. Toward an integration of landscape and food web ecology: the dynamics of spatially subsidized food webs. Annu. Rev. Ecol. Syst. 28, 289–316.

Potts, S.G., Imperatriz-Fonseca, V., Ngo, H.T., Aizen, M.A., Biesmeijer, J.C., Breeze, T.D., Dicks, L.V., Garibaldi, L.A., Hill, R., Settele, J., Vanbergen, A.J., 2016. Safeguarding pollinators and their values to human well-being. Nature 540, 220–229.

Roscher, C., Schumacher, J., Baade, J., Wilke, W.W., Gleixner, G., Weisser, W.W., Schmid, B., Schulze, E.D., 2004. The role of biodiversity for element cycling and trophic interactions: an experimental approach in a grassland community. Basic Appl. Ecol. 5, 107–121.

Roslin, T., et al., 2017. Higher predation risk for insect prey at low latitudes and elevations. Science 356, 742–744.

Rzanny, M., Voigt, W., 2012. Complexity of multitrophic interactions in a grassland ecosystem depends on plant species diversity. J. Anim. Ecol. 81, 614–627.

Scherber, C., Eisenhauer, N., Weisser, W.W., Schmid, B., Voigt, W., Fischer, M., Schulze, E.D., Roscher, C., Weigelt, A., Allan, E., Bessler, H., Bonkowski, M., Buchmann, N., Buscot, F., Clement, L.W., Ebeling, A., Engels, C., Halle, S., Kertscher, I., Klein, A.M., Koller, R., König, S., Kowalski, E., Kummer, V., Kuu, A., Lange, M., Lauterbach, D., Middelhoff, C., Migunova, V.D., Milcu, A., Müller, R., Partsch, S., Petermann, J.S., Renker, C., Rottstock, T., Sabais, A., Scheu, S., Schumacher, J., Temperton, V.M., Tscharntke, T., 2010. Bottom-up effects of plant diversity on multitrophic interactions in a biodiversity experiment. Nature 468, 553–556.

Schmitz, O.J., Wilmers, C.C., Leroux, S.J., Doughty, C.E., Atwood, T.B., Galetti, M., Davies, A.B., Goetz, S.J., 2018. Animals and the zoogeochemistry of the carbon cycle. Science 362, 1127.

Sharp, R., Tallis, H.T., Ricketts, T., Guerry, A.D., Wood, S.A., Chaplin-Kramer, R., Nelson, E., Ennaanay, D., Wolny, S., Olwero, N., Vigerstol, K., Pennington, D., Mendoza, G., Aukema, J., Foster, J., Forrest, J., Cameron, D., Arkema, K., Lonsdorf, E., Kennedy, C., Verutes, G., Kim, C.K., Guannel, G., Papenfus, M., Toft, J., Marsik, M., Bernhardt, J., Griffin, R., Glowinski, K., Chaumont, N., Perelman, A., Lacayo, M., Mandle, L., Hamel, P., Vogl, A.L., Rogers, L., Bierbower, W., 2015. InVEST user's guide. In: The Natural Capital Project. Stanford University, University of Minnesota, The Nature Conservancy, World Wildlife Fund.

Simons, N.K., Weisser, W.W., 2017. Agricultural intensification without biodiversity loss is possible in grassland landscapes. Nat. Ecol. Evol. 1, 1136–1145.

Soliveres, S., et al., 2016. Biodiversity at multiple trophic levels is needed for ecosystem multifunctionality. Nature 536, 456–459.

Stouffer, D.B., Camacho, J., Jiang, W., Amaral, L.A.N., 2007. Evidence for the existence of a robust pattern of prey selection in food webs. Proc. R. Soc. B 22, 1931–1940.

Suding, K.N., Higgs, E., Palmer, M.A., Callicott, J.B., Baker, M., Gutrich, J.J., Hondula, K.L., LaFevor, M.C., Larson, B.M.H., Randall, A., Ruhl, J.B., Schwartz, K.Z.S., 2015. Committing to ecological restoration. Science 348, 638–640.

TEEB Foundations, 2010. The Economics of Ecosystems and Biodiversity: Ecological and Economic Foundations. A Banson Production, London and Washington.

Thompson, R.M., Brose, U., Dunne, J.A., Hall, R.O.J., Hladyz, S., Kitching, R.L., Martinez, N.D., Rantala, H., Romanuk, T.N., Stouffer, D.B., Tylianakis, J.M., 2012. Food webs: reconciling the structure and function of biodiversity. Trends Ecol. Evol. 27, 689–697.

Tilman, D., Downing, J.A., 1994. Biodiversity and stability in grasslands. Nature 367, 363–365.

Tilman, D., Wedin, D., Knops, J., 1996. Productivity and sustainability influenced by biodiversity in grassland ecosystems. Nature 379, 718–720.
Tilman, D., Balzer, C., Hill, J., Befort, B.L., 2011. Global food demand and the sustainable intensification of agriculture. Proc. Natl. Acad. Sci. U. S. A. 108, 20260–20264.
Tscharntke, T., Klein, A.M., Kruess, A., Steffan-Dewenter, I., Thies, C., 2005. Landscape perspective on agricultural intensification and biodiversity-ecosystem service management. Ecol. Lett. 8, 857–874.
Tscharntke, T., Clough, Y., Wanger, T.C., Jackson, L., Motzke, I., Perfecto, I., Vandermeer, J., Whitbread, A., 2012a. Global food security, biodiversity conservation and the future of agricultural intensification. Biol. Conserv. 151, 53–59.
Tscharntke, T., Tylianakis, J.M., Rand, T.A., Didham, R.K., Fahrig, L., Batáry, P., Bengtsson, J., Clough, Y., Crist, T.O., Dormann, C.F., Ewers, R.W., Fründ, J., Holt, R.D., Holzschuh, A., Klein, A.M., Kleijn, D., Kremen, C., Landis, D.A., Laurance, W., Lindenmayer, D., Scherber, C., Sodhi, N., Steffan-Dewenter, I., Thies, C., van der Putten, W.H., Westphal, C., 2012b. Landscape moderation of biodiversity patterns and processes—eight hypotheses. Biol. Rev. 87, 661–685.
Tylianakis, J.M., Morris, R.J., 2017. Ecological networks across environmental gradients. Annu. Rev. Ecol. Evol. Syst. 48, 25–48.
Valiente-Banuet, A., Aizen, M.A., Alcántara, J.M., Arroyo, J., Cocucci, A., Galetti, M., García, M.B., García, D., Gómez, J.M., Jordano, P., Medel, R., Navarro, L., Obeso, J.R., Oviedo, R., Ramírez, N., Rey, P.J., Traveset, A., Verdú, M., Zamora, R., 2014. Beyond species loss: the extinction of ecological interactions in a changing world. Funct. Ecol. 29, 299–307.
Van Eck, N.J., Waltman, L., 2010. Software survey: VOS viewer, a computer program for bibliometric mapping. Scientometrics 84, 523–538.
Van Eck, J.J., Waltman, L., 2014. Visualizing bibliometric networks. In: Ding, Y., Rousseau, R., Wolfram, D. (Eds.), Measuring Scholarly Impact: Methods and Practice. Springer, pp. 285–320.
Verheyen, K., Vanhellemont, M., Auge, H., Baeten, L., Baraloto, C., Barsoum, N., Bilodeau-Gauthier, S., Bruelheide, H., Castagneyrol, B., Godbold, D., Haase, J., Hector, A., Jactel, H., Koricheva, J., Loreau, M., Mereu, S., Messier, C., Muys, B., Nolet, P., Paquette, A., Parker, J., Perring, M., Ponette, Q., Potvin, C., Reich, P., Smith, A., Weih, M., Scherer-Lorenzen, M., 2016. Contributions of a global network of tree diversity experiments to sustainable forest plantations. Ambio 45, 29–41.
Vitousek, P.M., Mooney, H.A., Lubchenco, J., Melillo, J.M., 1997. Human domination of earth's ecosystems. Science 277, 494–499.
Voigt, W., Perner, J., Davis, A.J., Eggers, T., Schmucher, J., Bahrmann, R., Fabian, B., Heinrich, W., Köhler, G., Lichter, D., Marstaller, R., Sander, F.W., 2003. Trophic levels are differentially sensitive to climate. Ecology 84, 244–2453.
Wardle, D., 2016. Do experiments exploring plant diversity-ecosystem functioning relationships inform how biodiversity loss impacts natural ecosystems? J. Veg. Sci. 27, 646–653.
Williams, R.J., Martinez, N.D., 2000. Simple rules yield complex food webs. Nature 404, 180–183.
Wilson, E.O., 1987. The little things that run the world (the importance and conservation of invertebrates). Conserv. Biol. 1, 344–346.
Wilson, E.O., 1999. The Diversity of Life. WW Norton & Company.
Worm, B., 2016. Averting a global fisheries disaster. Proc. Natl. Acad. Sci. U. S. A. 113, 4895–4897.
Worm, B., Barbier, E.B., Beaumont, N., Duffy, J.E., Folke, C., Halpern, B.S., Jackson, J.B.C., Lotze, H.K., Micheli, F., Palumbi, S.R., Sala, E., Selkoe, K.A., Stachowicz, J.J., Watson, R., 2006. Impact of biodiversity loss on ocean ecosystem services. Science 314, 787–790.

CHAPTER TEN

# Transferring biodiversity-ecosystem function research to the management of 'real-world' ecosystems

Peter Manning[a,*], Jacqueline Loos[b,c], Andrew D. Barnes[d,e,f],
Péter Batáry[g], Felix J.J.A. Bianchi[h], Nina Buchmann[i],
Gerlinde B. De Deyn[j], Anne Ebeling[k], Nico Eisenhauer[d,e],
Markus Fischer[l], Jochen Fründ[m], Ingo Grass[b], Johannes Isselstein[n],
Malte Jochum[d,e,l], Alexandra M. Klein[o], Esther O.F. Klingenberg[p],
Douglas A. Landis[q], Jan Lepš[r], Regina Lindborg[s], Sebastian T. Meyer[t],
Vicky M. Temperton[c], Catrin Westphal[u], Teja Tscharntke[b]

[a]Senckenberg Biodiversity and Climate Research Centre (SBiK-F), Frankfurt am Main, Germany
[b]Department of Crop Sciences, Division of Agroecology, University of Göttingen, Göttingen, Germany
[c]Faculty of Sustainability Science, Institute of Ecology, Leuphana University, Lüneburg, Germany
[d]German Centre for Integrative Biodiversity Research (iDiv) Halle-Jena-Leipzig, Leipzig, Germany
[e]Institute of Biology, Leipzig University, Leipzig, Germany
[f]School of Science, University of Waikato, Hamilton, New Zealand
[g]MTA Centre for Ecological Research, Institute of Ecology and Botany, Lendület Landscape and Conservation Ecology Research Group, Pest, Hungary
[h]Farming Systems Ecology, Wageningen University, Wageningen, Netherlands
[i]Department of Environmental Systems Science, ETH Zürich, Zürich, Switzerland
[j]Soil Biology Group, Wageningen University, Wageningen, Netherlands
[k]Institute of Ecology and Evolution, Friedrich Schiller University Jena, Jena, Germany
[l]Institute of Plant Sciences, University of Bern, Bern, Switzerland
[m]Department of Biometry and Environmental System Analysis, Albert-Ludwigs-University Freiburg, Freiburg, Germany
[n]Institute of Grassland Science, Georg-August-University Göttingen, Göttingen, Germany
[o]Nature Conservation and Landscape Ecology, Albert-Ludwigs-University Freiburg, Freiburg, Germany
[p]Department of Plant Ecology and Ecosystem Research, Georg-August University Göttingen, Göttingen, Germany
[q]Department of Entomology and Great Lakes Bioenergy Research Center, 204 Center for Integrated Plant System, Michigan State University, East Lansing, MI, United States
[r]Department of Botany, Faculty of Science, University of South Bohemia, Ceske Budejovice, Czech Republic
[s]Deptartment of Physical Geography, Stockholm University, Stockholm, Sweden
[t]Department of Ecology and Ecosystem Management, Technical University of Munich, Munich, Germany
[u]Functional Agrobiodiversity, Department of Crop Sciences, University of Göttingen, Göttingen, Germany
*Corresponding author: e-mail address: peter.manning@senckenberg.de

## Contents

1. Introduction — 324
2. Small-grain and highly-controlled experiments (Cluster A) — 326
   2.1 What can be transferred — 331

| | |
|---|---|
| 2.2 Barriers to transfer and directions for future research | 334 |
| 3. Small-grain studies with low experimental control (Cluster B) | 337 |
| 3.1 What can be transferred | 338 |
| 3.2 Barriers to transfer and directions for future research | 338 |
| 4. Large-grain studies without experimental control (Cluster C) | 340 |
| 4.1 What can be transferred | 341 |
| 4.2 Barriers to transfer and directions for future research | 342 |
| 5. Conclusion | 344 |
| Acknowledgements | 345 |
| References | 346 |

## Abstract

Biodiversity-ecosystem functioning (BEF) research grew rapidly following concerns that biodiversity loss would negatively affect ecosystem functions and the ecosystem services they underpin. However, despite evidence that biodiversity strongly affects ecosystem functioning, the influence of BEF research upon policy and the management of 'real-world' ecosystems, i.e., semi-natural habitats and agroecosystems, has been limited. Here, we address this issue by classifying BEF research into three clusters based on the degree of human control over species composition and the spatial scale, in terms of grain, of the study, and discussing how the research of each cluster is best suited to inform particular fields of ecosystem management. Research in the first cluster, small-grain highly controlled studies, is best able to provide general insights into mechanisms and to inform the management of species-poor and highly managed systems such as croplands, plantations, and the restoration of heavily degraded ecosystems. Research from the second cluster, small-grain observational studies, and species removal and addition studies, may allow for direct predictions of the impacts of species loss in specific semi-natural ecosystems. Research in the third cluster, large-grain uncontrolled studies, may best inform landscape-scale management and national-scale policy. We discuss barriers to transfer within each cluster and suggest how new research and knowledge exchange mechanisms may overcome these challenges. To meet the potential for BEF research to address global challenges, we recommend transdisciplinary research that goes beyond these current clusters and considers the social-ecological context of the ecosystems in which BEF knowledge is generated. This requires recognizing the social and economic value of biodiversity for ecosystem services at scales, and in units, that matter to land managers and policy makers.

## 1. Introduction

Widespread concerns over the consequences of global biodiversity loss led to an explosion of ecological research in the early 1990s into the relationship between biodiversity and the functioning of ecosystems (hereafter BEF research) (Schulze and Mooney, 1994; Loreau et al., 2001;

Hooper et al., 2005; Eisenhauer et al., 2019 this issue; Hines et al., 2019 this issue). Historically, most work in this field has been conducted in experimental settings, especially in grasslands, where extinction is simulated by randomly assembling plant communities differing in species and functional richness and where other environmental drivers of ecosystem function are controlled for (Hector et al., 1999; Tilman et al., 2001; Weisser et al., 2017). While this work has led to several robust conclusions regarding the form of biodiversity-function relationships and the mechanisms that drive them (Cardinale et al., 2012), there remain doubts regarding the capacity for experimental BEF research to inform the management of biodiversity and ecosystem functions and services in the 'real world' (i.e. ecosystems with communities that have not been experimentally manipulated) (Eisenhauer et al., 2016; Huston, 1997; Lepš, 2004; Srivastava and Vellend, 2005; Wardle, 2016). Much of this debate concerns the design of biodiversity experiments, which were established to investigate if biodiversity *could* affect function, and via what mechanisms (Loreau and Hector, 2001; Schmid et al., 2002; Tilman et al., 1996).

A more recent generation of BEF research has been conducted in non-experimental and naturally assembled real-world ecosystems such as natural and semi-natural (hereafter semi-natural) drylands, grasslands and forests (e.g. Maestre et al., 2012; Grace et al., 2016; Van Der Plas et al., 2016; Duffy et al., 2017; Fanin et al., 2018; Hautier et al., 2018, van der Plas, 2019). As they are performed in naturally assembled communities, shaped by both environmental drivers and global change factors, these studies are correlational and tend to rely upon statistical controls, thus limiting confident inference about the functional consequences of biodiversity loss in these systems. Removal experiments can help overcome this issue but, to date, relatively few have been conducted (Diaz et al., 2003; Fanin et al., 2018; Fry et al., 2013). While a lack of confident inference may limit transfer, many other knowledge gaps also limit the transferability of BEF research. For example, there is little consensus regarding on how strongly biodiversity loss affects ecosystem functioning, relative to other drivers (Duffy et al., 2017; Hooper et al., 2012; Srivastava and Vellend, 2005; van der Plas, 2019). Moreover, the functional consequences of the non-random extinction which occurs in semi-natural ecosystems have largely been estimated from correlational studies (Larsen et al., 2005; Duffy et al., 2017; van der Plas, 2019, but see Lyons and Schwartz, 2001 and Zavaleta and Hulvey, 2004). Further challenges in the knowledge transfer and application of BEF research emerge from a lack of information regarding the social and economic barriers to conserving biodiversity and promoting diversification (Fazey et al., 2013; Rosa-Schleich et al., 2019). Filling these knowledge gaps would help in providing reliable evidence to inform the management of the world's ecosystems, e.g., via the Intergovernmental Science-Policy Panel on Biodiversity and Ecosystem Services (IPBES) (Díaz et al., 2015; Díaz et al., 2018).

In this article, we review the current understanding of the BEF relationship and discuss how BEF research could inform the management of real-world ecosystems. We do this by assessing the suitability of current knowledge for transfer and how this is reflected in current applied research. We then identify barriers to transfer and expand on how these barriers can be overcome via future research and changes to knowledge exchange mechanisms. Throughout, we emphasize the transition of BEF research from a fundamental science to applied research that can inform management. By doing so we assume that the promotion of certain ecosystem services is desired (e.g. carbon storage or crop production).

To aid understanding of the potential transfer of BEF research, we classify it into three clusters based upon a) the degree of human control over the plant community, which in experiments manifests through removal of non-target species, and in real world ecosystems through management inputs, and b) the size of the study plots or area, i.e., grain (Fig. 1A). While these two axes represent continuous gradients, and some studies are difficult to classify, research within each cluster shares several features (described below), making a general critique possible. Furthermore, each of these clusters shares features with a subset of real-world ecosystems (e.g. similar levels of human control over plant community and the grain of management (Fig. 1B). Based on these similarities, we suggest possibilities and challenges for knowledge transfer and applications. We then identify future research needs (summarized in Table 1). Throughout our discussion, we focus on terrestrial ecosystems, particularly the role of plant diversity in grasslands and that of insects in agricultural landscapes. This focus is a result of our own expertise and the historical focus of much BEF research on these systems (Hines et al., 2019 this issue).

## 2. Small-grain and highly-controlled experiments (Cluster A)

Since the mid 90's, >600 experiments have been established to explore the causal relationship between biodiversity and ecosystem functioning (Cardinale et al., 2012), typically under field conditions (e.g. Hector et al., 1999; Roscher et al., 2004; Tilman et al., 1996). The primary goal of these

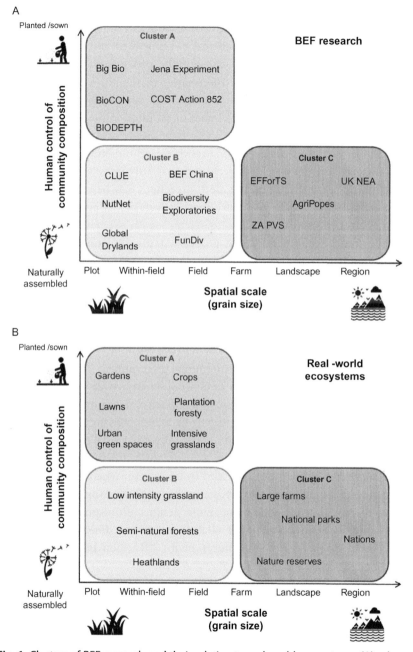

**Fig. 1** Clusters of BEF research and their relation to real world ecosystems. (A) selected research projects, (B) selected 'real-world' ecosystems. Note that, as spatial scale increases, the user of research findings changes from individual local scale managers to governments and institutions and the form of transfer changes from management
*(Continued)*

experiments was to establish whether biodiversity could affect ecosystem functioning, and so they controlled for potentially confounding effects of environmental conditions, functional composition, individual density, and non-random assembly and disassembly processes (Schmid et al., 2002, Schmid and Hector, 2004, Eisenhauer et al., 2019, this issue). To achieve this, BEF experiments apply a diversity treatment, where varying levels of plants species richness are sown or planted, and ecosystem functioning is measured (Schmid et al., 2002: Bruelheide et al., 2014). As such studies are highly controlled (e.g. via randomized blocking, weeding and the homogenisation of growing conditions), diversity effects may be ascribed with confidence and detailed inferences can be made regarding the identity of the mechanisms driving biodiversity effects (Loreau and Hector, 2001).

While these experiments act as model systems for BEF research, with generally applicable results to a wide range of systems (Eisenhauer et al., 2016; Schmid and Hector, 2004), the direct application of these insights in the management of real-world ecosystems could be limited for several reasons. First, the sown or planted community (and its species richness) is maintained through the repeated removal of non-target species, which typically does not occur in real-world systems. As a result, communities may be present that would not persist without human intervention. Second, the species richness gradient tends to span levels of diversity (typically 1- <20 plant species) that are much lower than many semi-natural communities (Wilson et al., 2012). Third, the studies tend to be conducted in replicated plots smaller than $500 \, m^2$ (Hector et al., 1999; Roscher et al., 2004; Tilman et al., 1996), with a median size of $3 \, m^2$ (Cardinale et al., 2012). As such studies are labour-intensive, they also tend to be unreplicated at the landscape scale (but see Hector et al., 1999; Kirwan et al., 2007). However, the large number of experiments with comparable designs allows meta-level, large extent analyses to be conducted (Balvanera et al., 2006; Isbell et al., 2015b; Lefcheck et al., 2015; Verheyen et al., 2016; Craven et al., 2018).

---

**Fig. 1—cont'd** practice recommendations to policy change, though these are clearly interrelated. Example references for the studies shown are: Jena experiment (Weisser et al., 2017), BigBio (Tilman et al., 2001), BioCON (Reich et al., 2001), COST Action 852 (Kirwan et al., 2007), BIODEPTH (Hector et al., 1999), BEF-China (Huang et al., 2018), CLUE (van der Putten et al., 2000), NutNet (Grace et al., 2016), Biodiversity Exploratories (Allan et al., 2015), Global Drylands (Maestre et al., 2012), FunDiv (van der Plas et al., 2016), EFForTS (Teuscher et al., 2016), AgriPopes (Emmerson et al., 2016), ZA PVS (Bretagnolle et al., 2018), UKNEA National Ecosystem Assessment (2011).

**Table 1** Research required to enable the real-world application of BEF research.

| Research need and approach | Potential benefit to transfer | Examples or foundational studies |
|---|---|---|
| Cluster A | | |
| Identify mechanistic general rules governing complementary species combinations in existing biodiversity experiments | Suggested combinations of species for restoration, intercropping and crop rotation, mixed plantations, etc. | Zuppinger-Dingley et al. (2014) and Brooker et al. (2015) |
| Demonstrate the biodiversity-multifunctionality relationship in sown or planted ecosystems, e.g., by identifying mixtures that provide multiple desired services | Could be used to design multifunctional species mixtures that provide benefits to a range of stakeholder groups | Baeten et al. (2019) and Finn et al. (2013) |
| Compare multispecies mixtures to the high performing species-poor systems of current management | Without realistic comparison to current management alternative option will not be adopted | Binder et al. (2018) |
| Perform BEF experiments with species pools that contain potentially useful and manageable species (e.g. self-sustaining mixtures) | High performing mixtures identified can be managed in a cost-effective manner | Kirwan et al. (2007) and Finn et al. (2013) |
| Generate measures of stability that are relevant to managers | To show relationship between biodiversity and the stability sought by stakeholders | Donohue et al. (2016) and Oliver et al. (2015) |
| Demonstrate the cost effectiveness of multispecies mixtures compared to existing management and develop technology that increases this (e.g. multicrop harvesters) | Unless clear benefits are demonstrated diversification may not be adopted | Finger and Buchmann (2015) and Blaauw and Isaacs (2014) |
| Cluster B | | |
| Form general predictions of how biodiversity and other drivers of ecosystem function changes in response to global change drivers | Accurate and general estimates and predictions of biodiversity loss are the foundation of accurate and general assessments of their impacts | Bjorkman et al. (2018) and Grace et al. (2016) |
| Develop mechanistic understanding of biodiversity in real world systems, e.g., by using new quantitative tools to disentangle biodiversity effects | Would increase confidence in correlational BEF relationships and allow their causes to be understood | Grace et al. (2016) |

*Continued*

**Table 1** Research required to enable the real-world application of BEF research.—cont'd

| Research need and approach | Potential benefit to transfer | Examples or foundational studies |
|---|---|---|
| Systematically assess the relative role of alpha and beta diversity, functional composition, abundance and other covariates including abiotic factors and understand the feedbacks and relationships between these drivers | Would lead to more precise estimates of the relative role of biodiversity in semi-natural systems and its relationship with other factors | Allan et al. (2015), Winfree et al. (2015) and van der Plas et al. (2016) |
| Establish a new generation of experiments that varies the above factors, across realistic gradients | Would allow causation to be inferred for the above relationships | Smith and Knapp (2003) and Manning et al. (2006) |
| Assess the role of biodiversity in species rich communities, including that of rare species | Most diversity loss occurs between high and intermediate levels and rare species are more likely to be lost | Soliveres et al. (2016a), Klein et al. (2003), and Lyons and Schwartz (2001) |
| Provide statistical estimates of where different components of biodiversity play their greatest role and test these estimates | Can be used in regional and global assessments and projections of the expected impacts of biodiversity loss | van der Plas (2019) |
| Explore the BEF relationship within the context of ecosystem restoration, and link this to community assembly mechanisms | The restoration of semi-natural habitats may be more effective if a high diversity of species is used | Bullock et al. (2007) and Weidlich et al. (2018) |
| Cluster C | | |
| Understand the strength and role of mechanisms linking biodiversity to ecosystem function at spatial and temporal scales (e.g. species matching to site conditions, dispersal processes) | Biodiversity may play a different role at large scales to that established in experiments | Loreau et al. (2003) and Mori et al. (2018) |
| Upscale ecosystem functions to large scales and link these to ecosystem services | Would allow the relationship between biodiversity, ecosystem functions and ecosystem services to be evaluated at management relevant scales | Clough et al. (2016), Lindborg et al. (2017) |
| Use upscaled measures to understand which taxa drive ecosystem services and disservices at landscape scales, and what factors drive the diversity of these taxa | Would allow important ecosystem service providers to be identified and managed appropriately | van der Plas et al. (2018), Winfree et al. (2018), and Grass et al. (2019) |

**Table 1** Research required to enable the real-world application of BEF research.—cont'd

| Research need and approach | Potential benefit to transfer | Examples or foundational studies |
| --- | --- | --- |
| Evaluate the role of biodiversity in driving landscape multifunctionality of ecosystem services (via upscaled measures) | Would allow the impact of biodiversity on a range of stakeholders and wider society to be communicated | van der Plas et al. (2018) and Manning et al. (2018) |
| Knowledge exchange (all clusters) | | |
| Disseminate research findings effectively (e.g. via web tools and demonstration sites). | Non-academic approaches are required for BEF research findings to reach potential end-users users | Activities of: Forum for the Future of Agriculture (FFA) (2019), European Landowners Organisation (ELO) (2019), F.R.A.N.Z. (2019), Conservation Evidence (2019) website, and RSPB Hope Farm (2019) |
| Work in collaboration with stakeholders to collect information on which ecosystem services are desired, at which different temporal and spatial scales, and their relative importance | This could inform applied BEF research, ensuring that it meets the needs of potential end-users | Geertsema et al. (2016) and Walter et al. (2017) |

## 2.1 What can be transferred

BEF experiments were designed to provide general mechanistic insights into the BEF relationship. Nevertheless, the close control of plant community composition and their low species diversity means that findings from BEF experiments are potentially transferable to highly managed ecosystems, e.g., intensive agricultural grasslands, plantation forestry, gardens, sown communities found in urban green spaces or ecosystems restored from a heavily degraded state (Fig. 1B). Such systems tend to be managed intensively and at small scales, e.g., via the application of selective herbicides, weeding and fertilisation. As these systems typically contain fewer species than most semi-natural ecosystems, we predict that cluster A BEF research is best able to inform work related to diversification, rather than the impacts of species loss. BEF experiment results suggest that diversification of such systems would lead to considerable gains in the supply of some ecosystem services, as numerous functions related to agricultural production and sustainability often increase with species diversity, including plant productivity, pollination, soil carbon storage and weed suppression (Isbell et al., 2017). Moreover, species-rich communities produce a more stable and constant yield (Craven et al., 2018; Isbell et al., 2015b), which may reduce risks to farmers (Finger and Buchmann, 2015).

Experimental results indicate that the benefits of diversification are greater when increasing diversity from low to intermediate levels (e.g. from 1 to 8 grassland species per m$^2$) than from medium to high (e.g. from 8 to 16), as the diversity-function relationship tends to saturate (Isbell et al., 2017). As species are typically grown in monocultures and in a wide range of low-diversity mixtures, data from these experiments can help to identify high performing species, but also high performing mixtures, for a range of ecosystem functions. Agronomists have conducted significant research on crop diversification for many years (Brooker et al., 2015; Vandermeer, 1992), and demonstrated that crop diversification can lead to various positive outcomes, such as increased primary crop yield and biocontrol (Iverson et al., 2014). Moreover, intercropping can improve yield stability (Raseduzzaman and Jensen, 2017), and more diverse mixtures of cover crops, especially those containing legumes, lead to multiple additional benefits (Blesh, 2018; Storkey et al., 2015), thus increasing their multifunctionality (defined here as ecosystem service multifunctionality, the co-supply of multiple ecosystem services relative to their human demand, Manning et al., 2018). Similarly, crop mixtures of multiple cultivars provide higher yields (Reiss and Drinkwater, 2018), and the mixing of rice varieties within a field reduces disease prevalence (Zhu et al., 2000). The frameworks and fundamental insights of BEF research may inform such research by identifying general rules governing complementary combinations of species and varieties (Brooker et al., 2015; Wright et al., 2017).

An additional benefit of BEF experiments is that they often provide information on a wider range of ecosystem services than many agricultural experiments and agronomic analyses, which tend to focus on yield and its sustainability, e.g., weed control and nutrient cycling (Meyer et al., 2018). Mixtures that promote the supply of multiple ecosystem services simultaneously may therefore be identified from BEF studies (Baeten et al., 2019; Storkey et al., 2015). Further evidence of existing BEF transfer comes from grassland studies, which indicate that there are multiple benefits of diversifying agroecosystems in terms of grass yield and reduced weed abundance (Finn et al., 2013). Studies have also shown that diverse grassland mixtures produce greater bioenergy yields (Khalsa et al., 2014; Tilman et al., 2006). However, another study of bioenergy production in grass mixtures showed that diverse mixtures were not more productive than currently used monocultures, thus showing that diversification might not always promote bioenergy production (Dickson and Gross, 2015). Even in the absence of positive impacts of diversity on productivity, other benefits may be realized; diverse bioenergy landscapes can promote the supply of other ecosystem services including greenhouse gas mitigation, pest suppression, pollination, and bird watching potential (Werling et al., 2014).

A number of other avenues of experimental BEF research have the capacity to inform the management of intensive systems. BEF experiments show that damage to plant growth and productivity from plant pathogens and pests is often weaker in more diverse communities, both aboveground (Civitello et al., 2015; Otway et al., 2005) and belowground (Maron et al., 2011; Schnitzer et al., 2011). Accordingly, information from BEF experiments on plant–soil feedbacks (e.g. Vogel et al., 2019 this issue) could potentially help to devise effective crop rotation sequences, e.g., by identifying consistent antagonistic or synergistic feedbacks between functional groups when grown together or in sequence (Barel et al., 2018; Ingerslew and Kaplan, 2018). The insights of BEF experiments are also applicable to gardens and green roof planting (Lundholm et al., 2010) and the restoration of highly degraded ecosystems. Here it may be possible to determine species mixtures or particular functional trait combinations, which, when sown or planted, deliver desired functions, such as soil aggregate stability and soil organic matter accumulation (Gould et al., 2016; Kollmann et al., 2016; Lange et al., 2015; Yang et al., 2019). In restoration, another promising approach would be to identify and sow mixtures of species that facilitate each other as this is a key mechanism underlying biodiversity effects in harsh environments (Wright et al., 2017). Finally, evidence from forests suggests that similar or higher amounts of timber production can be achieved in mixed plantations of native species compared to monocultures of plantation species, and that co-benefits, e.g., to biodiversity conservation, would also be realized (Gamfeldt et al., 2013; Huang et al., 2018; Hulvey et al., 2013; Pretzsch and Schütze, 2009). As with crops, the results of BEF studies can also be used to indicate the tree species mixtures that best achieve this multifunctionality (Baeten et al., 2019; Teuscher et al., 2016).

## 2.2 Barriers to transfer and directions for future research

While the plant communities of BEF experiments and human-dominated ecosystems share similarities, there are also marked differences. For instance, the species composition in BEF experiments is randomly assembled and they are usually performed in unfertilized, pesticide-free, unirrigated systems. In contrast, in intensively managed real-world systems, prior knowledge has led managers to select high performing, but often low diversity, mixtures by sowing and planting species that deliver high levels of desired services, and/or encouraging these via pesticide application, irrigation and fertilisation. The benefits of diversification therefore need to be demonstrated relative to these intensive low diversity communities, rather than the random low diversity assemblages found in BEF experiments. For example, in European grasslands farmers typically sow or maintain mixtures of a single grass, *Lolium perenne,* and a single legume, *Trifolium repens,* to which fertilizers are also applied (Peeters et al., 2014). Such a mixture clearly differs from the random species-poor mixtures of grassland biodiversity experiments. It is unclear if the relatively diverse and high-functioning communities of biodiversity experiments are generally able to deliver yield of a similar or higher quality, quantity and reliability. However, it has been demonstrated that diversification from 1–2 to 3–4 species provides significant increases in grassland yield and higher resistance to weed invasion (Finn et al., 2013; Kirwan et al., 2007; Nyfeler et al., 2009). We hypothesize that the species-poor communities found in intensively managed systems are more likely to resemble the high performing species-poor communities of BEF experiments (e.g. those dominated by tall grasses of fertile conditions) than the low performing communities, which may struggle to persist without regular weeding and close control (e.g. those containing only a few small herbs). In contrast, the low diversity situations found in experiments, where potentially dominant species are missing, could be relevant to isolated habitat patches, where species cannot disperse to potentially suitable conditions and the species pool is restricted.

As described above, current research suggests that links between BEF and agronomic research are beginning to emerge. However, current studies do not cover the wide range of situations in which diversification could be beneficial to agroecosystems. To the best of our knowledge, little work has yet made the transition to widespread adoption, an exception being the standard mixtures for forage production in Switzerland (see Fig. 2 for details), This lack of adoption highlights knowledge exchange as an important bottleneck

**Fig. 2** Swiss grassland diversification. In Switzerland species rich semi-natural grasslands (left) can decline to a more species-poor state (right) if fertilized and mown frequently. To counteract this loss many species rich sites are maintained via agri-environment policy schemes (Kampmann et al., 2012) and Swiss researchers have developed diversified seed mixtures suitable for a wide range of conditions that have been adopted by many Swiss farmers (Suter et al., 2017). We postulate that this adoption is likely to be attributable to a range of factors including: a strong cultural valuation of grassland, a clear mandate of agriculture to manage sustainably (in Swiss Constitution, article 104), generous agri-environment compensation schemes for many grassland types, and a strong focus on applied grassland research that has investigated which mixtures work over different time horizons (e.g. annual to permanent) and environmental conditions (moisture and elevational gradients) (e.g. Suter et al., 2015). Finally, there is effective communication from both researchers (e.g. Agroscope) and the Swiss grassland society (AGFF, 2019), which contains many farmers as members. Future BEF transfer work could investigate the role of such factors in successful transfer. Photo credits Peter Manning.

and another future need. To enable this, future BEF experiments could increase their relevance for management by drawing experimental communities from species pools that contain potentially useful and manageable species, and performing experiments in settings that are similar to those found in land use systems (e.g. fertilized or grazed grasslands). In this way, communities that are manageable and multifunctional may also be identified, and specific mixtures can be recommended (e.g. current policy in Switzerland). These should be cost-efficient and self-supporting and thus easily adapted and maintained by land managers.

Results on the relationship between biodiversity and the stability of ecosystem functions and services also require re-interpretation if they are to inform ecosystem management. While definitions of stability very greatly (Grimm and Wissel, 1997), BEF studies typically measure stability as the coefficient of variation (e.g. Craven et al., 2018; Knapp and van der Heijden, 2018), the resistance to perturbations, or the rate of recovery following these (Isbell et al., 2015b). In contrast, ecosystem managers often perceive stability differently (Donohue et al., 2016); while reliability is appreciated, and there are minimum levels of ecosystem service supply that are acceptable and over-performance (e.g. high productivity in favourable weather years, Wright et al., 2015) is often appreciated. Therefore, alternative measures of stability, e.g., that measure the number of years in which the supply of services exceed an acceptable threshold (Oliver et al., 2015), need to be employed if diversity-stability relationships are to be determined meaningfully for agroecosystems.

Finally, the transfer of BEF research findings to the real world may be limited by the uncertainties related to the profitability and management associated with diversifying species-poor communities and maintaining high species richness. For example, in many agricultural grasslands, plant species loss and dominance by a few nitrophilous species has occurred due to fertilisation (Gaujour et al., 2012; Gossner et al., 2016). Reducing nutrient availability and reversing these biodiversity declines can be difficult (Clark and Tilman, 2010; Smith et al., 2008; Storkey et al., 2015). Moreover, species-rich seed mixtures may prove expensive to create, and it remains to be seen if diverse and high functioning grasslands can be created and maintained cost-effectively over large areas. In croplands, multispecies mixtures might pose challenges to harvesting and sorting, as most modern agricultural machinery specializes in managing and cropping monocultures, and the harvesting of mixtures is relatively costly and labor-intensive (Magrini et al., 2018). We therefore need to know if, and under which conditions, encouraging diversity in agricultural systems is efficient and feasible, especially compared to management practices that deliver similar benefits (e.g. the promotion of productivity via diversification versus fertilisation) (Kleijn et al., 2018). A key part of this may be to acknowledge additional benefits of diversity (e.g. pest control, pollination or higher yield stability) and to factor this multifunctionality into comparisons. To better inform the management of agroecosystems and potentially lead to their diversification, a new generation of more applied and social-ecological BEF research is required (Geertsema et al., 2016). In this new work, comparisons should be made between the 'high performing low-diversity systems' that are the current norm and multifunctional 'sustainable high-diversity systems' that can be established and maintained at an equivalent cost to current systems, or which provide additional benefits that justify greater cost (e.g. carbon storage or avoided emissions) (Binder et al., 2018). Alternatively, evidence that high diversity systems can be intensified without negative environmental impacts could be sought, e.g., as demonstrated for biofuel grasslands (Yang et al., 2018). Clearly, such approaches require transdisciplinary research involving economic and/or multiple stakeholder-based assessments of the value of the diverse systems relative to current and future systems and practices (Jackson et al., 2012; Geertsema et al., 2016; Bretagnolle et al., 2018; Kleijn et al., 2018) (Table 1).

## 3. Small-grain studies with low experimental control (Cluster B)

The second cluster contains small-grain observational studies that investigate natural- or human-induced gradients of plant diversity in less intensively managed systems (e.g. Maestre et al., 2012; Soliveres et al., 2016b; van der Plas et al., 2016; Zhu et al., 2016) (Fig. 1). In this cluster, we also consider experiments in which particular species or functional groups are removed from intact ecosystems, often according to simulated global change scenarios (Cross and Harte, 2007; Fanin et al., 2018; Fry et al., 2013; Pan et al., 2016; Smith and Knapp, 2003; Suding et al., 2008), and those which boost diversity in established communities or disturbed sites, e.g., via seeding (Bullock et al., 2007; Stein et al., 2008; Van der Putten et al., 2000; Weidlich et al., 2018). Finally, we also consider global change driver experiments, where biodiversity change is treated as a co-variate and used to explain observed changes in function (e.g. Grace et al., 2016; Hautier et al., 2018). Plot sizes are similar to those in cluster A (i.e. $<500m^2$) and diversity levels vary greatly, from inherently species-poor ecosystems (e.g. Suding et al., 2008) to species-rich communities (Allan et al., 2015). Therefore, in contrast to most of the experiments of cluster A, studies from cluster B tend to contain more mature communities with higher species richness, fewer monocultures, less or no weeding, and species compositions and management regimes that are more similar to real-world low management intensity systems. In most of these studies, and in contrast to most BEF experiments that manipulate random community assembly, diversity loss occurs as non-random disassembly in response to environmental drivers. Observational studies of cluster B often statistically control for co-varying factors that may also drive ecosystem functions. These may include biotic covariates, such as functional composition and the abundance of different functional groups (Allan et al., 2015; Maestre et al., 2012; Soliveres et al., 2016a, 2016b; van der Plas et al., 2016), which strongly co-vary with diversity in many communities (Allan et al., 2015; Barnes et al., 2016; Soliveres et al., 2016a, 2016b).

The design of studies in this cluster limits interpretation about the cause of biodiversity effects as data for monoculture performances are usually unavailable, meaning that the mechanisms underlying biodiversity effects cannot be estimated (Loreau and Hector, 2001). This is unfortunate as these processes may differ in their strength compared to biodiversity experiments. For example, in mature communities, species may show higher levels of niche differentiation at both between and within species levels (Guimarães-Steinicke et al., 2019, this issue; Zuppinger-Dingley et al., 2014). A final property differentiating cluster B studies from those of cluster A is that variation in the diversity of other trophic levels is a complex product of responses to environmental drivers and concurrent changes in all trophic levels (Soliveres et al., 2016a, 2016b; Tscharntke et al., 2005), rather than primarily driven by variation in the diversity of primary producers (Scherber et al., 2010).

## 3.1 What can be transferred

Because they are conducted in unmanipulated real-world ecosystems, cluster B results are directly transferable to semi-natural ecosystems, which experience species loss and compositional change due to global environmental change. Cluster B studies provide direct estimates of the real-world impacts of global change drivers on diversity, and the corresponding impact of these changes on ecosystem function. However, most cluster B studies are observational, so patterns remain correlational, despite statistical controls. Nevertheless, due to their greater realism, syntheses of cluster B results (van der Plas, 2019), can provide statistical estimates of where different components of biodiversity play their greatest role, and estimates may be used as an evidence base for both local managers and in global assessments.

The experimental studies of cluster B can provide information on how diversification can boost ecosystem functioning in restored or enriched communities. For example, several studies show that sowing into intact communities can increase both species richness and ecosystem functioning, including community productivity and carbon storage (Bullock et al., 2007; Stein et al., 2008; Weidlich et al., 2018).

## 3.2 Barriers to transfer and directions for future research

For research in cluster B to become more directly transferable to the management of semi-natural ecosystems, greater confidence in the mechanisms underlying real-world BEF relationships is needed. While management recommendations may be drawn from selected case studies such as those presented above, a general understanding of the relative and interacting roles of environmental covariates, direct effects of global change drivers and various facets of diversity and compositional change is lacking (van der Plas, 2019). Biodiversity could play an important role in maintaining ecosystem function in real world ecosystems. Yet, whether loss of a few species at this scale makes a strong contribution to function, relative to these other drivers, has been only been tested in a limited number of cases (e.g. Allan et al., 2015; Grace et al., 2016; Manning et al., 2006; Winfree et al., 2015), and inconsistently, making generalisation difficult (van der Plas, 2019). To address this issue, observational studies need to ensure that factors such as abundance and functional composition are properly controlled for statistically. Predictions of the impacts of drivers on ecosystem services can be made by combining (a) estimates of expected biodiversity change according to different global change drivers across a range of conditions (e.g. Bjorkman et al., 2018; Grace et al., 2016; Hautier et al., 2018), (b) knowledge of how great a difference to functions and services such changes will make (e.g. Craven et al., 2018), and (c) ecosystem service production functions (Isbell et al., 2015a). This in turn allows for estimates of where ecosystem service-based arguments for conservation are strongest. Such predictions, if verified, could then form a sound basis for management decisions.

Transfer would also be enabled by a new generation of experiments. These could include a wider range of non-random extinction scenarios, assessments of the relative importance of abiotic drivers of function and biodiversity (e.g. Isbell et al., 2013; Manning et al., 2006), and the reduction of diversity from high to intermediate levels (Zobel et al., 1994), in order to verify, or refute the results of observational studies. To do this, manipulations such as the manipulation of dominance and functional composition, trait dissimilarity, or other aspects of biodiversity could be employed (Cross and Harte, 2007; Manning et al., 2006; Smith and Knapp, 2003). Manipulations that simulate the homogenisation of biota (i.e. the loss of beta diversity, while alpha diversity remains unchanged), may also prove informative, as this may be as, or more, common than alpha diversity loss in real-world ecosystems (Flohre et al., 2011; Vellend et al., 2013; Dornelas et al., 2014; Gossner et al., 2016; Wardle, 2016). Finally, it may be possible to link community assembly mechanisms (e.g. founder effects and habitat filtering) and functional BEF research to identify how to increase species richness and promote certain ecosystem functions, information that would be particularly useful in ecosystem restoration (Bullock et al., 2007; Kirmer et al., 2012; Stein et al., 2008; Weidlich et al., 2018) (Table 1).

Work is also needed in converting the measures of ecosystem function commonly taken in ecological studies into measures of ecosystem services that are of relevance to stakeholders (Mace et al., 2012; Kleijn et al., 2018). This requires the development of new metrics, e.g., trait measures that link to nutritional quality or cultural services such as aesthetic appeal. Applied studies could explicitly measure relevant ecosystem services, e.g., by involving stakeholders, assessing which services are most important to them, and adapting function measures to quantify these (King et al., 2015; Manning et al., 2018; Martín-López et al., 2012). This approach, and many of the others outlined above requires inter- and transdisciplinary research involving stakeholders and researchers from other disciplines, e.g., with farmers, local governments, agronomists and economists.

## 4. Large-grain studies without experimental control (Cluster C)

The third cluster (C) contains BEF studies that cover large areas (from 100 m$^2$ to landscapes) (e.g. Garibaldi et al., 2013; Larsen et al., 2005; Winfree et al., 2018). Due to the huge efforts required to manipulate diversity at a large spatial and temporal grain (Teuscher et al., 2016), such studies tend to be observational, comparative, and of low replication, although the large number of such studies has allowed for meta-level analyses to be conducted (Lichtenberg et al., 2017). The focal study organisms also tend to be invertebrates, particularly pollinators, instead of plants. The measurement of biodiversity (e.g. species richness and functional diversity) is also often limited in these studies due to the effort required to measure it directly over large areas. As a result, it is often landscape variables, such as landscape configuration and the proportion of different land uses that are related to function, rather than diversity (e.g. Bosem Baillod et al., 2017; Hass et al., 2018). These landscape properties may influence the dispersal, abundance and diversity of organisms within the landscape, and may also correlate with management factors and abiotic drivers of ecosystem function (Dominik et al., 2018; Gámez-Virués et al., 2015; Lindborg et al., 2017). As a result of these covariances, the role of biodiversity in driving ecosystem functioning cannot always be confidently ascribed (Tscharntke et al., 2016).

Within this cluster, we also place remote sensing studies (e.g. Oehri et al., 2017) and national and regional correlational studies (e.g. Anderson et al., 2009). In these, biodiversity can only be measured using proxies or with presence/absence data within large grid cells (e.g. 10 × 10 km), e.g., from national monitoring schemes. These coarse biodiversity measures are then correlated with ecosystem service proxy measures such as carbon storage and recreational use. These studies often lack a strong mechanistic basis, and focus instead on how biodiversity co-varies with ecosystem services (e.g. Anderson et al., 2009; Maskell et al., 2013). Even where covariates are included and mechanistic relationships postulated (e.g. Oehri et al., 2017; Duffy et al., 2017), causal links are hard to infer due to the strong covariance between biodiversity and other drivers, and the high probability of missing, or improperly measuring, important covariates.

Another common type of BEF study at this scale are those showing that functional biodiversity co-varies or differs across environmental gradients and management regimes (Gámez-Virués et al., 2015; Rader et al., 2014). While there is significant evidence that functional traits do relate to ecosystem processes and properties at landscape and national scales (e.g. Garibaldi et al., 2015; Lavorel et al., 2011; Manning et al., 2015), evidence for a mechanistic link between the functional diversity of traits to the supply of ecosystem services at these scales is generally limited.

## 4.1 What can be transferred

As the studies of cluster C are performed in real landscapes, and as management is often conducted at large scales (e.g. by farmers or foresters), research findings from this cluster are potentially of high relevance to policy and large-scale management, e.g., via payments for ecosystem service schemes. In recent years, a number of studies have demonstrated large-scale benefits of landscapes with high diversity of crops and non-crop habitats, which support higher biodiversity (Gardiner et al., 2009; Redlich et al., 2018). These benefits include more effective pollination and biological pest control (Garibaldi et al., 2013; Winfree et al., 2018). By showing how diversity and diversification practices influence ecosystem service delivery, these practices can then be incorporated into agronomic considerations (Rosa-Schleich et al., 2019) and into agri-environment policy (Garibaldi et al., 2014). Studies at this scale also complement those of the other clusters by showing that biodiversity not only promotes ecosystem function and services at the plot scale but also via spillover effects into the surrounding landscape, with ecosystem service benefits including pest suppression, pollination, and bird watching potential (Blitzer et al., 2012; Werling et al., 2014). However, biodiversity does not always promote function at these scales. For example, natural enemy diversity does not always relate to pest abundance, nor higher crop yields (Tscharntke et al., 2016), and in some cases biodiversity does not control pests as effectively as pesticides (Samnegard et al., 2019).

## 4.2 Barriers to transfer and directions for future research

The observational nature of most research in this cluster means that the exact role of diversity in driving ecosystem function and providing ecosystem services at these scales is hard to ascertain. This general limitation is compounded by several other barriers which can prevent transfer to landscape management and policy. First, several processes could drive BEF relationships at landscape scales that do not operate at the smaller grain size of clusters A and B, and as a result are little acknowledged in BEF research, outside of theory (Lindborg et al., 2017; Loreau et al., 2003; Tscharntke et al., 2012). These include the spatial processes that maintain diversity, the matching between species and environmental conditions in which they perform well (Leibold et al., 2017; Mori et al., 2018), and the potential for different species to provide different functions and services in different patches of the landscape, thus boosting landscape multifunctionality (van der Plas et al., 2016; van der Plas et al., 2019). The strength and role of such mechanisms clearly needs to be demonstrated. Another key problem in transferring BEF research to large scales is that landscape managers typically seek to simultaneously promote multiple ecosystem services, i.e., the multifunctionality of landscapes, not single ecosystem functions at the plot scale (Kremen and Merenlender, 2018; Manning et al., 2018). A focus on single functions is problematic if they trade-off and the components of diversity that boost some ecosystem services diminish others. For example, the maintenance of biodiversity-rich habitats may add resilience to multiple ecosystem functions at the landscape scale, but also occupies land that could be used for crop production.

New research approaches are required to overcome the difficulties in identifying how biodiversity controls ecosystem functioning at large scales, and how biodiversity may be conserved and promoted to increase the supply of ecosystem services. First, to ensure that service measures are of relevance to stakeholders, we require a better understanding of which services are demanded by different stakeholders, and at which different temporal and spatial scales, so that relevant indicator variables or ecosystem service production functions can be used (Tallis, 2011). A more holistic approach, which accounts for the relative demand for different ecosystem services and how this changes with socio-economic context, is therefore required, e.g., to assess how much land can be returned to a high biodiversity condition while maintaining desired levels of food production and other ecosystem services (Clough et al., 2011; Kremen and Merenlender, 2018; Manning et al., 2018). Such studies should also identify what drives patterns of land use and management and hence biodiversity loss, so that appropriate interventions can be identified (Grass et al., 2019).

To consider landscape multifunctionality and its dependence on biodiversity, multiple ecosystem services need to be scaled up in space and time, which is challenging. Some of the functions that can be measured at the plot scale can be 'linearly' scaled up, e.g., by using remote sensing proxies of diversity and functional traits, and interpolated maps, e.g., of climate and soil properties (Manning et al., 2015; van der Plas et al., 2018). Others, however, require an understanding of spatial interactions that makes their upscaling more complex, e.g., pollination and nutrient leaching (Koh et al., 2016; Lindborg et al., 2017). Furthermore, some services that operate at large scales (e.g. flood control, landscape aesthetics) cannot be predicted and scaled up from small-scale measures. Therefore, new procedures and methods are needed to quantify large-scale multifunctionality and the role of biodiversity in driving it. There have been calls for landscape-scale experiments to address these issues (Koh et al., 2009; Landis, 2017). One example is the recent EFForTS project in which "tree islands" of varying size and tree diversity (0–6 species) have been planted in oil-palm clearings (Teuscher et al., 2016). Initial results indicate no economic trade-off: the islands generate yield gains which compensate for the reduced number of oil palms (Gérard et al., 2017). However, the high financial cost and/or logistical effort of such experiments means it may be more realistic to use biophysical models in most cases. Unfortunately, such models do not currently fully represent the complexity of biodiversity or its relationship with ecosystem functions and services (Lavorel et al., 2017).

To understand biodiversity-landscape multifunctionality relationships, a greater knowledge of which aspects of diversity underpin different ecosystem services is also required. While knowledge exists regarding the drivers of many ecosystem service provider groups at the landscape scale (e.g. plants, birds, butterflies and pollinators, Roschewitz et al., 2005; Rösch et al., 2015; Kormann et al., 2015; Grab et al., 2019), this understanding needs to be extended to other groups, including soil microbes and soil fauna. Similarly, understanding of how spatial biodiversity dynamics affect functions and the services they underpin needs to be extended to taxa involved in services other than pest control and pollination (Table 1). In some cases, there may be trade-offs between services, e.g., if the conditions that maximize the diversity of one taxa do not favour another (van der Plas et al., 2019). This research may also demonstrate that when it comes to real-world ecosystem services and landscape-level multifunctionality, biodiversity effects are not easily generalizable, but depend on the context. Thus, the rules of this context-dependency need to be identified (Allan et al., 2015; Birkhofer et al., 2018; Samnegard et al., 2019). Doing this will limit uncertainty; managers could be less reluctant to manage for biodiversity when the degree to which it provides ecosystem service benefits at larger scales has been clearly demonstrated. In semi-natural ecosystems the promotion of the biodiversity components underpinning ecosystem services are most likely to be achieved via management options that are simple and effective over large areas, and so the practices that would promote the desired facets of biodiversity, e.g., mowing or the introduction of selective grazers, may need to be identified.

## 5. Conclusion

A vast array of BEF studies has taught us much about the complex relationship between biodiversity and ecosystem functioning. In this article, we argue that with some re-analysis and re-interpretation, some of this research could be directly transferred to policy and management, where practitioners could use its insights to guide the diversification of agricultural and other human-dominated ecosystems, and inform the conservation of biodiversity in semi-natural ecosystems. However, there are numerous challenges to the transfer of BEF research to more applied research and practice, and we argue that these challenges differ depending on the spatial grain of the study and the degree of community manipulation. While acknowledging the differences in transferability between these clusters of BEF research may help resolve the ongoing debate about relevance of BEF findings a new generation of BEF research is also required. This would involve the merging and connecting research between the current clusters, e.g., the setup of a new generation of biodiversity experiments that bridge the gap between current BEF experiments and observational studies. These should be complemented by new observational studies which more comprehensively account for covarying factors and which better acknowledge the link between ecosystem function and ecosystem services (Table 1).

It should be noted that the main message transferred from BEF research may simply be a stronger and more confident argument that it is important to conserve the diversity that is already present in semi-natural systems. In some cases BEF research may also show that not every species plays a positive or strong role in driving certain ecosystem functions, and that a small number of species dominate the supply of certain services (Kleijn et al., 2015). In such cases, acknowledging the non-market benefits of species and returning to more traditional ethical arguments will help promote biodiversity conservation (e.g. Hill et al., 2019).

Finally, to make BEF research more applied, large-scale studies that utilize novel approaches to investigate the role of diversity in providing the desired ecosystem services at the landscape scale are required (Table 1). Accordingly, key considerations in applied BEF research are to acknowledge when research is fundamental or applied, and to clarify when services, rather than functions, are being considered, thus making it transparent which services and functions are focal and why, and acknowledging which stakeholder groups may benefit. In many respects, the technical solutions to the challenges addressed in this article are already being investigated. However, if the potential for BEF research to address global challenges is to be fully realized, future BEF must also be transdisciplinary, and include the main stakeholders of the ecosystem collaboratively from their inception. By considering social-ecological context, BEF research should be better able to demonstrate the social and economic value of biodiversity at the scales that matter to land managers and policy makers.

## Acknowledgements

This work was funded by Deutsche Forschungsgemeinschaft; DFG, German Research Foundation Grant Ei 862/13 to MF, NB, AK, NE and TT. The Jena Experiment is funded by the Deutsche Forschungsgemeinschaft (DFG, German Research Foundation; FOR 1451), the Friedrich Schiller University Jena, the Max Planck Institute for Biogeochemistry in Jena, and the Swiss National Science Foundation. NE ADB and MJ acknowledge support by the German Centre for Integrative Biodiversity Research (iDiv) Halle-Jena-Leipzig (DFG FZT 118). DAL acknowledges support from Great Lakes Bioenergy Research Center, U.S. Department of Energy, Office of Science, Office of Biological and Environmental Research (Awards DE-SC0018409 and DE-FC02-07ER64494), by the National Science Foundation Long-term Ecological Research Program (DEB 1637653) at the Kellogg Biological Station, and by Michigan State University AgBioResearch. CW is grateful for funding by the Deutsche Forschungsgemeinschaft (DFG) (Project number 405945293).

## References

Arbeitsgemeinschaft zur Förderung des Futterbaues (AGFF), 2019. http://www.agff.ch/deutsch/aktuell.html.

Allan, E., Manning, P., Alt, F., Binkenstein, J., Blaser, S., Blüthgen, N., Böhm, S., Grassein, F., Hölzel, N., Klaus, V.H., Kleinebecker, T., 2015. Land use intensification alters ecosystem multifunctionality via loss of biodiversity and changes to functional composition. Ecol. Lett. 18, 834–843.

Anderson, B.J., Armsworth, P.R., Eigenbrod, F., Thomas, C.D., Gillings, S., Heinemeyer, A., Roy, D.B., Gaston, K.J., 2009. Spatial covariance between biodiversity and other ecosystem service priorities. J. Appl. Ecol. 46, 888–896.

Baeten, L., Bruelheide, H., van der Plas, F., Kambach, S., Ratcliffe, S., Jucker, T., Allan, E., Ampoorter, E., Barbaro, L., Bastias, C.C., Bauhus, J., 2019. Identifying the tree species compositions that maximize ecosystem functioning in European forests. J. Appl. Ecol. 56, 733–744.

Balvanera, P., Pfisterer, A.B., Buchmann, N., He, J.S., Nakashizuka, T., Raffaelli, D., Schmid, B., 2006. Quantifying the evidence for biodiversity effects on ecosystem functioning and services. Ecol. Lett. 9, 1146–1156.

Barel, J.M., Kuyper, T.W., de Boer, W., Douma, J.C., De Deyn, G.B., 2018. Legacy effects of diversity in space and time driven by winter cover crop biomass and nitrogen concentration. J. Appl. Ecol. 55, 299–310.

Barnes, A.D., Weigelt, P., Jochum, M., Ott, D., Hodapp, D., Haneda, N.F., Brose, U., 2016. Species richness and biomass explain spatial turnover in ecosystem functioning across tropical and temperate ecosystems. Philos. Trans. R. Soc., B 371, 20150279.

Binder, S., Isbell, F., Polasky, S., Catford, J.A., Tilman, D., 2018. Grassland biodiversity can pay. Proc. Natl. Acad. Sci. 115, 3876–3881.

Birkhofer, K., Andersson, G.K., Bengtsson, J., Bommarco, R., Dänhardt, J., Ekbom, B., Ekroos, J., Hahn, T., Hedlund, K., Jönsson, A.M., Lindborg, R., 2018. Relationships between multiple biodiversity components and ecosystem services along a landscape complexity gradient. Biol. Conserv. 218, 247–253.

Bjorkman, A.D., Myers-Smith, I.H., Elmendorf, S.C., Normand, S., Rüger, N., Beck, P.S., Blach-Overgaard, A., Blok, D., Cornelissen, J.H.C., Forbes, B.C., Georges, D., et al., 2018. Plant functional trait change across a warming tundra biome. Nature 562, 57.

Blaauw, B.R., Isaacs, R., 2014. Flower plantings increase wild bee abundance and the pollination services provided to a pollination-dependent crop. J. Appl. Ecol. 51, 890–898.

Blesh, J., 2018. Functional traits in cover crop mixtures: biological nitrogen fixation and multifunctionality. J. Appl. Ecol. 55, 38–48.

Blitzer, E.J., Dormann, C.F., Holzschuh, A., Klein, A.M., Rand, T.A., Tscharntke, T., 2012. Spillover of functionally important organisms between managed and natural habitats. Agr. Ecosyst. Environ. 146, 34–43.

Bosem Baillod, A., Tscharntke, T., Clough, Y., Batáry, P., 2017. Landscape-scale interactions of spatial and temporal cropland heterogeneity drive biological control of cereal aphids. J. Appl. Ecol. 54, 1804–1813.

Bretagnolle, V., Berthet, E., Gross, N., Gauffre, B., Plumejeaud, C., Houte, S., Badenhausser, I., Monceau, K., Allier, F., Monestiez, P., Gaba, S., 2018. Towards sustainable and multifunctional agriculture in farmland landscapes: lessons from the integrative approach of a French LTSER platform. Sci. Total Environ. 627, 822–834.

Brooker, R.W., Bennett, A.E., Cong, W.F., Daniell, T.J., George, T.S., Hallett, P.D., Hawes, C., Iannetta, P.P., Jones, H.G., Karley, A.J., Li, L., 2015. Improving intercropping: a synthesis of research in agronomy, plant physiology and ecology. New Phytol. 206, 107–117.

Bruelheide, H., Nadrowski, K., Assmann, T., Bauhus, J., Both, S., Buscot, F., Chen, X.Y., Ding, B., Durka, W., Erfmeier, A., Gutknecht, J.L., 2014. Designing forest biodiversity experiments: general considerations illustrated by a new large experiment in subtropical China. Methods Ecol. Evol. 5, 74–89.

Bullock, J.M., Pywell, R.F., Walker, K.J., 2007. Long-term enhancement of agricultural production by restoration of biodiversity. J. Appl. Ecol. 44, 6–12.

Cardinale, B.J., Duffy, J.E., Gonzalez, A., Hooper, D.U., Perrings, C., Venail, P., Narwani, A., Mace, G.M., Tilman, D., Wardle, D.A., Kinzig, A.P., 2012. Biodiversity loss and its impact on humanity. Nature 486, 59.

Civitello, D.J., Cohen, J., Fatima, H., Halstead, N.T., Liriano, J., McMahon, T.A., Ortega, C.N., Sauer, E.L., Sehgal, T., Young, S., Rohr, J.R., 2015. Biodiversity inhibits parasites: broad evidence for the dilution effect. Proc. Natl. Acad. Sci. 112, 8667–8671.

Clark, C.M., Tilman, D., 2010. Recovery of plant diversity following N cessation: effects of recruitment, litter, and elevated N cycling. Ecology 91, 3620–3630.

Clough, Y., Barkmann, J., Juhrbandt, J., Kessler, M., Wanger, T.C., Anshary, A., Buchori, D., Cicuzza, D., Darras, K., Putra, D.D., Erasmi, S., et al., 2011. Combining high biodiversity with high yields in tropical agroforests. Proc. Natl. Acad. Sci. 108, 8311–8316.

Clough, Y., Krishna, V.V., Corre, M.D., Darras, K., Denmead, L.H., Meijide, A., Moser, S., Musshoff, O., Steinebach, S., Veldkamp, E., Allen, K., et al., 2016. Land-use choices follow profitability at the expense of ecological functions in Indonesian smallholder landscapes. Nat. Commun. 7, 13137.

Conservation Evidence, 2019. https://www.conservationevidence.com/.

Craven, D., Eisenhauer, N., Pearse, W.D., Hautier, Y., Roscher, C., Isbell, F., Bahn, M., Beierkuhnlein, C., Bönisch, G., Buchmann, N., Byun, C., et al., 2018. Multiple facets of biodiversity drive the diversity-stability relationship. Nat Ecol Evol 2, 1.

Cross, M.S., Harte, J., 2007. Compensatory responses to loss of warming-sensitive plant species. Ecology 88, 740–748.

Díaz, S., Symstad, A.J., Chapin III, F.S., Wardle, D.A., Huenneke, L.F., 2003. Functional diversity revealed by removal experiments. Trends Ecol. Evol. 18, 140–146.

Díaz, S., Demissew, S., Carabias, J., Joly, C., Lonsdale, M., Ash, N., Larigauderie, A., Adhikari, J.R., Arico, S., Báldi, A., Bartuska, A., et al., 2015. The IPBES conceptual framework—connecting nature and people. Curr. Opin. Environ. Sustain. 14, 1–16.

Díaz, S., Pascual, U., Stenseke, M., Martín-López, B., Watson, R.T., Molnár, Z., Hill, R., Chan, K.M., Baste, I.A., Brauman, K.A., Polasky, S., 2018. Assessing nature's contributions to people. Science 359, 270–272.

Dickson, T.L., Gross, K.L., 2015. Can the results of biodiversity-ecosystem productivity studies be translated to bioenergy production? PLoS One 10, e0135253.

Dominik, C., Seppelt, R., Horgan, F.G., Settele, J., Václavík, T., 2018. Landscape composition, configuration, and trophic interactions shape arthropod communities in rice agroecosystems. J. Appl. Ecol. 55, 2461–2472.

Donohue, I., Hillebrand, H., Montoya, J.M., Petchey, O.L., Pimm, S.L., Fowler, M.S., Healy, K., Jackson, A.L., Lurgi, M., McClean, D., O'Connor, N.E., O'Gorman, E.J., Yang, Q., Adler, F., 2016. Navigating the complexity of ecological stability. Ecol. Lett. 19, 1172–1185.

Dornelas, M., Gotelli, N.J., McGill, B., Shimadzu, H., Moyes, F., Sievers, C., Magurran, A.E., 2014. Assemblage time series reveal biodiversity change but not systematic loss. Science 344, 296–299.

Duffy, J.E., Godwin, C.M., Cardinale, B.J., 2017. Biodiversity effects in the wild are common and as strong as key drivers of productivity. Nature 549, 261.

Eisenhauer, N., Barnes, A.D., Cesarz, S., Craven, D., Ferlian, O., Gottschall, F., Hines, J., Sendek, A., Siebert, J., Thakur, M.P., Türke, M., 2016. Biodiversity–ecosystem function experiments reveal the mechanisms underlying the consequences of biodiversity change in real world ecosystems. J. Veg. Sci. 27, 1061–1070.

Eisenhauer, N., Schielzeth, H., Barnes, A.D., Barry, K.E., Bonn, A., Brose, U., Bruelheide, H., Buchmann, N., Buscot, F., Ebeling, A., et al., 2019. A multitrophic perspective on biodiversity–ecosystem functioning research. Adv. Ecol. Res. 61, 1–54.

Emmerson, M., Morales, M.B., Oñate, J.J., Batáry, P., Berendse, F., Liira, J., Aavik, T., Guerrero, I., Bommarco, R., Eggers, S., Pärt, T., 2016. How agricultural intensification affects biodiversity and ecosystem services. In: Advances in Ecological Research. 55, Academic Press, pp. 43–97.

European Landowners Organisation (ELO), 2019. https://www.europeanlandowners.org/.

Fanin, N., Gundale, M.J., Farrell, M., Ciobanu, M., Baldock, J.A., Nilsson, M.C., Kardol, P., Wardle, D.A., 2018. Consistent effects of biodiversity loss on multifunctionality across contrasting ecosystems. Nat. Ecol. Evol. 2, 269.

Fazey, I., Evely, A.C., Reed, M.S., Stringer, L.C., Kruijsen, J., White, P.C., Newsham, A., Jin, L., Cortazzi, M., Phillipson, J., Blackstock, K., 2013. Knowledge exchange: a review and research agenda for environmental management. Environ. Conserv. 40, 19–36.

Finger, R., Buchmann, N., 2015. An ecological economic assessment of risk-reducing effects of species diversity in managed grasslands. Ecol. Econ. 110, 89–97.

Finn, J.A., Kirwan, L., Connolly, J., Sebastià, M.T., Helgadottir, A., Baadshaug, O.H., Bélanger, G., Black, A., Brophy, C., Collins, R.P., Čop, J., 2013. Ecosystem function enhanced by combining four functional types of plant species in intensively managed grassland mixtures: a 3-year continental-scale field experiment. J. Appl. Ecol. 50, 365–375.

Flohre, A., Fischer, C., Aavik, T., Bengtsson, J., Berendse, F., Bommarco, R., Ceryngier, P., Clement, L.W., Dennis, C., Eggers, S., Emmerson, M., 2011. Agricultural intensification and biodiversity partitioning in European landscapes comparing plants, carabids, and birds. Ecol. Appl. 21, 1772–1781.

Forum for the Future of Agriculture (FFA), 2019. http://www.forumforagriculture.com/.

Fry, E.L., Manning, P., Allen, D.G., Hurst, A., Everwand, G., Rimmler, M., Power, S.A., 2013. Plant functional group composition modifies the effects of precipitation change on grassland ecosystem function. PLoS One 8, e57027.

Für Ressourcen, Agrarwirtschaft & Naturschutz mit Zukunft (F.R.A.N.Z.), 2019, F.R.A.N.Z. www.franz-projekt.de.

Gámez-Virués, S., Perović, D.J., Gossner, M.M., Börschig, C., Blüthgen, N., De Jong, H., Simons, N.K., Klein, A.M., Krauss, J., Maier, G., Scherber, C., et al., 2015. Landscape simplification filters species traits and drives biotic homogenization. Nat. Commun. 6, 8568.

Gamfeldt, L., Snäll, T., Bagchi, R., Jonsson, M., Gustafsson, L., Kjellander, P., Ruiz-Jaen, M.C., Fröberg, M., Stendahl, J., Philipson, C.D., Mikusiński, G., 2013. Higher levels of multiple ecosystem services are found in forests with more tree species. Nat. Commun. 4, 1340.

Gardiner, M.M., Landis, D.A., Gratton, C., DiFonzo, C.D., O'neal, M., Chacon, J.M., Wayo, M.T., Schmidt, N.P., Mueller, E.E., Heimpel, G.E., 2009. Landscape diversity enhances biological control of an introduced crop pest in the north-Central USA. Ecol. Appl. 19, 143–154.

Garibaldi, L.A., Steffan-Dewenter, I., Winfree, R., Aizen, M.A., Bommarco, R., Cunningham, S.A., Kremen, C., Carvalheiro, L.G., Harder, L.D., Afik, O., Bartomeus, I., et al., 2013. Wild pollinators enhance fruit set of crops regardless of honey bee abundance. Science 339, 1608–1611.

Garibaldi, L.A., Carvalheiro, L.G., Leonhardt, S.D., Aizen, M.A., Blaauw, B.R., Isaacs, R., Kuhlmann, M., Kleijn, D., Klein, A.M., Kremen, C., Morandin, L., et al., 2014. From research to action: enhancing crop yield through wild pollinators. Front. Ecol. Environ. 12, 439–447.

Garibaldi, L.A., Bartomeus, I., Bommarco, R., Klein, A.M., Cunningham, S.A., Aizen, M.A., Boreux, V., Garratt, M.P., Carvalheiro, L.G., Kremen, C., Morales, C.L., 2015.

Trait matching of flower visitors and crops predicts fruit set better than trait diversity. J. Appl. Ecol. 52, 1436–1444.

Gaujour, E., Amiaud, B., Mignolet, C., Plantureux, S., 2012. Factors and processes affecting plant biodiversity in permanent grasslands. A review. Agron. Sustain. Dev. 32, 133–160.

Geertsema, W., Rossing, W.A., Landis, D.A., Bianchi, F.J., Van Rijn, P.C., Schaminée, J.H., Tscharntke, T., Van Der Werf, W., 2016. Actionable knowledge for ecological intensification of agriculture. Front. Ecol. Environ. 14, 209–216.

Gérard, A., Wollni, M., Hölscher, D., Irawan, B., Sundawati, L., Teuscher, M., Kreft, H., 2017. Oil-palm yields in diversified plantations: initial results from a biodiversity enrichment experiment in Sumatra, Indonesia. Agr. Ecosyst. Environ. 240, 253–260.

Gossner, M.M., Lewinsohn, T.M., Kahl, T., Grassein, F., Boch, S., Prati, D., Birkhofer, K., Renner, S.C., Sikorski, J., Wubet, T., et al., 2016. Land-use intensification causes multitrophic homogenization of grassland communities. Nature 540, 266.

Gould, I.J., Quinton, J.N., Weigelt, A., De Deyn, G.B., Bardgett, R.D., 2016. Plant diversity and root traits benefit physical properties key to soil function in grasslands. Ecol. Lett. 19, 1140–1149.

Grab, H., Branstetter, M.G., Amon, N., Urban-Mead, K.R., Park, M.G., Gibbs, J., Blitzer, E.J., Poveda, K., Loeb, G., Danforth, B.N., et al., 2019. Agriculturally dominated landscapes reduce bee phylogenetic diversity and pollination services. Science 363, 282–284.

Grace, J.B., Anderson, T.M., Seabloom, E.W., Borer, E.T., Adler, P.B., Harpole, W.S., Hautier, Y., Hillebrand, H., Lind, E.M., Pärtel, M., et al., 2016. Integrative modelling reveals mechanisms linking productivity and plant species richness. Nature 529, 390.

Grass, I., Loos, J., Baensch, S., Batáry, P., Librán-Embid, F., Ficiciyan, A., Klaus, F., Riechers, M., Rosa, J., Tiede, J., Udy, K., 2019. Land-sharing/−sparing connectivity landscapes for ecosystem services and biodiversity conservation. People and Nature 1, 262–272.

Grimm, V., Wissel, C., 1997. Babel, or the ecological stability discussions: an inventory and analysis of terminology and a guide for avoiding confusion. Oecologia 109, 323–334.

Guimarães-Steinicke, C., Weigelt, A., Ebeling, A., Eisenhauer, N., Duque-Lazo, J., Reu, B., Roscher, C., Schumacher, J., Wagg, C., Wirth, C., 2019. Terrestrial laser scanning reveals temporal changes in biodiversity mechanisms driving grassland productivity. Adv. Ecol. Res. 61, 133–161.

Hass, A.L., Kormann, U.G., Tscharntke, T., Clough, Y., Baillod, A.B., Sirami, C., Fahrig, L., Martin, J.L., Baudry, J., Bertrand, C., Bosch, J., 2018. Landscape configurational heterogeneity by small-scale agriculture, not crop diversity, maintains pollinators and plant reproduction in western Europe. Proc. R. Soc. B Biol. Sci. 285, 20172242.

Hautier, Y., Isbell, F., Borer, E.T., Seabloom, E.W., Harpole, W.S., Lind, E.M., MacDougall, A.S., Stevens, C.J., Adler, P.B., Alberti, J., Bakker, J.D., et al., 2018. Local loss and spatial homogenization of plant diversity reduce ecosystem multifunctionality. Nat. Ecol. Evol. 2, 50.

Hector, A., Schmid, B., Beierkuhnlein, C., Caldeira, M., Diemer, M., Dimitrakopoulos, P., Finn, J., Freitas, H., Giller, P., Good, J., et al., 1999. Plant diversity and productivity experiments in European grasslands. Science 286, 1123–1127.

Hill, R., Nates-Parra, G., Quezada-Euán, J.J.G., Buchori, D., LeBuhn, G., Maués, M.M., Pert, P.L., Kwapong, P.K., Saeed, S., Breslow, S.J., da Cunha, M.C., et al., 2019. Biocultural approaches to pollinator conservation. Nature Sustainability 2, 214.

Hines, J., Ebeling, A., Barnes, A., Brose, U., Scherber, C., Scheu, S., Tscharntke, T., Weisser, W.W., Giling, D.P., Klein, A.M., Eisenhauer, N., 2019. Mapping change in biodiversity and ecosystem function research: food webs foster integration of experiments and science policy. Adv. Ecol. Res. 61, 297–322.

Hooper, D.U., Chapin, F., Ewel, J., Hector, A., Inchausti, P., Lavorel, S., Lawton, J., Lodge, D., Loreau, M., Naeem, S., et al., 2005. Effects of biodiversity on ecosystem functioning: a consensus of current knowledge. Ecol. Monogr. 75, 3–35.

Hooper, D.U., Adair, E.C., Cardinale, B.J., Byrnes, J.E., Hungate, B.A., Matulich, K.L., Gonzalez, A., Duffy, J.E., Gamfeldt, L., O'Connor, M.I., 2012. A global synthesis reveals biodiversity loss as a major driver of ecosystem change. Nature 486, 105.

Huang, Y., Chen, Y., Castro-Izaguirre, N., Baruffol, M., Brezzi, M., Lang, A., Li, Y., Härdtle, W., von Oheimb, G., Yang, X., Liu, X., et al., 2018. Impacts of species richness on productivity in a large-scale subtropical forest experiment. Science 362, 80–83.

Hulvey, K.B., Hobbs, R.J., Standish, R.J., Lindenmayer, D.B., Lach, L., Perring, M.P., 2013. Benefits of tree mixes in carbon plantings. Nat. Clim. Chang. 3, 869.

Huston, M.A., 1997. Hidden treatments in ecological experiments: re-evaluating the ecosystem function of biodiversity. Oecologia 110, 449–460.

Ingerslew, K.S., Kaplan, I., 2018. Distantly related crops are not better rotation partners for tomato. J. Appl. Ecol. 55, 2506–2516.

Isbell, F., Reich, P.B., Tilman, D., Hobbie, S.E., Polasky, S., Binder, S., 2013. Nutrient enrichment, biodiversity loss, and consequent declines in ecosystem productivity. Proc. Natl. Acad. Sci. 110, 11911–11916.

Isbell, F., Tilman, D., Polasky, S., Loreau, M., 2015a. The biodiversity-dependent ecosystem service debt. Ecol. Lett. 18, 119–134.

Isbell, F., Craven, D., Connolly, J., Loreau, M., Schmid, B., Beierkuhnlein, C., Bezemer, T.M., Bonin, C., Bruelheide, H., De Luca, E., et al., 2015b. Biodiversity increases the resistance of ecosystem productivity to climate extremes. Nature 526, 574–577.

Isbell, F., Adler, P.R., Eisenhauer, N., Fornara, D., Kimmel, K., Kremen, C., Letourneau, D.K., Liebman, M., Polley, H.W., Quijas, S., 2017. Benefits of increasing plant diversity in sustainable agroecosystems. J. Ecol. 105, 871–879.

Iverson, A.L., Marín, L.E., Ennis, K.K., Gonthier, D.J., Connor-Barrie, B.T., Remfert, J.L., Cardinale, B.J., Perfecto, I., 2014. Do polycultures promote win-wins or trade-offs in agricultural ecosystem services? A meta-analysis. J. Appl. Ecol. 51, 1593–1602.

Jackson, L.E., Pulleman, M.M., Brussaard, L., Bawa, K.S., Brown, G.G., Cardoso, I.M., De Ruiter, P.C., García-Barrios, L., Hollander, A.D., Lavelle, P., Ouédraogo, E., 2012. Social-ecological and regional adaptation of agrobiodiversity management across a global set of research regions. Glob. Environ. Chang. 22, 623–639.

Kampmann, D., Lüscher, A., Konold, W., Herzog, F., 2012. Agri-environment scheme protects diversity of mountain grassland species. Land Use Policy 29, 569–576.

Khalsa, J., Fricke, T., Weigelt, A., Wachendorf, M., 2014. Effects of species richness and functional groups on chemical constituents relevant for methane yields from anaerobic digestion: results from a grassland diversity experiment. Grass Forage Sci. 69, 49–63.

King, E., Cavender-Bares, J., Balvanera, P., Mwampamba, T., Polasky, S., 2015. Trade-offs in ecosystem services and varying stakeholder preferences: evaluating conflicts, obstacles, and opportunities. Ecol. Soc. 20, 25.

Kirmer, A., Baasch, A., Tischew, S., 2012. Sowing of low and high diversity seed mixtures in ecological restoration of surface mined-land. Appl. Veg. Sci. 15, 198–207.

Kirwan, L., Lüscher, A., Sebastià, M.T., Finn, J.A., Collins, R.P., Porqueddu, C., Helgadottir, A., Baadshaug, O.H., Brophy, C., Coran, C., Dalmannsdóttir, S., et al., 2007. Evenness drives consistent diversity effects in intensive grassland systems across 28 European sites. J. Ecol. 95, 530–539.

Kleijn, D., Winfree, R., Bartomeus, I., Carvalheiro, L.G., Henry, M., Isaacs, R., Klein, A.M., Kremen, C., M'gonigle, L.K., Rader, R., Ricketts, T.H., 2015. Delivery of crop pollination services is an insufficient argument for wild pollinator conservation. Nat. Commun. 6, 7414.

Kleijn, D., Bommarco, R., Fijen, T.P., Garibaldi, L.A., Potts, S.G., van der Putten, W.H., 2018. Ecological intensification: bridging the gap between science and practice. Trends Ecol. Evol. 34, 154–166.

Klein, A.M., Steffan-Dewenter, I., Tscharntke, T., 2003. Fruit set of highland coffee increases with the diversity of pollinating bees. Proc. R. Soc. Lond. B Biol. Sci. 270, 955–961.

Knapp, S., van der Heijden, M.G., 2018. A global meta-analysis of yield stability in organic and conservation agriculture. Nat. Commun. 9, 3632.

Koh, L.P., Levang, P., Ghazoul, J., 2009. Designer landscapes for sustainable biofuels. Trends Ecol. Evol. 24, 431–438.

Koh, I., Lonsdorf, E.V., Williams, N., Brittain, C., Isaacs, R., Gibbs, J., Ricketts, T.H., 2016. Modeling the status, trends, and impacts of wild bee abundance in the United States. Proc. Natl. Acad. Sci. 113, 140–145.

Kollmann, J., Meyer, S.T., Bateman, R., Conradi, T., Gossner, M.M., de Souza Mendonça Jr, M., Fernandes, G.W., Hermann, J.M., Koch, C., Müller, S.C., Oki, Y., 2016. Integrating ecosystem functions into restoration ecology—recent advances and future directions. Restor. Ecol. 24, 722–730.

Kormann, U., Rösch, V., Batáry, P., Tscharntke, T., Orci, K.M., Samu, F., Scherber, C., 2015. Local and landscape management drive trait-mediated biodiversity of nine taxa on small grassland fragments. Divers. Distrib. 21, 1204–1217.

Kremen, C., Merenlender, A.M., 2018. Landscapes that work for biodiversity and people. Science 362, eaau6020.

Landis, D.A., 2017. Designing agricultural landscapes for biodiversity-based ecosystem services. Basic Appl. Ecol. 18, 1–12.

Lange, M., Eisenhauer, N., Sierra, C.A., Bessler, H., Engels, C., Griffiths, R.I., Mellado-Vázquez, P.G., Malik, A.A., Roy, J., Scheu, S., Steinbeiss, S., et al., 2015. Plant diversity increases soil microbial activity and soil carbon storage. Nat. Commun. 6, 6707.

Larsen, T.H., Williams, N.M., Kremen, C., 2005. Extinction order and altered community structure rapidly disrupt ecosystem functioning. Ecol. Lett. 8, 538–547.

Lavorel, S., Grigulis, K., Lamarque, P., Colace, M.P., Garden, D., Girel, J., Pellet, G., Douzet, R., 2011. Using plant functional traits to understand the landscape distribution of multiple ecosystem services. J. Ecol. 99, 135–147.

Lavorel, S., Bayer, A., Bondeau, A., Lautenbach, S., Ruiz-Frau, A., Schulp, N., Seppelt, R., Verburg, P., van Teeffelen, A., Vannier, C., Arneth, A., 2017. Pathways to bridge the biophysical realism gap in ecosystem services mapping approaches. Ecol. Indic. 74, 241–260.

Lefcheck, J.S., Byrnes, J.E., Isbell, F., Gamfeldt, L., Griffin, J.N., Eisenhauer, N., Hensel, M.J., Hector, A., Cardinale, B.J., Duffy, J.E., 2015. Biodiversity enhances ecosystem multifunctionality across trophic levels and habitats. Nat. Commun. 6, 6936.

Leibold, M.A., Chase, J.M., Ernest, S.M., 2017. Community assembly and the functioning of ecosystems: how metacommunity processes alter ecosystems attributes. Ecology 98, 909–919.

Lepš, J., 2004. What do the biodiversity experiments tell us about consequences of plant species loss in the real world? Basic Appl. Ecol. 5, 529–534.

Lichtenberg, E.M., Kennedy, C.M., Kremen, C., Batáry, P., Berendse, F., Bommarco, R., Bosque-Pérez, N.A., Carvalheiro, L.G., Snyder, W.E., Williams, N.M., Winfree, R., 2017. A global synthesis of the effects of diversified farming systems on arthropod diversity within fields and across agricultural landscapes. Glob. Chang. Biol. 23, 4946–4957.

Lindborg, R., Gordon, L.J., Malinga, R., Bengtsson, J., Peterson, G., Bommarco, R., Deutsch, L., Gren, A., Rundlöf, M., Smith, H.G., 2017. How spatial scale shapes the generation and management of multiple ecosystem services. Ecosphere 8, e01741. https://doi.org/10.1002/ecs2.1741.

Loreau, M., Hector, A., 2001. Partitioning selection and complementarity in biodiversity experiments. Nature 413, 548.

Loreau, M., Naeem, S., Inchausti, P., Bengtsson, J., Grime, J.P., Hector, A., Hooper, D.U., Huston, M.A., Raffaelli, D., Schmid, B., Tilman, D., Wardle, D.A., 2001. Biodiversity and ecosystem functioning: current knowledge and future challenges. Science 294, 804–808.

Loreau, M., Mouquet, N., Gonzalez, A., 2003. Biodiversity as spatial insurance in heterogeneous landscapes. Proc. Natl. Acad. Sci. 100, 12765–12770.

Lundholm, J., MacIvor, J.S., MacDougall, Z., Ranalli, M., 2010. Plant species and functional group combinations affect green roof ecosystem functions. PLoS One 5, e9677.

Lyons, K.G., Schwartz, M.W., 2001. Rare species loss alters ecosystem function–invasion resistance. Ecol. Lett. 4, 358–365.

Mace, G.M., Norris, K., Fitter, A.H., 2012. Biodiversity and ecosystem services: a multilayered relationship. Trends Ecol. Evol. 27, 19–26.

Maestre, F.T., Quero, J.L., Gotelli, N.J., Escudero, A., Ochoa, V., Delgado-Baquerizo, M., García-Gómez, M., Bowker, M.A., Soliveres, S., Escolar, C., 2012. Plant species richness and ecosystem multifunctionality in global drylands. Science 335, 214–218.

Magrini, M.B., Anton, M., Chardigny, J.M., Duc, G., 2018. Pulses for sustainability: breaking agriculture and food sectors out of lock-in. Front. Sust. Food Syst. 2, 64.

Manning, P., Newington, J.E., Robson, H.R., Saunders, M., Eggers, T., Bradford, M.A., Bardgett, R.D., Bonkowski, M., Ellis, R.J., Gange, A.C., 2006. Decoupling the direct and indirect effects of nitrogen deposition on ecosystem function. Ecol. Lett. 9, 1015–1024.

Manning, P., Vries, F.T., Tallowin, J.R., Smith, R., Mortimer, S.R., Pilgrim, E.S., Harrison, K.A., Wright, D.G., Quirk, H., Benson, J., Shipley, B., et al., 2015. Simple measures of climate, soil properties and plant traits predict national-scale grassland soil carbon stocks. J. Appl. Ecol. 52, 1188–1196.

Manning, P., Plas, F., Soliveres, S., Allan, E., Maestre, F.T., Mace, G., Whittingham, M.J., Fischer, M., 2018. Redefining ecosystem multifunctionality. Nat. Ecol. Evol. 2, 427.

Maron, J.L., Marler, M., Klironomos, J.N., Cleveland, C.C., 2011. Soil fungal pathogens and the relationship between plant diversity and productivity. Ecol. Lett. 14, 36–41.

Martín-López, B., Iniesta-Arandia, I., García-Llorente, M., Palomo, I., Casado-Arzuaga, I., Del Amo, D.G., Gómez-Baggethun, E., Oteros-Rozas, E., Palacios-Agundez, I., Willaarts, B., González, J.A., 2012. Uncovering ecosystem service bundles through social preferences. PLoS One 7, e38970.

Maskell, L.C., Crowe, A., Dunbar, M.J., Emmett, B., Henrys, P., Keith, A.M., Norton, L.R., Scholefield, P., Clark, D.B., Simpson, I.C., Smart, S.M., Clough, Y., 2013. Exploring the ecological constraints to multiple ecosystem service delivery and biodiversity. J. Appl. Ecol. 50, 561–571.

Meyer, S.T., Ptacnik, R., Hillebrand, H., Bessler, H., Buchmann, N., Ebeling, A., Eisenhauer, N., Engels, C., Fischer, M., Halle, S., Klein, A.M., et al., 2018. Biodiversity–multifunctionality relationships depend on identity and number of measured functions. Nat. Ecol. Evol. 2, 44.

Mori, A.S., Isbell, I., Seidl, R., 2018. β-diversity, community assembly, and ecosystem functioning. Trends Ecol. Evol. 33, 549–564.

Nyfeler, D., Huguenin-Elie, O., Suter, M., Frossard, E., Connolly, J., Lüscher, A., 2009. Strong mixture effects among four species in fertilized agricultural grassland led to persistent and consistent transgressive overyielding. J. Appl. Ecol. 46, 683–691.

Oehri, J., Schmid, B., Schaepman-Strub, G., Niklaus, P.A., 2017. Biodiversity promotes primary productivity and growing season lengthening at the landscape scale. Proc. Nat. Acad. Sci. 114, 10160–10165.

Oliver, T.H., Heard, M.S., Isaac, N.J., Roy, D.B., Procter, D., Eigenbrod, F., Freckleton, R., Hector, A., Orme, C.D.L., Petchey, O.L., 2015. Biodiversity and resilience of ecosystem functions. Trends Ecol. Evol. 30, 673–684.

Otway, S.J., Hector, A., Lawton, J.H., 2005. Resource dilution effects on specialist insect herbivores in a grassland biodiversity experiment. J. Anim. Ecol. 74, 234–240.
Pan, Q., Tian, D., Naeem, S., Auerswald, K., Elser, J.J., Bai, Y., Huang, J., Wang, Q., Wang, H., Wu, J., Han, X., 2016. Effects of functional diversity loss on ecosystem functions are influenced by compensation. Ecology 97, 2293–2302.
Peeters, A., Beaufoy, G., Canals, R.M., de Vliegher, A., Huyghe, C., Isselstein, J., Jones, G., Kessler, W., Kirilov, A., Mosquera-Losada, M.R., et al., 2014. Grassland term definitions and and classifications adapted to the delivery of European grassland-based systems. Grassl. Sci. Eur. 19, 743–750.
Pretzsch, H., Schütze, G., 2009. Transgressive overyielding in mixed compared with pure stands of Norway spruce and European beech in Central Europe: evidence on stand level and explanation on individual tree level. Eur. J. For. Res. 128, 183–204.
Rader, R., Birkhofer, K., Schmucki, R., Smith, H.G., Stjernman, M., Lindborg, R., 2014. Organic farming and heterogeneous landscapes positively affect different measures of plant diversity. J. Appl. Ecol. 51, 1544–1553.
Raseduzzaman, M., Jensen, E.S., 2017. Does intercropping enhance yield stability in arable crop production? A meta-analysis. Eur. J. Agron. 91, 25–33.
Redlich, S., Martin, E.A., Steffan-Dewenter, I., 2018. Landscape-level crop diversity benefits biological pest control. J. Appl. Ecol. 55, 2419–2428.
Reich, P.B., Knops, J., Tilman, D., Craine, J., Ellsworth, D., Tjoelker, M., Lee, T., Wedink, D., Naeem, S., Bahauddin, D., et al., 2001. Plant diversity enhances ecosystem responses to elevated $CO_2$ and nitrogen deposition. Nature 410, 809–812.
Reiss, E.R., Drinkwater, L.E., 2018. Cultivar mixtures: a meta-analysis of the effect of intraspecific diversity on crop yield. Ecol. Appl. 28, 62–77.
Rosa-Scleich, J., Loos, J., Musshoff, O., Tscharntke, T., 2019. Ecological-economic trade-offs of diversified farming systems—a review. Ecol. Econ. 160, 251–263.
Rösch, V., Tscharntke, T., Scherber, C., Batáry, P., 2015. Biodiversity conservation across taxa and landscapes requires many small as well as single large habitat fragments. Oecologia 179, 209–222.
Roscher, C., Schumacher, J., Baade, J., Wilcke, W., Gleixner, G., Weisser, W.W., Schmid, B., Schulze, E.-D., 2004. The role of biodiversity for element cycling and trophic interactions: an experimental approach in a grassland community. Basic Appl. Ecol. 5, 107–121.
Roschewitz, I., Gabriel, D., Tscharntke, T., Thies, C., 2005. The effects of landscape complexity on arable weed species diversity in organic and conventional farming. J. Appl. Ecol. 2005 (42), 873–882.
Royal Society for the Protection of Birds (RSPB), Hope Farm, 2019. https://www.rspb.org.uk/our-work/conservation/conservation-and-sustainability/farming/hope-farm/.
Samnegard, U., Alins, G., Boreux, V., Bosch, J., García, D., Happe, A.-K., Klein, A.M., Miñarro, M., Mody, K., Porcel, M., et al., 2019. Management trade-offs on ecosystem services in apple orchards across Europe: direct and indirect effects of organic production. J. Appl. Ecol. 56, 802–811.
Scherber, C., Eisenhauer, N., Weisser, W.W., Schmid, B., Voigt, W., Fischer, M., Schulze, E.-D., Roscher, C., Weigelt, A., Allan, E., et al., 2010. Bottom-up effects of plant diversity on multitrophic interactions in a biodiversity experiment. Nature 468, 553.
Schmid, B., Hector, A., 2004. The value of biodiversity experiments. Basic Appl. Ecol. 5, 535–542.
Schmid, B., Hector, A., Huston, M.A., Inchausti, P., Nijs, I., Leadley, P.W., Tilman, D., 2002. The design and analysis of biodiversity experiments. In: Biodiversity and Ecosystem Functioning: Synthesis and Perspectives. Oxford University Press, Oxford, pp. 61–75.

Schnitzer, S.A., Klironomos, J.N., HilleRisLambers, J., Kinkel, L.L., Reich, P.B., Xiao, K., Rillig, M.C., Sikes, B.A., Callaway, R.M., Mangan, S.A., 2011. Soil microbes drive the classic plant diversity–productivity pattern. Ecology 92, 296–303.

Schulze, E.-D., Mooney, H.A., 1994. Ecosystem function of biodiversity: a summary. In: Biodiversity and Ecosystem Function. Springer, pp. 497–510.

Smith, M.D., Knapp, A.K., 2003. Dominant species maintain ecosystem function with non-random species loss. Ecol. Lett. 6, 509–517.

Smith, R., Shiel, R., Bardgett, R.D., Millward, D., Corkhill, P., Evans, P., Quirk, H., Hobbs, P., Kometa, S., 2008. Long-term change in vegetation and soil microbial communities during the phased restoration of traditional meadow grassland. J. Appl. Ecol. 45, 670–679.

Soliveres, S., Manning, P., Prati, D., Gossner, M.M., Alt, F., Arndt, H., Baumgartner, V., Binkenstein, J., Birkhofer, K., Blaser, S., et al., 2016a. Locally rare species influence grassland ecosystem multifunctionality. Philos. Trans. R. Soc. B 371, 20150269.

Soliveres, S., Van Der Plas, F., Manning, P., Prati, D., Gossner, M.M., Renner, S.C., Alt, F., Arndt, H., Baumgartner, V., Binkenstein, J., et al., 2016b. Biodiversity at multiple trophic levels is needed for ecosystem multifunctionality. Nature 536, 456.

Srivastava, D.S., Vellend, M., 2005. Biodiversity-ecosystem function research: is it relevant to conservation? Annu. Rev. Ecol. Evol. Syst. 36, 267–294.

Stein, C., Auge, H., Fischer, M., Weisser, W.W., Prati, D., 2008. Dispersal and seed limitation affect diversity and productivity of montane grasslands. Oikos 117, 1469–1478.

Storkey, J., Döring, T., Baddeley, J., Collins, R., Roderick, S., Jones, H., Watson, C., 2015. Engineering a plant community to deliver multiple ecosystem services. Ecol. Appl. 25, 1034–1043.

Suding, K.N., Ashton, I.W., Bechtold, H., Bowman, W.D., Mobley, M.L., Winkleman, R., 2008. Plant and microbe contribution to community resilience in a directionally changing environment. Ecol. Monogr. 78, 313–329.

Suter, M., Connolly, J., Finn, J.A., Loges, R., Kirwan, L., Sebastià, M.T., Lüscher, A., 2015. Nitrogen yield advantage from grass–legume mixtures is robust over a wide range of legume proportions and environmental conditions. Glob. Chang. Biol. 21, 2424–2438.

Suter, D., Rosenberg, E., Mosimann, E., Frick, R., 2017. Standardmischungen für den Futterbau Revision 2017–2020. Agrarforschung Schweiz 8, 1–16.

Tallis, H., 2011. Natural Capital: Theory and Practice of Mapping Ecosystem Services. Oxford University Press.

Teuscher, M., Gérard, A., Brose, U., Buchori, D., Clough, Y., Ehbrecht, M., Hölscher, D., Irawan, B., Sundawati, L., Wollni, M., Kreft, H., 2016. Experimental biodiversity enrichment in oil-palm-dominated landscapes in Indonesia. Front. Plant Sci. 7, 1538.

Tilman, D., Wedin, D., Knops, J., 1996. Productivity and sustainability influenced by biodiversity in grassland ecosystems. Nature 379, 718.

Tilman, D., Reich, P.B., Knops, J., Wedin, D., Mielke, T., Lehman, C., 2001. Diversity and productivity in a long-term grassland experiment. Science 294, 843–845.

Tilman, D., Hill, J., Lehman, C., 2006. Carbon-negative biofuels from low-input high-diversity grassland biomass. Science 314, 1598–1600.

Tscharntke, T., Klein, A.M., Kruess, A., Steffan-Dewenter, I., Thies, C., 2005. Landscape perspectives on agricultural intensification and biodiversity–ecosystem service management. Ecol. Lett. 8, 857–874.

Tscharntke, T., Tylianakis, J.M., Rand, T.A., Didham, R.K., Fahrig, L., Batáry, P., Bengtsson, J., Clough, Y., Crist, T.O., Dormann, C.F., Ewers, R.M., 2012. Landscape moderation of biodiversity patterns and processes-eight hypotheses. Biol. Rev. 87, 661–685.

Tscharntke, T., Karp, D.S., Chaplin-Kramer, R., Batáry, P., DeClerck, F., Gratton, C., Hunt, L., Ives, A., Jonsson, M., Larsen, A., Martin, E.A., 2016. When natural habitat fails to enhance biological pest control–five hypotheses. Biol. Conserv. 204, 449–458.

UK National Ecosystem Assessment, 2011. The UK National Ecosystem Assessment: Synthesis of the Key Findings. UNEP-WCMC, Cambridge.

van der Plas, F., 2019. Biodiversity and ecosystem functioning in naturally assembled communities. Biol. Rev. https://doi.org/10.1111/brv.12499.

van der Plas, F., Manning, P., Soliveres, S., Allan, E., Scherer-Lorenzen, M., Verheyen, K., Wirth, C., Zavala, M.A., Ampoorter, E., Baeten, L., Barbaro, L., et al., 2016. Biotic homogenization can decrease landscape-scale forest multifunctionality. Proc. Natl. Acad. Sci. 113, 3557–3562.

van der Plas, F., Ratcliffe, S., Ruiz-Benito, P., Scherer-Lorenzen, M., Verheyen, K., Wirth, C., Zavala, M.A., Ampoorter, E., Baeten, L., Barbaro, L., Bastias, C.C., et al., 2018. Continental mapping of forest ecosystem functions reveals a high but unrealised potential for forest multifunctionality. Ecol. Lett. 21, 31–42.

van der Plas, F., Allan, E., Fischer, M., Alt, F., Arndt, H., Binkenstein, J., Blaser, S., Blüthgen, N., Böhm, S., Hölzel, N., Klaus, V.H., et al., 2019. Towards the development of general rules describing landscape heterogeneity–multifunctionality relationships. J. Appl. Ecol. 56, 168–179.

Van der Putten, W.H., Mortimer, S.R., Hedlund, K., Van Dijk, C., Brown, V.K., Lepä, J., Rodriguez-Barrueco, C., Roy, J., Len, T.D., Gormsen, D., Korthals, G.W., 2000. Plant species diversity as a driver of early succession in abandoned fields: a multi-site approach. Oecologia 124, 91–99.

Vandermeer, J.H., 1992. The Ecology of Intercropping. Cambridge University Press.

Vellend, M., Baeten, L., Myers-Smith, I.H., Elmendorf, S.C., Beauséjour, R., Brown, C.D., De Frenne, P., Verheyen, K., Wipf, S., 2013. Global meta-analysis reveals no net change in local-scale plant biodiversity over time. Proc. Natl. Acad. Sci. 110, 19456–19459.

Verheyen, K., Vanhellemont, M., Auge, H., Baeten, L., Baraloto, C., Barsoum, N., Bilodeau-Gauthier, S., Bruelheide, H., Castagneyrol, B., Godbold, D., Haase, J., 2016. Contributions of a global network of tree diversity experiments to sustainable forest plantations. Ambio 45, 29–41.

Vogel, A., Ebeling, A., Gleixner, G., Roscher, C., Scheu, S., Ciobanu, M., Koller-France, E., Lange, M., Lochner, A., Meyer, S.T., et al., 2019. A new experimental approach to test why biodiversity effects strengthen as ecosystems age. Adv. Ecol. Res. 61, 221–264 (Chapter 7).

Walter, A., Finger, R., Huber, R., Buchmann, N., 2017. Opinion: smart farming is key to developing sustainable agriculture. Proc. Natl. Acad. Sci. 114, 6148–6150.

Wardle, D.A., 2016. Do experiments exploring plant diversity–ecosystem functioning relationships inform how biodiversity loss impacts natural ecosystems? J. Veg. Sci. 27, 646–653.

Weidlich, E.W., von Gillhaussen, P., Max, J.F., Delory, B.M., Jablonowski, N.D., Rascher, U., Temperton, V.M., 2018. Priority effects caused by plant order of arrival affect below-ground productivity. J. Ecol. 106, 774–780.

Weisser, W.W., Roscher, C., Meyer, S.T., Ebeling, A., Luo, G., Allan, E., Beßler, H., Barnard, R., Buchmann, N., Buscot, F., Engels, C., et al., 2017. Biodiversity effects on ecosystem functioning in a 15-year grassland experiment: patterns, mechanisms, and open questions. Basic Appl. Ecol. 23, 1–73.

Werling, B.P., Dickson, T.L., Isaacs, R., Gaines, H., Gratton, C., Gross, K.L., Liere, H., Malmstrom, C.M., Meehan, T.D., Ruan, L., Robertson, B.A., et al., 2014. Perennial grasslands enhance biodiversity and multiple ecosystem services in bioenergy landscapes. Proc. Natl. Acad. Sci. 111, 41652–41657.

Wilson, J.B., Peet, R.K., Dengler, J., Pärtel, M., 2012. Plant species richness: the world records. J. Veg. Sci. 23, 796–802.

Winfree, R., Jeremy, W.F., Williams, N.M., Reilly, J.R., Cariveau, D.P., 2015. Abundance of common species, not species richness, drives delivery of a real-world ecosystem service. Ecol. Lett. 18, 626–635.

Winfree, R., Reilly, J.R., Bartomeus, I., Cariveau, D.P., Williams, N.M., Gibbs, J., 2018. Species turnover promotes the importance of bee diversity for crop pollination at regional scales. Science 359, 791–793.

Wright, A.J., Ebeling, A., De Kroon, H., Roscher, C., Weigelt, A., Buchmann, N., Buchmann, T., Fischer, C., Hacker, N., Hildebrandt, A., Leimer, S., et al., 2015. Flooding disturbances increase resource availability and productivity but reduce stability in diverse plant communities. Nat. Commun. 6, 6092.

Wright, A.J., Wardle, D.A., Callaway, R., Gaxiola, A., 2017. The overlooked role of facilitation in biodiversity experiments. Trends Ecol. Evol. 32, 383–390.

Yang, Y., Tilman, D., Lehman, C., Trost, J.J., 2018. Sustainable intensification of high-diversity biomass production for optimal biofuel benefits. Nature Sustainability 1, 686.

Yang, Y., Tilman, D., Furey, G., Lehman, C., 2019. Soil carbon sequestration accelerated by restoration of grassland biodiversity. Nat. Commun. 10, 718.

Zavaleta, E.S., Hulvey, K.B., 2004. Realistic species losses disproportionately reduce grassland resistance to biological invaders. Science 306, 1175–1177.

Zhu, Y., Chen, H., Fan, J., Wang, Y., Li, Y., Chen, J., Fan, J., Yang, S., Hu, L., Leung, H., 2000. Genetic diversity and disease control in rice. Nature 406, 718.

Zhu, J., Jiang, L., Zhang, Y., 2016. Relationships between functional diversity and aboveground biomass production in the northern Tibetan alpine grasslands. Sci. Rep. 6, 34105.

Zobel, K., Zobel, M., Rosén, E., 1994. An experimental test of diversity maintenance mechanisms, by a species removal experiment in a species-rich wooded meadow. Folia Geobotanica et Phytotaxonomica 29, 449–457.

Zuppinger-Dingley, D., Schmid, B., Petermann, J.S., Yadav, V., De Deyn, G.B., Flynn, D.F., 2014. Selection for niche differentiation in plant communities increases biodiversity effects. Nature 515, 108.